Graphene as Energy Storage Material for Supercapacitors

Edited by

Inamuddin[1,2,3], Rajender Boddula[4], Mohammad Faraz Ahmer[5] and Abdullah M. Asiri[1,2]

[1]Chemistry Department, Faculty of Science, King Abdulaziz University, Jeddah 21589, Saudi Arabia

[2]Centre of Excellence for Advanced Materials Research, King Abdulaziz University, Jeddah 21589, Saudi Arabia

[3]Department of Applied Chemistry, Faculty of Engineering and Technology, Aligarh Muslim University, Aligarh-202 002, India

[4]CAS Key Laboratory of Nanosystem and Hierarchical Fabrication, National Center for Nanoscience and Technology, Beijing 100190, PR China

[5]Department of Electrical Engineering, Mewat College of Engineering and Technology, Mewat-122103, India

Published by **Materials Research Forum LLC**
Millersville, PA 17551, USA

Published as part of the book series
Materials Research Foundations
Volume 64 (2020)
ISSN 2471-8890 (Print)
ISSN 2471-8904 (Online)

Print ISBN 978-1-64490-054-3
eBook ISBN 978-1-64490-055-0

Distributed worldwide by

Materials Research Forum LLC
105 Springdale Lane
Millersville, PA 17551
USA
http://www.mrforum.com

Manufactured in the United States of America
10 9 8 7 6 5 4 3 2 1

Table of Contents

Preface

Currently, supercapacitor energy storage devices are highly demanding due to their cost-effectiveness, eco-friendly nature, high power density, moderate energy density, and long-term cycle stability. These qualities are meeting the interest for supercapacitor applications in electric power systems of portable electronics, hybrid-electric vehicles, mobile phones, etc. However, energy storage materials play a key role in supercapacitor devices. Among the various materials used in supercapacitor devices, graphene has turned into a primary research focus for the development of elite supercapacitors because of its excellent electrical conductivity, high surface area, and mechanical properties. The combinations of graphene with various materials forming composite materials of different shapes and structures have pulled significant attention in the development of supercapacitor devices. However, constraints to the power density of supercapacitors brought about by electrode kinetics. Therefore, the development of energy storage devices with higher explicit power has turned out to be appealing research innovation themes.

The book is devoted to providing a comprehensive review of graphene-based supercapacitor technology with a wide scope of discussion related to materials, mechanisms, processes, and device architectures. This invaluable guide will fascinate researchers and electrochemists working in the area of materials science, nanotechnology, and surface electrochemistry. This book aims to provide knowledge of essential aspects of graphene material synthesis and characterization, fundamental electrochemical properties, and the most promising applications. It also provides knowledge of the fundamentals, fabrication strategies, engineering pathways, and graphene material strutures (1D, 2D, 3D, holey, heteroatom-doped, aerogels, monoliths) used in supercapacitor technology. This book is the result of commitments by leading researchers working in the field of graphene having various backgrounds and expertises. The book is well structured and essential resource for scientists, undergraduate and postgraduate students, faculty, research and development professionals, energy chemists, and industrial experts.

Inamuddin[1,2,3], Rajender Boddula[4], Mohammad Faraz Ahmer[5] and Abdullah M. Asiri[1,2]

[1]Chemistry Department, Faculty of Science, King Abdulaziz University, Jeddah 21589, Saudi Arabia

[2]Centre of Excellence for Advanced Materials Research, King Abdulaziz University, Jeddah 21589, Saudi Arabia

[3]Department of Applied Chemistry, Faculty of Engineering and Technology, Aligarh Muslim University, Aligarh-202 002, India

[4]CAS Key Laboratory of Nanosystem and Hierarchical Fabrication, National Center for Nanoscience and Technology, Beijing 100190, PR China

[5]Department of Electrical Engineering, Mewat College of Engineering and Technology, Mewat-122103, India

Graphene as Energy Storage Material for Supercapacitors   Materials Research Forum LLC
Materials Research Foundations **64** (2020) 1-24   https://doi.org/10.21741/9781644900550-1

Chapter 1

Fabrication Approaches for Graphene in Supercapacitors

Riya Thomas*, B. Manoj

CHRIST (Deemed to be University), Bengaluru-560029, India

*riyapooppillil@gmail.com

Abstract

Supercapacitors are promising candidates with tremendous potential to replace conventional batteries because of their remarkable properties. The exceptional physical, mechanical and chemical features of 2D materials, such as graphene help to utilize the close relationship between the structure and its functional properties to deliver a high performance supercapacitor. The possibilities are vast, if a compatible synthesis route for graphene is feasible. In this chapter, the synthesis approach for graphene, outcomes and challenges involved are discussed.

Keywords

Fabrication, Electric Double Layer Capacitor, Advanced Electrode, Energy Storage, Energy Conversion, Free-Standing Graphene

Contents

1. Introduction

The evolution of the capacitor set off from electrostatic to electrolyte and finally to electric double layer capacitors (EDLCs). Extensive research on improving the performance and reduce cost led to the production of so-called supercapacitors from mid-90s till date. The supercapacitor is an emerging proficient electrochemical device for energy storage with remarkable specific capacitance, elevated power and energy densities, high charge-discharge performance rates, safe operation and long cycle life stability [1-2]. It is identified to be a potential candidate to replace conventional batteries in wearable and foldable devices [3], hybrid electrical vehicles [4], light-weight power sources [5] and smart textiles [6].

A supercapacitor is fabricated with up to two electrodes with a separator in between, an electrolyte and a current collector. Based on the electrode material, supercapacitors can be categorized as electrochemical double layer capacitors (EDLCs) and pseudo-capacitors. EDLCs have carbon electrodes and their capacitances originate from the charge accumulation or interfacial polarization at electrode/electrolyte layer along with the fast exchange kinetics of ions. The control over surface area, porosity and electrical conductivity can significantly enhance the performance of EDLCs. The pseudo-capacitors have the electrodes made out of conductive polymers which yield higher specific capacitance and energy density than that of EDLCs. But the poor long term stability hinders its electrochemical performance.

There is a close relationship between the structure of a molecule and its functional properties. The synthesis of these structures is very demanding, because through proper design the related properties can be tuned from single-molecule level to nanoscale and can achieve multifunctional materials. The 2D material, graphene is well known for its imperative properties like tunable surface area, improved electrical and thermal conductivities, high mechanical strength, excellent chemical stability and light weight. These exceptional physical, chemical and mechanical features which are very essential for the electrochemical energy storage make graphene, a most suitable electrode material for supercapacitors. Though, the perfect graphene sheet has theoretical specific capacitance of $\sim$21 uF cm^{-2}, the practical capacitance is lesser due to the agglomeration of graphene layers during the synthesis [7]. But, the graphene-based nanostructured electrodes with varying dimensions such as 0-D, 1-D, 2-D and 3-D can in turn benefit the electrochemical performance.

This chapter is aimed to analyze the synthesis approaches in achieving graphene-based nanostructures for advanced supercapacitor electrodes, the successful outcomes, challenges involved and the future prospects. The promises are immense if a controllable fabrication route for graphene electrode is feasible.

2. Synthesis of graphene electrodes

Graphene, the monolayer of graphite with sp^2 hybridized carbon assembled in the honeycomb lattice has got wide spectrum of applications from electronics to electrodes due to its impressive inherent characteristics [8,9]. This multi-purpose utilization of graphene has invoked various fabrication routes which can be approached in different outlooks. Based on the viability for the energy storage and generation, the preparation methods of graphene sheets can be classified into the following categories (Fig.1).

   i.   Mechanical exfoliation of graphite

   ii. Chemical exfoliation of graphite

   iii. Reduction of graphene oxide (GO)

   iv. Epitaxial growth

   v.   Chemical vapour deposition (CVD) of graphene on substrates

   vi. Multilayered graphene by arc-discharge method

   vii. Graphene platelets by microwave reactor

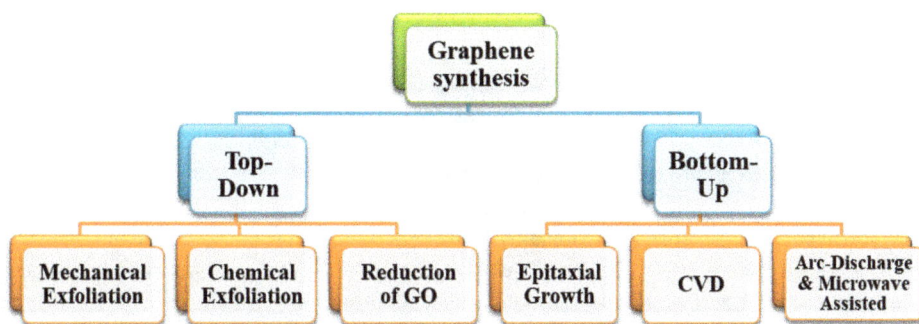

Fig. 1 Synthesis techniques of Graphene

2.1   Mechanical exfoliation

Graphene layers which are piled up by van der Waals forces can be extracted from the bulk material graphite by means of some mechanical methods. It is the first recognized top-down technology for the cleavage of graphene flakes either from a single crystal graphite or natural graphite [10]. An external force of 300 nN/μm^2 necessitates to peel-off graphene layers having interlayer bond energy of 2 eV/nm^2 [11]. The exfoliation is achieved by creating a longitudinal or traverse stress on the exterior of layered material using different agents like atomic force microscopy (AFM) cantilever or scotch tape, electric field and by some printing techniques.

The mechanical exfoliation by simple scotch or (adhesive) tape fabricates few layer (30~layer) graphene [12], offering linear current-voltage (I-V) traits, and enormous sensible currents ($>10^8$ A/cm^2), however it is associated with low yield, thickness and lack of sustainable graphene flakes. As prepared graphene flakes with fewer defects are used for FET devices (Fig.2).

Fig.2 Scanning electron microscope image of Few layer graphene and schematic view of the device.

The electrostatic force assisted exfoliation of graphene (EFEG) detaches the layers of pre-patterned highly ordered pyrolytic graphene (HOPG) by means of high voltage [13]. The features are patterned on the surface by lithographic techniques and are brought into contact with a Si substrate. Exfoliation of pre-patterned layer is then obtained by applying a voltage across HOPG and the substrate. [Fig.3]. The peeled-off graphene layers get stuck on the surface of the substrate without the help of an adhesive layer. But the obtained single graphene layers were of the size ranging from nanometer to few micrometers and chances of contamination due to the material handling are also anticipated [14].

Materials Research Forum LLC
https://doi.org/10.21741/9781644900550-1

Minor scale production and intensive labour invlovement arouse the need for the development of other synthesis techniques.

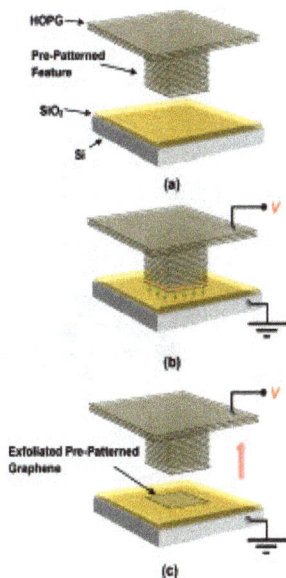

Fig 3 Schematic flowchart of Electrostatic force Assisted exfoliation of graphene

2.2 Chemical exfoliation

A quest for low cost, high yield and purity led to the development of chemical methods for the exfoliation of graphene from graphite. Primarily it forms graphene-intercalated compounds with increased interlayer spacing by reducing interlayer vander Waals force [15]. Then the layers are seperated out either by heating or ultrasonication.

Chemical method is a scalable approach for synthesizing chemically derived graphenes. With the help of Hummers [16] or modified Hummers method [17], graphite is initially oxidised to graphite oxide. But the disadvantage is that the harsh oxidation conditions may cause structural defects [18] and deteriorate electronic structure [19] of graphene as such. The less hydrophobic graphite oxide can easily be exfoliated into thin unifom graphene oxide (GO) layers by using water. The measurement of the zeta potential of GO shows negative charges [20], implying the probability of attaining stable aqeous colloidal dispersions out of them without the need of a foreign stabilizer because of their favourable electrostatic repulsions. Solvents for the homogenous suspensions were decided on the basis of surface energy of graphene and cohesive energy of solvent. The dried GO with layered structure can produce free standing paper-like materials [21-23] suitable for flexible electrodes [Fig.4].

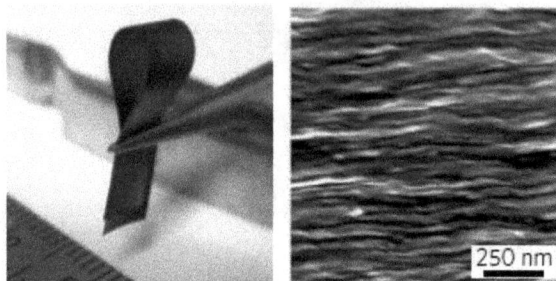

Fig. 4 Folded graphene oxide paper and the SEM image of its layered structure.

Small domains of differently sized and shaped graphene could also be synthesized through cyclodehydrogenation of polycyclic aromatic hydrocarbons [24,25]. Mono or poly dispersed hyperbranched polyphenylenes were identified as ideal precursor for chemical graphene formation (Fig. 5) [26]. Dispersion of these graphene flakes was achieved in various solvents such as orthodichloro benzene [27], chloroform [28], perfluorinated aromatic solvents [29] etc.

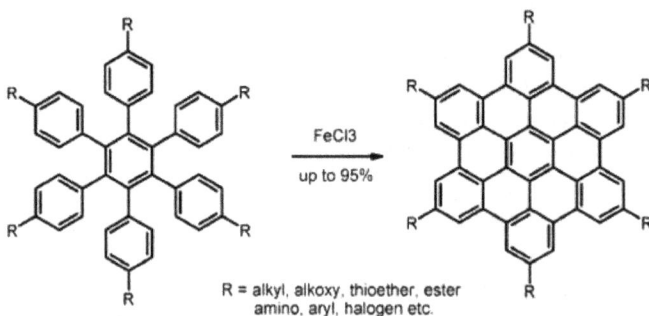

R = alkyl, alkoxy, thioether, ester
amino, aryl, halogen etc.

Fig. 5 Cyclohydrogenation of Polyphenylenes to graphene.

Choice of precursor, the chemical reactions in solvents to obtain graphene and its interactions with functional groups are still a challenge.

2.3   Reduction of GO into graphene

The chemically derived graphene can be obtained by the reduction of GO by using thermal method and reducing agents [30,31] as demonstrated in Fig.6.

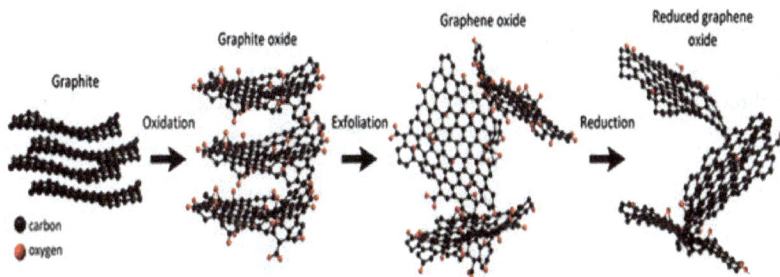

Fig. 6 Synthesis route from graphite to graphene oxide

Reduction of GO by hydrazine is a low-cost and effective method, with improved charge transport mechanism but it produces agglomerated graphene platelets [32-34]. As chemically derived graphene shows good electrical conductivity and large surface area, a solution based reduction was evolved with specific capacitances of 135 and 99 Fg^{-1} in aqueous and organic electrolytes respectively [35]. In order to further lessen the functional groups, the two-step reduction process was extended: deoxygenation with $NaBH_4$ and then dehydration with sulphuric acid. It involves inconsiderate reduction which can destruct the comb-like structure [36].

To reduce the toxicity and restore the graphene structure, green materials like sugar, melatonin, vitamin C, polyphenols of green tea, etc. [37-39] are used as reducing agents. In addition to that green methods like hydrothermal reduction, catalytic and photocatalytic reduction were also employed [40,41]. Reduced graphene oxide by hydrothermal approach yielded a high conductance of 5×10^{-3} S cm^{-1} and a high specific capacitance of 175 F g^{-1} in aqueous electrolyte.

As it is very difficult to dispose off all oxygen from the graphene lattice, there will remain a remanant of C/O in the ratio of 10 [41].

2.4   Epitaxial growth

Epitaxial growth is a different approach of growing graphene on the substrate from the carbon already exists in the substrate. The final layers are reliant only on the factors like temperature, heating rate and pressure, but not on the size of the initial graphitic crystal. A simple heating and cooling of SiC crystal could generate graphene whose growth direction depend essentially on the specific polarity of the crystal [42]. Growth of the graphene is quicker along the C-face than the Si-face, although growth constitute larger disordered domains (Fig.7). Subsequently with the help of the diffusion method, a thin layer of Ni was deposited on the surface of SiC. Depending on the heating rates, carbon diffuse through the Ni layer to form graphene layers and make it easier to detach the layers in contrast to the one grown on simple SiC crystal [43].

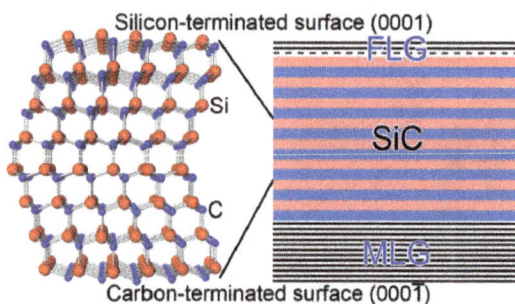

Fig. 7 Graphene grown on the SiC surface. On Si face, few layer graphenes (FLG) and On C surface, multilayer graphenes are deposited.

The feeble interface with the substrate helps to grow high quality graphene wafers on commercially available semi-insulating substrate [44]. The graphene on SiC substrate guarantees high electron mobility and quantum hall effect [45] that confers high performance electronic device applications. As the growth of graphene initiated simultaneously at different locations, layers were not homogenously good for massive production. The need of high temperature annealing at ultrahigh vacuum (UHV) limits its wide opportunities and make the accessibility quite expensive.

2.5   Chemical vapour deposition

Chemical vapour deposition (CVD) is an amenable way for the massive manufacturing of single or few layer graphene. In this process, the gaseous compound which is exposed to the metal substrate gradually decomposes on the surface of substrate to grow a thin film and allowing the by-products to evaporate. The major concern about the thickness of the layers could be then resolved by the possible chemical etching of the substrate in order to transfer high quality graphene from one substrate to other [46]. The turbostratic graphite doesn't have the Bernal stacking and thus the relative motion of layers during etching hardly affect their electronic properties as the layers interrelate marginally. The stamping of these graphene layers on the substrate also lessens the resistance [47]. Thus it reduced the complexity and added the quality of the synthesis technique.

Some of the viable routes for achieving vapour deposition involve heating the sample either with filament or by plasma. Growth of theses graphene layers on a metal substrate is a three-step process [48] as shown schematically in Fig.8.

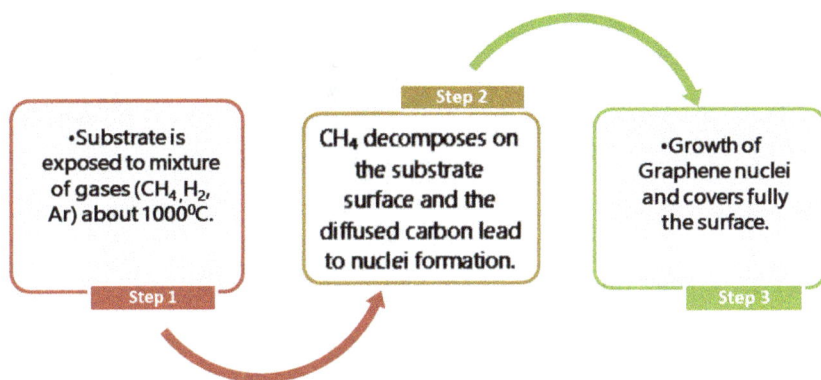

	Step 2	
•Substrate is exposed to mixture of gases (CH$_4$, H$_2$, Ar) about 1000^{0}C.	CH$_4$ decomposes on the substrate surface and the diffused carbon lead to nuclei formation.	•Growth of Graphene nuclei and covers fully the surface.
Step 1		Step 3

Fig. 8 Schematic representation of the steps involved in CVD

Few layer graphene was grown on the Ni surface by decomposition of methane at higher temperature and followed by cooling in argon atmosphere. The number of layers developed was a consequence of the thickness of growing substrate, while the shape of the graphene can be mastered through the preformed pattern on the Ni surface [49]. Graphene layers with substantial surface area have also been fabricated on other metal substrates like Co [50], Pd [51], Pt [52], Ru [53], Re [54], Rh [55], Ir [56] and Au [57]. To get a single layer graphene, a substrate with low solubility of carbon atoms would be of great significance as it offers a path for growth based on the self-limiting process [58-60]. Copper has been identified as an idyllic substrate for the production of single crystal monolayer graphene with large surface area and high quality (Fig.9). Centimeter sized graphene monolayer has been attained on the oxygen passivated Cu foil, which was the largest single crystal of graphene reported so far [61]. The pretreatment of Cu surface with oxygen reduced the graphene nucleation density to depress polycrystallinity. Conventional CVD has got challenges with scaled production of graphene due to convection and radiation loss of heat from the reactor. By connecting the heat source directly to the Cu foil which is termed as rapid heating CVD (RHCVD), it helps to accomplish larger Cu grains for an extensive production for graphene in a shorter time frame [62].

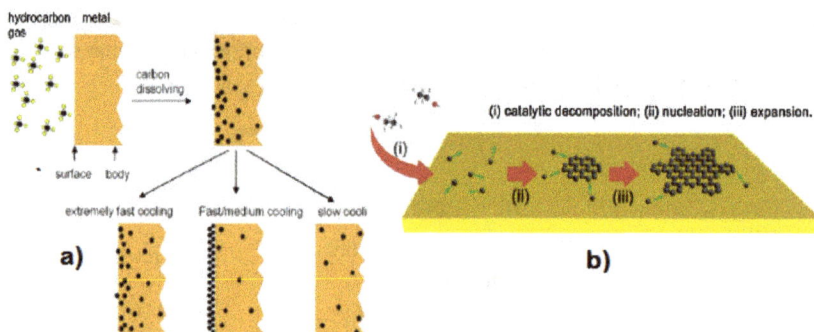

Fig.9 CVD grown graphene a) on a metal substrate b) Cu foil

Binary alloys are another class of substrate that permits uniform formation of graphene layer through controllable carbon solubility by tailoring the atomic fraction of alloy. Cu-Ni alloy [63] and Ni-Mo alloy [64] are well known for the production of chemically stable graphene film in the elevated temperature growth process. But the use of metal catalyst makes the segregation of metal species a complicated and time consuming mechanism. This inadequacy can be overcome by using either non-metal substrate or shielding the catalytic activity of underlying metal. Graphene grown on the insulating substrates such as MgO [65], SiO_2 [66], sapphire [67] and Cu foil shielded with an insulating layer of hexagonal boron-nitride (h-BN) [68] pursued smooth topological surface and progressed quality compared to the one grown on a metal substrate (Fig. 10).

Large crystal domains and fewer defects by the CVD technique, crafts the graphene sheets for electronic and photonic device applications. But, synthesis of graphene at wafer-scale and precise control over the number of layer are yet to be achieved.

Materials Research Foundations **64** (2020) 1-24   https://doi.org/10.21741/9781644900550-1

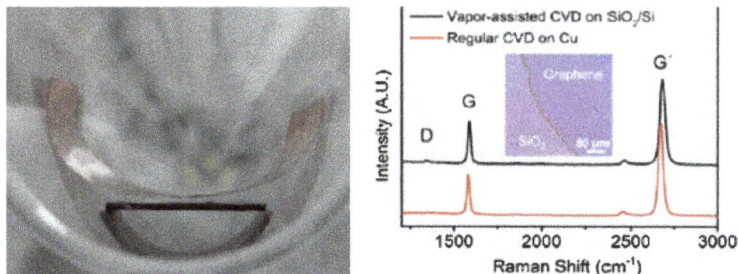

Fig. 10i) View of installed Cu foil and Si/SiO₂ substrate ii) Raman spectra of monolayer Graphene

2.6 Arc discharge method

The electric arc discharge technique is based on the principle that the carbon is evaporated by implementing DC arc voltage across two graphite electrodes in an inert gas atmosphere. The inert gas in the chamber acts as a quencher to promote nucleation and growth of graphene. When the rods are brought together, discharge occurs by consumption of anode leading to plasma formation. Once the discharge ceases, soot is collected from the inner wall of the chamber (Fig.11).

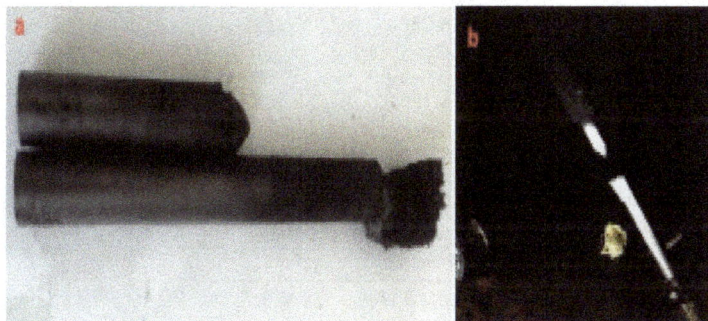

Fig. 11 a) Consumption of anode and formation of Slag b) Inner wall of the reaction chamber.

When the diameter of the electrode increases, the purity increases and the number of layer decreases which is similar to the commercially available graphene [70].

Sublimation of carbon electrode and its deposition on a heated substrate has a correlation with the crystallinity. Graphene synthesized on the Cu substrate at very high temperature near the melting point of copper has relatively high mobility of electron necessary for the production of nanoplates. While, low temperature delivery at the Cu surface yields amorphous carbon structures [71]. The arc discharge method has the sole advantage of producing large scale high quality graphene at low cost. But it is hard to control which is essential for characterization and application.

2.7   Microwave assisted synthesis

Microwave irradiation is a facile processing technique for the rapid reduction of GO into graphene sheets without the need of high temperature and pressure. The difference in the dielectric constants of the reactant and reducing solvent, provide an instantaneous temperature rise of the reactant to initiate exfoliation. GO prepared by Hummers method is mixed with a reducing agent (Ethylene diamine, Hydrazine hydrate) and heated in a conventional microwave to obtain controlled sized and shaped graphene platelets [72]. The use of these highly toxic and carcinogenic reducing agents was expelled by dry microwave treatment of GO to synthesize reduced graphene (rG) (Fig.12) [73].

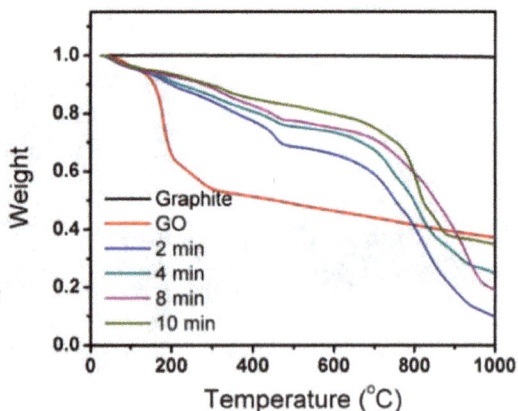

Fig.12 TGA characteristics of sample by microwave irradiation

But preparation of GO via the Hummers method is a time consuming process that also involves environment hazardous oxidizing agents and deadly gases. Microwave irradiation of graphite is an ease, scalable and eco-friendly synthesis operation of graphene in a short span of reaction time [74]. Graphene can directly have been synthesized from graphite with a single reduction of graphene oxide by using microwave assisted solvothermal technique [75]. The graphite was initially treated with a dicarboxylic acid (glutaric acid) to intercalate the graphitic layers. Then the graphene is dispersed in methanol and introduced into the microwave. The produced monolayer retains the original structure without much crystal defects in it. In a similar way, few layer graphene was synthesized using NH_3 as intercalation molecule and dibasic ether (DBE) as the solvent [76].

From the economic and commercial point of view, microwave assisted growth is a rapid and forthcoming method for the production of graphene. Some solvents absorb microwave and thus the careful selection of solvent is a crucial criterion. Heating beyond the boiling point of the solvent can cause evolution of gases to buildup pressure within the system and that requires special attention as well.

Table 1. Overview of the different methods

Method	Quality	Quantity	Size	Complexity	Control
Mechanical Exfoliation	x	✓	x	x	x
Chemical Exfoliation	✓	✓	x	✓	x
Reduction of GO	✓	✓	x	✓	x
Epitaxial Growth	✓	x	x	✓	x
CVD	x	✓	✓	✓	✓
Arc Discharge	✓	✓	✓	x	x
Microwave Assisted	✓	✓	✓	x	✓

2.8　Other methods

There are still further synthesis techniques for graphene which is been evolved during the course of time. It includes Pyrolysis of Sodium Ethoxide [77] and PMMA (Poly Methyl Metacrylate) [78], unzipping of carbon nanotubes [79], hydrothermal synthesis [80], chemical exfoliation of coal [81-82] and conversion of nanodiamond [83].

High purity graphene sheets were easily prepared by pyrolyzing sodium ethoxide [77]. Similarly, gram-scale of multi-layer graphene was fabricated through pyrolization of PMMA [78]. But the quality of graphene was uncertain due to the defects introduced in it. Single layer or multilayer graphene can be etched from single walled or multi-walled carbon nanotubes respectively. Chemical [84], plasma [85] and electrical [86] etching methods are employed for the controlled and defect-free synthesis. Hydrothermal route is a one-pot synthesis mechanism for graphene. Graphene oxide is found to be an efficient precursor for the hydrothermal etching [87], which make it important in the electrical and optical field.

Coal has got preformed graphitic crystalline planes, which can be easily exfoliated compared to that of pure allotropes of carbon. It is a proficient way of producing large magnitude of graphene at low-cost [88]. Graphene that possess desirable properties can be derived from nanodiamond through thermal conversion [83]. Graphene sample own large surface area of $2600 \ m^2 g^{-1}$ and thus exhibit specific capacitance of $35Fg^{-1}$.

Conclusions

The immense demand for next-generation energy storage devices provoked the research community to an intensive research for innovative and advanced electrode materials. The incomparable characteristics of the 2D material, Graphene emphasized its potential purpose of being an electrode in supercapacitors. The macrostructural complexity of graphene led to the progression of numerous synthesis techniques that comprise quantity and quality of the product. Despite the existing method, fabrication of cost-effective and free-standing graphene in larger mass is still a challenge. A clear perception of its chemical structure, physical aspects and reaction mechanism is very essential for the revision of energy storage and conversion in accordance with the current research trends and global demand.

References

[1]   T. Purkait, G. Singh, D. Kumar, M. Singh, R.S. Dey, High-performance flexible supercapacitors based on electrochemically tailored three-dimensional reduced graphene oxide networks, Sci. Rep. 8 (2018) 640. https://doi.org/10.1038/s41598-017-18593-3.

[2]   H. Yang, S. Kannappan, A.S. Pandian, J.H. Jang, Y.S. Lee, W. Lu, Graphene supercapacitor with both high power and energy density, Nanotechnology 44 (2017) 445401. https://doi.org/10.1088/1361-6528/aa8948.

[3]   M. Kaempgen, C.K. Chan, J. Ma, Y. Cui, G. Gruner, Printable thin film supercapacitors using single-walled carbon nanotubes, Nano Lett. 5 (2009) 1872-1876. https://doi.org/10.1021/nl8038579

[4]   Y. Lei, Z.H. Huang, Y. Yang, W. Shen, Y. Zheng, H. Sun, F. Kang, Porous mesocarbon microbeads with graphitic shells: constructing a high-rate, high-capacity cathode for hybrid supercapacitor, Sci. Rep. 3 (2013) 2477. https://doi.org/10.1038/srep02477

[5]   D. Yu, K. Goh, H. Wang, L. Wei, W. Jiang, Q. Zhang, L. Dai, Y. Chen, Scalable synthesis of hierarchically structured carbon nanotube–graphene fibres for capacitive energy storage, Nat. Nanotechnol. 7 (2014) 555-562. https://doi.org/10.1038/nnano.2014.93

[6]   S. Zhai, W. Jiang, L. Wei, H.E. Karahan, Y. Yuan, A.K. Ng, Y. Chen, All-carbon solid-state yarn supercapacitors from activated carbon and carbon fibers for smart textiles, Mater. Horiz. 6 (2015) 598-605. https://doi.org/10.1039/C5MH00108K

[7]   J. Xia, F. Chen, J. Li, N. Tao, Measurement of the quantum capacitance of graphene, Nat. Nanotechnol. 8 (2009) 505-509. https://doi.org/10.1038/nnano.2009.177

[8]   T. Upreti, V. Gupta, S. Chand, Graphene-Based Solar Cells, In Graph. Sci. Handb, CRC press (2016) 376-340.

[9]   X. Wang, I. Zhi, K. Müllen, Transparent, conductive graphene electrodes for dye-sensitized solar cells, Nano Lett. 1 (2008) 323-327. https://doi.org/10.1021/nl072838r

[10] K.S. Novoselov, D. Jiang, F. Schedin, T.J. Booth, V.V. Khotkevich, S.V. Morozov, A.K. Geim, Two-dimensional atomic crystals, PNAS 30 (2005) 10451-10453. https://doi.org/10.1073/pnas.0502848102

[11] Y. Zhang, J.P. Small, W.V. Pontius, P. Kim, Fabrication and electric-field dependent transport measurements of mesoscopic graphite devices, Appl. Phys. Lett. 86 (2005) 073104. https://doi.org/10.1063/1.1862334

[12] K.S. Novoselov, A.K. Geim, S.V. Morozov, D. Jiang, Y. Zhang, S.V. Dubonos, I.V. Grigorieva, A.A. Firsov, Electric field effect in atomically thin carbon films, Science 306 (2004) 666–669. DOI: 10.1126/science.1102896

[13] N.A. Sidorov, M.M. Yazdanpanah, R. Jalilian, P.J. Ouseph, R.W. Cohn, G.U. Sumanasekera, Electrostatic deposition of graphene, Nanotechnology 13 (2007) 135301. https://doi.org/10.1088/0957-4484/18/13/135301

[14] X. Liang, A.S.P. Chang, Y. Zhang, B.D. Harteneck, H. Choo, D.L. Olynick, S. Cabrini, Electrostatic force assisted exfoliation of prepatterned few-layer graphenes into device sites, Nano Lett. 1 (2008) 467-472. https://doi.org/10.1021/nl803512z

[15] Y.H. Wu, T. Yu, Z.X. Shen, Two-dimensional carbon nanostructures: fundamental properties, synthesis, characterization, and potential applications. J. Appl. Phys. 108, (2010) 071301. https://doi.org/10.1063/1.3460809

[16] W.S. Hummers, R.E. Offeman, Preparation of graphitic oxide, J. Am. Chem. Soc. 6 (1958) 1339-1339. https://doi.org/10.1021/ja01539a017

[17] N. I. Kovtyukhova, P.J. Ollivier, B.R. Martin, T.E. Mallouk, S.A. Chizhik, E.V. Buzaneva, A.D. Gorchinskiy, Layer-by-layer assembly of ultrathin composite films from micron-sized graphite oxide sheets and polycations, Chem. Mater. 11 (1999) 771–778. https://doi.org/10.1021/cm981085u

[18] G. Eda, G. Fanchini, M. Chhowalla, Large-area ultrathin films of reduced graphene oxide as a transparent and flexible electronic material, Nat. Nanotechnol. 5 (2008) 270-274. https://doi.org/10.1038/nnano.2008.83

[19] I. Jung, M. Pelton, R. Piner, D.A. Dikin, S. Stankovich, S. Watcharotone, M. Hausner, R.S. Ruoff, Simple approach for high-contrast optical imaging and characterization of graphene-based sheets, Nano Lett. 12 (2007) 3569-3575. https://doi.org/10.1021/nl0714177

[20] D. Li, M.B. Müller, S. Gilje, R.B. Kaner, G.G. Wallace. Processable aqueous dispersions of graphene Nanosheets, Nat. Nanotechnol. 2 (2008) 101-105. https://doi.org/10.1038/nnano.2007.451

[21] S. Park, J. An, R.D. Piner, I. Jung, D. Yang, A. Velamakanni, S.T. Nguyen, R.S. Ruoff, Aqueous suspension and characterization of chemically modified graphene sheets, Chem. Mater. 21 (2008) 6592-6594. https://doi.org/10.1021/cm801932u

[22] H. Chen, M.B. Müller, K.J. Gilmore, G.G. Wallace, D. Li. Mechanically strong, electrically conductive, and biocompatible graphene paper, Adv. Mater. 18 (2008) 3557-3561. https://doi.org/10.1002/adma.200800757

[23] D.A. Dikin, S. Stankovich, E.J. Zimney, R.D. Piner, G.H.B. Dommett, G. Evmenenko, S.T. Nguyen, R.S. Ruoff, Preparation and characterization of graphene oxide paper, Nature 7152 (2007) 457-460. https://doi.org/10.1038/nature06016

[24] X. Fan, W. Peng, Y. Li, X. Li, S. Wang, G. Zhang, F. Zhang, Deoxygenation of exfoliated graphite oxide under alkaline conditions: a green route to graphene preparation, Adv. Mater. 23 (2008) 4490-4493.https://doi.org/10.1002/adma.200801306

[25] J. Wu, W. Pisula, K. Mu"llen, Graphenes as potential material for electronics, Chem. Rev. 107 (2007) 718–747. https://doi.org/10.1021/cr068010r

[26] J. Cai, P. Ruffieux, R.B. Jaafar, T. Braun, S. Blankenburg, M. Matthias, A.P. Seitsonen, S. Moussa, X. Feng, K. Mu"llen, R. Fasel, Atomically precise bottom-up fabrication of graphene nanoribbons, Nature 466 (2010) 470–473. https://doi.org/10.1038/nature09211

[27] L. Zhi, K. Müllen, A bottom-up approach from molecular nanographenes to unconventional carbon materials, J. Mater. Chem. 13 (2008) 1472-1484. https://doi.org/10.1039/B717585J

[28] C.E. Hamilton, J. R. Lomeda, Z. Sun, J. M. Tour, A.R. Barron, High-yield organic dispersions of unfunctionalized graphene, Nano Lett. 10 (2009) 3460-3462. https://doi.org/10.1021/nl9016623

[29] A. O'Neill, U. Khan, P.N. Nirmalraj, J. Boland, J. N. Coleman, Graphene dispersion and exfoliation in low boiling point solvents, J. Phys. Chem.13 (2011) 5422-5428. https://doi.org/10.1021/jp110942e

Materials Research Forum LLC
https://doi.org/10.21741/9781644900550-1

[30] A.B. Bourlinos, V. Georgakilas, R. Zboril, T.A. Steriotis, A.K. Stubos, Liquid-phase exfoliation of graphite towards solubilized graphenes, Small 16 (2009) 1841-1845. https://doi.org/10.1002/smll.200900242

[31] Q. Ke, Y. Liu, H. Liu, Y. Zhang, Y. Hu, J. Wang, Surfactant-modified chemically reduced graphene oxide for electrochemical supercapacitors, RSC Adv. 50 (2014) 26398-26406. https://doi.org/10.1039/C4RA03826F

[32] S. Park, R.S. Ruoff, Chemical methods for the production of graphenes, Nat. Nanotechnol. 4 (2009) 217-224. https://doi.org/10.1038/nnano.2009.58

[33] E.J.C Amieva, J.L. Barroso, A.L.M. Hernández, C.V. Santos, Graphene-based materials functionalization with natural polymeric biomolecules, In recent advances in graphene research. InTech, 2016.DOI:10.5772/64001

[34] S. Park, J. An, J.R. Potts, A. Velamakanni, S. Murali, R.S. Ruoff, Hydrazine-reduction of graphite-and graphene oxide, Carbon 9 (2011) 3019-3023. https://doi.org/10.1016/j.carbon.2011.02.071

[35] R. Kumar, A. Kaur, Charge transport mechanism of hydrazine hydrate reduced graphene oxide, IET Circuits Devices & Systems 6 (2015) 392-396. https://doi.org/10.1049/iet-cds.2015.0034

[36] M.D. Stoller, S. Park, Y. Zhu, J. An, R.S. Ruoff, Graphene-based ultracapacitors, Nano Lett. 10 (2008) 3498-3502. https://doi.org/10.1021/nl802558y

[37] W. Gao, L.B. Alemany, L. Ci, P.M. Ajayan, New insights into the structure and reduction of graphite oxide, Nat. Chem. 5 (2009) 403-408. https://doi.org/10.1038/nchem.281

[38] C. Zhu, S. Guo, Y. Fang, S. Dong, Reducing sugar: new functional molecules for the green synthesis of graphene Nanosheets, ACS Nano 4 (2010) 2429-2437. https://doi.org/10.1021/nn1002387

[39] M.J. Fernández-Merino, L. Guardia, J.I. Paredes, S. Villar-Rodil, P. Solís-Fernández, A. Martínez-Alonso, J.M.D. Tascon, Vitamin C is an ideal substitute for hydrazine in the reduction of graphene oxide suspensions, J. Phys. Chem. C 14 (2010) 6426-6432. https://doi.org/10.1021/jp100603h

[40] O. Akhavan, M. Kalaee, Z.S. Alavi, S.M.A. Ghiasi, A. Esfandiar, Increasing the antioxidant activity of green tea polyphenols in the presence of iron for the reduction of graphene oxide, Carbon 50 (2012) 3015-3025. https://doi.org/10.1016/j.carbon.2012.02.087

[41] G. Williams, B. Seger, P.V. Kamat, TiO_2-graphene nanocomposites UV assisted photocatalytic reduction of graphene oxide, ACS Nano 2 (2008) 1487-1491. https://doi.org/10.1021/nn800251f

[42] W. Shi, J. Zhu, D.H. Sim, H.H. Tay, Z. Lu, X. Zhang, Y. Sharma, M. Srinivasan, H. Zhang, H.H. Hnga, Q. Yan, Achieving high specific charge capacitances in Fe3O4/reduced graphene oxide nanocomposites, J Mater Chem. 21 (2011) 3422-3427. https://doi.org/10.1039/C0JM03175E

[43] I. Forbeaux, J.M. Themlin, J.M. Debever, Heteroepitaxial graphite on 6 H−SiC (0001): Interface formation through conduction-band electronic structure, Phys. Rev. B 24 (1998) 16396. https://doi.org/10.1103/PhysRevB.58.16396

[44] C. Enderlein, Graphene and its interaction with different substrates studied by angular-resolved photoemission spectroscopy, PhD diss., Freie Universität Berlin, 2010.

[45] W. Norimatsu, M. Kusunoki, Epitaxial graphene on SiC {0001}: Advances and perspectives, Phys. Chem. Chem. Phys. 8 (2014) 3501-3511. https://doi.org/10.1039/C3CP54523G

[46] S. Tanabe, Y. Sekine, H. Kageshima, M. Nagase, H. Hibino, Carrier transport mechanism in graphene on SiC (0001), Phys. Rev. 11 (2011) 115458-115463. https://doi.org/10.1103/PhysRevB.84.115458

[47] H.R. Byon, S.W. Lee, S. Chen, P.T. Hammond, S.Y. Horn, Thin films of carbon nanotubes and chemically reduced graphenes for electrochemical microcapacitors, Carbon 49 (2011) 457-467. https://doi.org/10.1016/j.carbon.2010.09.042

[48] S. Bae, H. Kim, Y. Lee, X. Xu, J.S. Park, Y. Zheng, J. Balakrishnan, T. Lei, R. Kim, Y.I. Song, Y.J. Kim, Roll-to-roll production of 30-inch graphene films for transparent electrodes, Nat. Nanotechnol. 5 (2010) 574-578. https://doi.org/10.1038/nnano.2010.132

[49] W.K. Chee, H.N. Lim, Z. Zainal, N.M. Huang, I. Harrison, Y. Andou, Flexible graphene based supercapacitors: a review, J. Phys. Chem. C 120 (2016) 4153-4172.https://doi.org/10.1021/acs.jpcc.5b10187

[50] K.S. Kim, Y. Zhao, H. Jang, S. Y. Lee, J. M. Kim, K.S. Kim, J.H. Ahn, P. Kim, J.Y. Choi, B.H. Hong, Large-scale pattern growth of graphene films for stretchable transparent electrodes, Nature 7230 (2009) 706-710. https://doi.org/10.1038/nature07719

[51] M.E. Ramón, A. Gupta, C. Corbet, D. A. Ferrer, H.C.P Movva, G. Carpenter, L. Colombo, CMOS-compatible synthesis of large-area, high-mobility graphene by chemical vapor deposition of acetylene on cobalt thin films, ACS Nano 9 (2011) 7198-7204. https://doi.org/10.1021/nn202012m

[52] S.Y. Kwon, C.V. Ciobanu, V. Petrova, V.B. Shenoy, J. Bareno, V. Gambin, I. Petrov, S. Kodambaka, Growth of semiconducting graphene on palladium, Nano Lett. 12 (2009) 3985-3990. https://doi.org/10.1021/nl902140j

[53] B.J. Kang, J.H. Mun, C.Y. Hwang, B.J. Cho, Monolayer graphene growth on sputtered thin film platinum, J. App. Phys. 10 (2009)104309. https://doi.org/10.1063/1.3254193

[54] Y. Pan, H. Zhang, D. Shi, J. Sun, S. Du, F. Liu, H.J. Gao, Highly ordered, millimeter scale, continuous, single-crystalline graphene monolayer formed on Ru (0001), Adv. Mater. 27 (2009) 2777-2780. https://doi.org/10.1002/adma.200800761

[55] E. Miniussi, M. Pozzo, T.O. Menteş, M.A. Niño, A. Locatelli, E. Vesselli, G. Comelli, S. Lizzit, D. Alfè, A. Baraldi, The competition for graphene formation on Re (0 0 0 1): A complex interplay between carbon segregation, dissolution and carburization, Carbon 73 (2014) 389-402. https://doi.org/10.1016/j.carbon.2014.02.081

[56] A.B. Preobrajenski, M.L. Ng, A.S. Vinogradov, N. Mårtensson, Controlling graphene corrugation on lattice-mismatched substrates, Phys. Rev. B 78 (2008) 073401-173405. https://doi.org/10.1103/PhysRevB.78.073401

[57] J. Coraux, T.N. Alpha 'Diaye, C. Busse, T. Michely, Structural coherency of graphene on Ir (111), Nano Lett. 2 (2008) 565-570. https://doi.org/10.1021/nl0728874

[58] T. Oznuluer, E. Pince, E. O. Polat, O. Balci, O. Salihoglu, C. Kocabas, Synthesis of graphene on gold, App. Phys. Lett. 18 (2011) 183101. https://doi.org/10.1063/1.3584006

[59] X. Li, W. Cai, J. An, S. Kim, J. Nah, D. Yang, R. Piner, Large-area synthesis of high-quality and uniform graphene films on copper foils, Science 5932 (2009) 1312-1314. https://doi.org/10.1126/science.1171245

[60] Y. Hao, M.S. Bharathi, L. Wang, Y. Liu, H. Chen, S. Nie, X. Wang, The role of surface oxygen in the growth of large single-crystal graphene on copper, Science 6159 (2013) 720-723. https://doi.org/10.1126/science.1243879

[61] Q. Yu, J. Lian, S. Siriponglert, H. Li, Y.P. Chen, S.S. Pei, Graphene segregated on Ni surfaces and transferred to insulators, App. Phys. Lett. 11 (2008) 113103. https://doi.org/10.1063/1.2982585

[62] P. Zhao, A. Kumamoto, S. Kim, X. Chen, B. Hou, S. Chiashi, E. Einarsson, Y. Ikuhara, S. Maruyama, Self-limiting chemical vapor deposition growth of monolayer graphene from ethanol, J. Phys. Chem. C 20 (2013) 10755-10763. https://doi.org/10.1021/jp400996s

[63] S. M. Kim, J.H. Kim, K.S. Kim, Y. Hwangbo, J.H. Yoon, E.K. Lee, J. Ryu, H. Joo Lee, S. Cho, S.M. Lee, Synthesis of CVD-graphene on rapidly heated copper foils, Nanoscale 9 (2014) 4728-4734. https://doi.org/10.1039/C3NR06434D

[64] X. Liu, L. Fu, N. Liu, T. Gao, Y. Zhang, L. Liao, Z. Liu, Segregation growth of graphene on Cu–Ni alloy for precise layer control, J. Phys. Chem. C 24 (2011) 11976-11982. https://doi.org/10.1021/jp202933u

[65] B. Dai, L. Fu, Z. Zou, M. Wang, H. Xu, S. Wang, Z. Liu, Rational design of a binary metal alloy for chemical vapour deposition growth of uniform single-layer graphene, Nat. Commun. 2 (2011) 522. https://doi.org/10.1038/ncomms1539

[66] M.H. Rummeli, A. Bachmatiuk, A. Scott, F. Borrnert, J.H. Warner, V. Hoffman, J.H. Lin, G. Cuniberti, B. Buchner, Direct low-temperature nanographene CVD synthesis over a dielectric insulator, ACS Nano 7 (2010) 4206-4210. https://doi.org/10.1021/nn100971s

[67] H. Medina, Y.C. Lin, C. Jin, C.C Lu, C.H. Yeh, K.P. Huang, K. Suenaga, J. Robertson, P.W. Chiu, Metal free growth of nanographene on silicon oxides for

transparent conducting applications, Adv. Funct. Mater. 10 (2012) 2123-2128. https://doi.org/10.1002/adfm.201102423

[68] H.J. Song, M. Son, C. Park, H. Lim, M.P. Levendorf, A.W. Tsen, J. Park, H.C. Choi, Large scale metal-free synthesis of graphene on sapphire and transfer-free device fabrication, Nanoscale 10 (2012) 3050-3054. https://doi.org/10.1039/C2NR30330B

[69] M. Wang, S.K. Jang, W.J. Jang, M. Kim, S.Y. Park, S.W. Kim, S.J. Kahng, A Platform for Large scale graphene electronics–CVD growth of single layer graphene on CVD grown hexagonal boron nitride, Adv. Mater. 19 (2013) 2746-2752. https://doi.org/10.1002/adma.201204904

[70] H. Kim, I. Song, C. Park, M. Son, M. Hong, Y. Kim, J.S. Kim, H.J. Shin, J. Baik, H.C. Choi, Copper-vapor-assisted chemical vapor deposition for high-quality and metal-free single-layer graphene on amorphous SiO_2 substrate, ACS Nano 8 (2013) 6575-6582. https://doi.org/10.1021/nn402847w

[71] U. Cotul, E.D.S. Parmak, C. Kaykilarli, O. Saray, O. Colak, D. Uzunsoy, Development of High Purity, Few-Layer Graphene synthesis by electric arc discharge technique, Acta Phys. Polo.134 (1) (2018). https://doi.org/10.12693/APhysPolA.134.289

[72] X. Fang, A. Shashurin, M. Keidar, Role of substrate temperature at graphene synthesis in an arc discharge, J. App. Phys. 10 (2015) 103304. https://doi.org/10.1063/1.4930177

[73] H.M.A. Hassan, V. Abdelsayed, S.K.A. El Rahman, K.M. AbouZeid, J. Terner, M. S. El-Shall, S.I. Al-Resayes, A.A. El-Azhary, Microwave synthesis of graphene sheets supporting metal nanocrystals in aqueous and organic media, J. Mater. Chem. 23 (2009) 3832-3837. https://doi.org/10.1039/B906253J

[74] Z. Li, Y. Yao, Z. Lin, K.S. Moon, W. Lin, C. Wong, Ultrafast dry microwave synthesis of graphene sheets, J. Mater. Chem. 23 (2010) 4781-4783. https://doi.org/10.1039/C0JM00168F

[75] X. Wang, H. Tang, S. Huang, L. Zhu, Fast and facile microwave-assisted synthesis of graphene oxide Nanosheets, RSC Adv. 104 (2014) 60102-60105. https://doi.org/10.1039/C4RA12022A

[76] F.S. Al Hazmi, G.H. Al-Harbi, G.W. Beall, A.A. Al-Ghamdi, A.Y. Obaid, W.E. Mahmoud, One pot synthesis of graphene based on microwave assisted solvothermal

22

technique, Synth. Met. 200 (2015) 54-57.
https://doi.org/10.1016/j.synthmet.2014.12.028

[77] F. Jiang, Y. Yu, Y. Wang, A. Feng, L. Song, A novel synthesis route of graphene via microwave assisted intercalation-exfoliation of graphite, Mater. Lett. 200 (2017) 39-42. https://doi.org/10.1016/j.matlet.2017.04.048

[78] M. Choucair, P. Thordarson, J.A. Stride, Gram-scale production of graphene based on solvothermal synthesis and sonication, Nat. Nanotechnol. 4 (2009) 30–33. https://doi.org/10.1038/nnano.2008.365

[79] N. Hong, W. Yang, C. Bao, S. Jiang, L. Song, Y. Hu, Facile synthesis of graphene by pyrolysis of poly (methyl methacrylate) on nickel particles in the confined microzones, Mater. Res. Bull.12 (2012) 4082-4088. https://doi.org/10.1016/j.materresbull.2012.08.049

[80] L. Jiao, L. Zhang, X. Wang, G. Diankov, H. Dai, Narrow graphene nanoribbons from carbon nanotubes, Nature 7240 (2009) 877-880.https://doi.org/10.1038/nature07919

[81] J. Shen, Y. Zhu, X. Yang, J. Zong, J. Zhang, C. Li, One-pot hydrothermal synthesis of graphene quantum dots surface-passivated by polyethylene glycol and their photoelectric conversion under near-infrared light, New J. Chem. 1 (2012) 97-101. https://doi.org/10.1039/C1NJ20658C

[82] R. Thomas, E. Jayaseeli, N.M.S. Sharma, B. Manoj, Opto-electric property relationship in phosphorus embedded nanocarbon, Results Phys. 10 (2018) 633-639. https://doi.org/10.1016/j.rinp.2018.07.018

[83] R. Ye, C. Xiang, J. Lin, Z. Peng, K. Huang, Z. Yan, N.P. Cook, Coal as an abundant source of graphene quantum dots, Nat. Commun. 4 (2013) 2943. https://doi.org/10.1038/ncomms3943

[84] K.S. Subrahmanyam, S.R.C. Vivekchand, A. Govindaraj, C.N.R. Rao, A study of graphenes prepared by different methods: characterization, properties and solubilization, J. Mater. Chem. 13 (2008) 1517-1523. https://doi.org/10.1039/B716536F

[85] A.G. Cano-Marquez, F.J. Rodríguez-Macías, J.C. Delgado, C.G. Espinosa-González, F. Tristán-López, D. Ramírez-González, D.A. Cullen, D.J. Smith, M. Terrones, Y.I. Vega-Cantú, Ex-MWNTs: graphene sheets and ribbons produced by

Materials Research Forum LLC
https://doi.org/10.21741/9781644900550-1

lithium intercalation and exfoliation of carbon nanotubes, Nano Lett. 4 (2009) 1527-1533. https://doi.org/10.1021/nl803585s

[86] S. Mohammadi, Z. Kolahdouz, S. Darbari, S. Mohajerzadeh, N. Masoumi, Graphene formation by unzipping carbon nanotubes using a sequential plasma assisted processing, Carbon 52 (2013) 451-463.https://doi.org/10.1016/j.carbon.2012.09.056

[87] K. Kim, A. Sussman, A. Zettl, Graphene nanoribbons obtained by electrically unwrapping carbon nanotubes, ACS Nano 3 (2010) 1362-1366.https://doi.org/10.1021/nn901782g

[88] Y. Zhu, G. Wang, H. Jiang, L. Chen, X. Zhang, One-step ultrasonic synthesis of graphene quantum dots with high quantum yield and their application in sensing alkaline phosphatase, Chem. Comm. 5 (2015) 948-951. https://doi.org/10.1039/C4CC07449A

Graphene as Energy Storage Material for Supercapacitors   Materials Research Forum LLC
Materials Research Foundations **64** (2020) 25-62   https://doi.org/10.21741/9781644900550-2

Chapter 2

Correlation between Synthesis and Properties of Graphene

Praveen Mishra, Badekai Ramachandra Bhat*

Catalysis and Materials Laboratory, Department of Chemistry, National Institute of Technology Karnataka, Surathkal, Mangalore - 575025, Karnataka, India

*ram@nitk.edu.in

Abstract

The discovery of a modest way to obtain graphene has led to an exponential rise in the development of materials for the electrodes of supercapacitors. The fabrication of graphene since then has come a long way. The synthesis of graphene using physical and chemical methods completes the top down and bottom up sets for this wonder material. In this chapter, we will review the various methods for the preparation and their implications on the properties of graphene and subsequently the effects on the parameters of fabricated device.

Keywords

Graphene, 2D Nanoparticles, Top-down Method, Bottom up Method, Supercapacitor

Contents

1.   Introduction

Energy is a vital resource for human development. As of today, the energy consumption and production are heavily relying on fossil fuels. This dependence if not addressed in time, will severely affect the world economy and ecology. Therefore, there is a high demand to fabricate high performance renewable energy storage and generation devices which are environment-friendly. Electrochemical energy is an obvious part of the clean energy group. Energy conversion and storage devices such as electrochemical cells, fuel cells, and supercapacitors are among the popular ones to harness the workings of the principle of electrochemical energy conversion. Among these devices, supercapacitors owing to exceptionally high specific capacitance and longevity are gaining widespread attention for electronic applications[1]. One of the salient features of the supercapacitors is their ability to connect the large voids which exist between the power and energy densities of conventional capacitors (high power density), and batteries and fuel cells (low power density) [2] as represented in the Ragone plot (Fig. 1). The figure clearly depicts the overlap of the power density of the supercapacitors with conventional capacitors while their energy density coincides with batteries and fuel cells. However, as per current research progress, there is no electrochemical device which can contend an internal combustion engine in terms of energy density. Therefore, the energy and power densities of electrochemical devices need to be increased to compete with combustion engines [2]. Other noted advantages of supercapacitors include very low maintenance operation, absence of memory effect, and non-explosive nature (unlike Li-ion batteries) [3-5].

Graphene as Energy Storage Material for Supercapacitors   Materials Research Forum LLC

Materials Research Foundations **64** (2020) 25-62   https://doi.org/10.21741/9781644900550-2

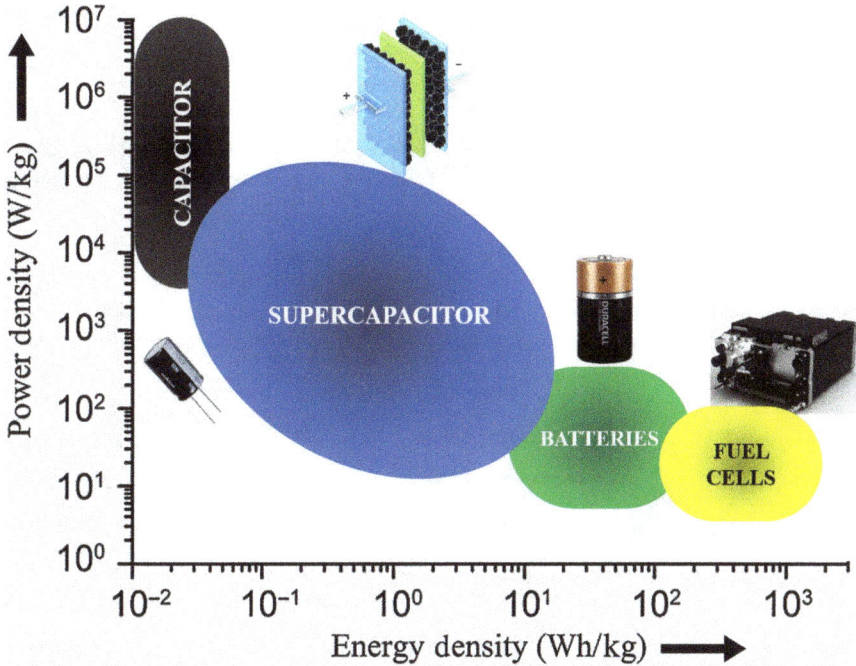

Figure 1. The positions of various electrochemical energy conversion systems on a Ragone plot [6].

Supercapacitors are largely aimed to be employed as power sources for various personal electronic gadgets like cellphones, notebooks, cameras, etc. [7] Supercapacitors are ambitiously looked upon as an alternative solution for energy need of rural section of the country where power grids are either not available or carry heavy cost of installations [1]. Additionally, the high power densities of the supercapacitors make them an ideal component for electric and hybrid vehicles [8]. A figurative depiction of the applications of the supercapacitors is shown in Figure 2.

The initiation of the capacitor technology is credited to Ewald Georg von Kleist and Musschenbroek for inventing the Leyden jar in 1745. However, the modern electrolytic capacitors may trace their origins in the patent filed by Samuel Ruben in 1925 [9]. The first supercapacitor was patented in the year 1957 by General Electric using activated carbon as electrodes [10]. This was the first report on the electric double layer capacitors

(EDLC) which is the continuing architecture of most of the reported supercapacitors made with carbonaceous materials. The electrostatic or non-Faradic charge storage which involves no migration of charge between electrodes and within the electrolyte is the primary mechanism involved in an EDLC [11]. This phenomenon makes them reversible with significantly high cyclic stability as compared to conventional batteries. Recently, the carbon nanomaterials such as activated carbons, carbon nanotubes (CNT), carbon aerogels, carbide-derived carbon, and graphene are the most preferred structures for EDLC [12]. These materials are beneficial due to their huge specific surface area (SSA), excellent electrical conductivity, and superior mechanical properties and chemical stability which make them suitable for electrode construction. The specific capacitance of the supercapacitors has been increased by developing new electrochemically active materials to fabricate the devices classified as pseudocapacitors and based on the Faradaic charge transfer [13].

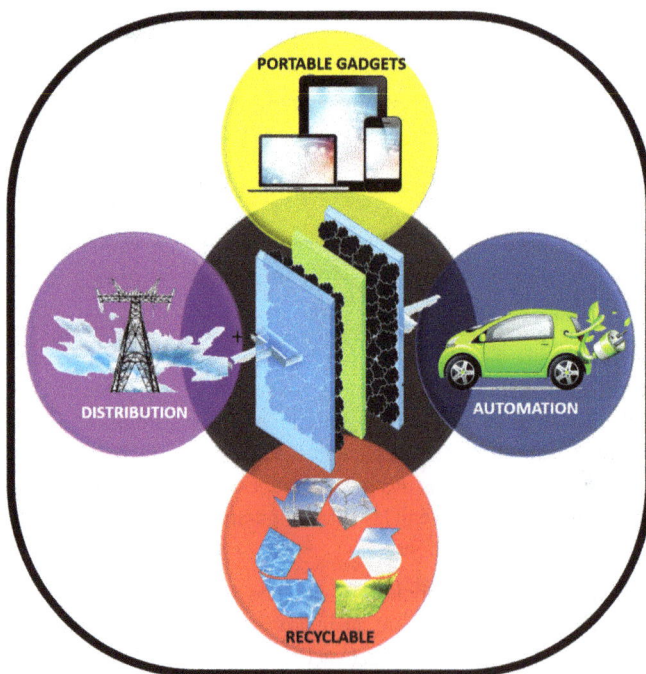

Figure 2.   Applications of Supercapacitors

Among various materials explored for supercapacitors, graphene is arguably the most interesting one. It is 2-dimensional stretch of sp^2-hybridized carbon with single atomic layer thickness [14]. Few of the fascinating properties of graphene are lightweight, high thermal as well as electrical conductivities, strong mechanical strength (~1 TPa), chemical resistance, and easily tunable surface area [15-18]. Therefore, graphene and graphene-based materials are considered as high-performance structural nanocomposites and materials for electronics, mimetics, environmental protection, and energy devices [19-25]. Moreover, these exceptional properties have led the use of graphene and its composites prevalent in sustainable electrochemical energy storage and generation devices such as Li-ion batteries, fuel cells, supercapacitors, and photovoltaics [11, 26-28].

Single layer graphene is pitched to have a theoretical capacitance as high as 550 Fg^{-1} [18, 29]. However, the capacitive behavior of graphene and graphene-based materials when evaluated for usable device construction has been found much inferior than the expected value because of the self-accumulation and restacking of graphene sheets [30-33]. Therefore, enhancing the overall electrochemical properties of graphene-based materials still remains one of the obstacles to achieve desirable device performance. However, hybrid route for achieving higher capacitance with graphene-based materials has been tried by incorporation of materials to induce pseudocapacitive behavior [34]. Such graphene-based materials as conducting network have been prepared by the introduction of redox behavior of metal oxides, hydroxides, hydrides, and conducting polymers for better specific capacitance [35]. These hybrid electrodes exhibit superior electrochemical properties as a result of the synergistic effect of graphene and dispersed nanoparticles used as filler to avail desired pseudo capacitance.

In this present chapter, we would focus on numerous ways to prepare graphene and related materials to be used in a supercapacitor device.

2.   Synthesis of Graphene

The interest in producing a single atomic layer of graphite or now called "Graphene" can be traced as early as 1947 when P. Wallace gave a theoretical accord of the properties of a monolayer graphene [36]. An year later, Ruess and Vogt published the first transmission electron micrograph of graphene, prepared by the thermal reduction of graphene oxide in a stream of hydrogen [37]. Foundation of this may be assumed to be now famous Hummer's method which is used to prepare graphene oxide from graphite [38]. In another report by Hofmann and co-workers in 1962, they reported isolated free graphene by reducing graphene oxide with hydrazine [39, 40]. The term graphene, however was coined by Mouras et al. in 1987 [41, 42]. Graphene later was rediscovered by Geim and Novoselov in 2004 [14] after which the materials have taken over the interest of researchers for a wide array of applications like a fish to the water [23].

As is the case of any nanomaterials, synthesis of graphene may be classified into two general categories (Fig. 3).

 1. Top Down Approach: Synthesis of graphene from larger materials such as graphite.

 2. Bottom Up Approach: Synthesis of graphene from simple reactants like acetylene.

Following sections discuss each of the synthetic routes in detail.

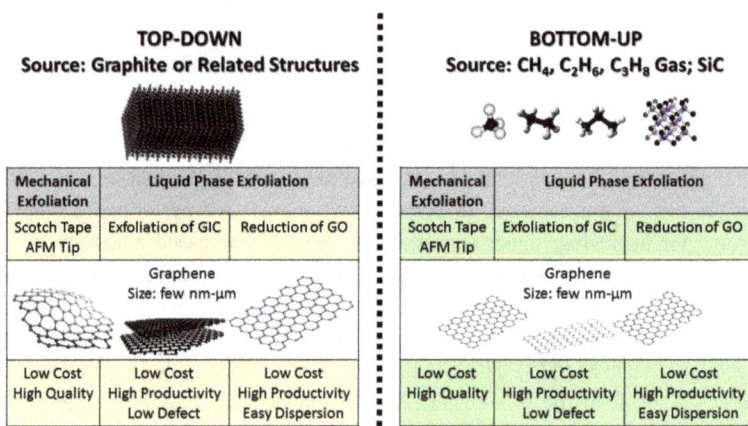

Figure 3. Schematic depiction of the top-down approach and bottom-up approach for synthesis of graphene[43].

2.1 Top-down approach

Graphite, arguably the most used precursor for synthesizing graphene is a piled-up arrangement of graphene layers held by the Van-der Waals forces. Therefore, it is imperative to overcome these Van-der Waals forces in order to separate these layers [44]. This is a simple example of a top-down approach. This approach is associated with several intrinsic challenges such as the occurrence of surface defects accompanied during the separation of sheet and the re-agglomerating of the product. Few of the top-down approaches also produce graphene in very low yields and often have cumbersome procedures [44].

2.1.1 Micromechanical cleavage

The micromechanical cleavage also commonly known as the 'scotch tape method' or the 'peel-off method' is a facile exfoliation of highly oriented pyrolytic graphite (HOPG) using adhesive tape (hence named scotch tape method) to pull the layers apart. This is the first popularized method to obtain high purity graphene [14]. The successive peeling of a HOPG yielded mono-layer, bi-layer, or few-layer graphene which could be identified by an optical microscope over a SiO_2/Si substrate. This identification is done by being benefited with the change in refractive index of light when passing between graphene and SiO_2 substrate. This method of obtaining graphene is advantageous due to the high purity of graphene obtained because of very limited processing requirement. Micromechanical cleavage also found its use in producing other two-dimensional atomic crystals including boron nitride and MoS_2 [45]. However, due to the method being slow, highly labor intensive, and yielding very small amount of graphene, the product may be reserved only to study its fundamental properties over its applications such as preparation of electrodes in large quantity for supercapacitors [46].

2.1.2 Exfoliation of graphite in solvent

The ultrasonic exfoliation of natural flakes of graphite in variety of solvents is a rather recent method to produce graphene [47, 48]. The ability of graphene to be able to disperse in numerous solvents was reported to conclude that the best solvents for such use should be the one which is having the parameter of Hildebrand solubility (δT) ~ 23 $MPa^{1/2}$ and the parameter of Hansen solubility i.e., δD, δP, and δH as ~ 18 $MPa^{1/2}$, ~ 9.3 $MPa^{1/2}$, and ~ 23 $MPa^{1/2}$, respectively. The surface tension of such solvents should be nearly 40 $mJ.m^{-2}$ [49]. With the given parameters, N-Methyl-2-Pyrrolidone (NMP) qualifies as the most suitable solvent which practically yield highest percentage of the monolayer graphene in the dispersion. However, cyclopentanone gives highest absolute concertation of mono- to few-layer graphene with the solubility of 0.0085 - 0.0012 mg mL^{-1}. Solvent based exfoliation of unmodified graphite is considered advantageous due the use of lesser volumes of solvents. An easy attempt to increase the concentration of graphite has been to rise the sonication time or the use of a sonic probe for the sonication [50, 51]. This is so because the sonic probes deliver higher sonication power as compared to the sonic bath whereas effect of increase in time is obvious. However, the suitable solvents for dispersing graphite such as NMP should have higher boiling point (> 200 C), and hence removal of solvents from the composite materials where films or coating have been formed as in the case of electrodes for supercapacitors is time consuming and difficult. Sonication for prolonged period in low boiling point solvents has shown to give concentration of graphene as high as 0.5 mg mL^{-1}[52]. In comparison, the sonication of graphite in NMP for 460 hours reportedly yielded a concentration of 1mg/mL [50]. It is worth mentioning that the lengthy sonication period improves the concentration of graphene in the dispersion, however with smaller and more defective flakes [50]. Another strategy involves the sonication of graphite in aqueous surfactant solutions for exfoliation [53, 54]. An additional advantage of both the methods is the inhibition of the reaggregation of the graphene caused by the repulsive potential barrier induced between the surfactant coated graphene sheets [55]. Sodium cholate solutions used to sonicate graphite for ~400 hours results in the dispersion strength of up to 0.3 mg mL^{-1} [54]. However, the removal of surfactants is difficult and affects the properties of graphene.

2.1.3 Graphene exfoliated from the graphite intercalation compounds (GIC)

Exfoliation of GIC is probably the most commonly attempted route to synthesize graphene or the oxide of graphene called graphene oxide (GO). Most commonly known attempt was made as early as 1958 by Hummers and Offeman to prepare graphitic oxide from graphite [38]. This method involved intercalation of graphite using concentrated H_2SO_4 in presence of $NaNO_3$ and $KMnO_3$. This method had a serious problem of emissions of hazardous gases like NO_2/N_2O_4 during the synthesis. Later the method was perfected by omitting the use of $NaNO_3$ as a reagent which made a more health friendly synthesis of GO possible [56, 57]. The prepared GO can be further reduced to give something called reduced graphene oxide (rGO) [58]. Both GO and rGO are widely used for the fabrication of supercapacitors [59-61]. Viculis et al. described the use of potassium to intercalate graphite sheets followed by exfoliation in ethanol to obtain the dispersion of graphene [62]. It was observed that during sonication, the exfoliated carbon nanosheets attained a rolled structure like a scroll therefore named as nanoscrolls. Transmission electron microscopy (TEM) analysis revealed that there were $\sim 40 \pm 15$ layers in each sheet. These experiments showed the possibilities for the separation of the layers of graphene from graphite. Researchers have fine-tuned the technique of intercalation graphite compounds to provide crystallites of ~100 μm in size [15].

Among various methods to exfoliate GIC for the preparation of graphene, solvent-assisted and thermal exfoliation have been most popular [44]. In a solvent-assisted method, GIC are generally sonicated as dispersion in a suitable solvent to aid exfoliation. Alkali metal GIC may be spontaneously exfoliated in NMP [63]. As the solvent molecules intercalating the GIC expand the interlayer distance of the contained graphene sheets [64], the subsequent evolution of gases to further aid the exfoliation. Few common examples for such gas assisted exfoliation are lithium GIC being sonicated in water [65], and ethanol being used as a solvent for the sonication of alkali metal GIC [66], which results in the formation of alkali metal ethoxide and hydrogen gas as by products.

Thermal expansion of GIC using bromide was probably first described in 1916 [44]. In the late 1960s, the foils of exfoliated graphite prepared by such methods were used in the manufacturing of gasket and sealant materials [67]. When GIC are heated, the thermal decomposition of intercalates produces gases which expand the graphite, called as expanded graphite along with the commonly known exfoliated graphite. Exfoliated graphite, being an excellent precursor for the preparation of graphene, may also be industrially utilized as composite filler material and thermal insulator [67]. The reaction of graphite with strong acids such as H_2SO_4 and HNO_3 is commonly applied to obtain GIC for thermal expansion. These GIC are then exfoliated by means of rapid heating or by microwave irradiation [68]. Graphene of monolayer to few atomic layer thickness has been reportedly obtained by the expanded graphite when dispersed and sonicated in ethanol [69], or NMP [70]. Repetitive cycles of intercalation and exfoliation followed by the sonication in Dimethylformamide (DMF) may yield in excess of 50% monolayer and bi-layer graphene [71]. GIC for thermal exfoliation may also be obtained by the intercalation of graphite with iron chloride ($FeCl_3$) and nitromethane (CH_3NO_2) [72], or ionic liquid crystals [73]. In the former method, $FeCl_3$ promotes the intercalation of CH_3NO_2, which decomposes at ~100 °C under the irradiance of microwave. Latter involves slight heating to invoke the intercalation of graphite by ionic liquid due to the reduced viscosity of ionic liquids at higher temperatures. The thermal decomposition however is achieved by heating the GIC to ~ 700 °C to obtain expanded graphite. Exfoliation of graphite brought about with the help of supercritical CO_2 involves the intercalation of the graphitic layers by the CO_2 in the supercritical phase followed by rapid expansion when the container is depressurized releasing CO_2 thereby forcing the graphitic layers apart [74].

2.1.4 Preparation of graphene by the exfoliation of graphite oxide

GO is one of the most popular products obtained from GIC reduction which can be used for the formation of rGO as seen in section 2.1.2. Apart from the aforementioned Hummers method for the preparation of graphitic oxide, other noted methods to synthesize graphite oxide by oxidation of graphite using concentrated acids have been reported by Straudenmaier [75] and Brodie [76]. Both GO and rGO are often classified as 'functionalized graphene' rather than 'graphene' as they carry oxygen rich functional groups and complete reduction of oxide form of graphene is yet to be achieved. The exfoliation of graphite oxide is readily achieved and often return higher yield than graphite by means of thermal treatments or under aqueous sonication [77]. This is because the oxidation of graphene leads to the formation of intercalated oxide bonds within the layers of graphite thereby reducing the effect of Van der Waal's force. Hence graphite oxide is considered to be a GIC. Furthermore, the exfoliation of graphite oxide results in GO which may be used, as obtained for supercapacitor applications or after being reduced to rGO by various chemical or thermal approaches [78, 79]. The most common methods for the chemical reduction of GO involve the use of reducing agents like hydrazine hydrate, sodium borohydride, Vitamin C, pyrogallol, potassium hydroxide and hydroiodic acid etc. [58]. Table 1 compiles various methods used for reducing GO to rGO.

Table 1. *Reduction of GO using various methods [58]*

Mode of Reduction	C/O ratio	σ (S/cm)	Ref.
Reduction using Hydrazine hydrate ($N_2H_4.H_2O$)	10.3	2	[32]
Reduction using hydrazine (N_2H_4)	--	72	[80]
Reduction using $NaBH_4$	8.6	0.045	[81]
Reduction using vapor of N_2H_4	8.8	--	[82]
UHV assisted Thermal reduction @ 900 °C	14.1		
UHV assisted Thermal reduction @ 1100 °C	--	~103	[83]
Thermal reduction @ 1100 °C under Ar/H_2	--	727	[84]
(a) $NaBH_4$	(a) 4.78	(a) 0.823	[85]
(b) Concentrated H_2SO_4 180 °C, 12 h	(b) 8.57	(b) 16.6	
(c) Thermal reduction @ 1100 °C in Ar/H_2	(c) >246	(c) 202	
(a) Vit. C	(a) 12.5	(a) 77	[86]
(b) $N_2H_4.H_2O$	(b) 12.5	(b) 99.6	
(c) Pyrogallol	(c) --	(c) 4.8	
(d) KOH	(d) --	(d) 1.9×10^{-3}	
55% HI reduction	>14.9	298	[87]

UHV: ultra-high vacuum.

GO is an electrical insulator due to the oxygen rich sp^3 hybridized C disrupting the long-range order of sp^2 hybridization of C present in the graphene layer. Hence, the reduction of GO is desired to restore its electrical properties. The chemical reduction of GO makes rGO hydrophobic resulting in the agglomerate precipitating out of the dispersion, thereby inhibiting further reduction [32]. However, in an unlikely case of complete reduction of

GO, the large number of defects brought about by harsh oxidation process, means that the properties of rGO will be different than the pristine graphene. Therefore, these materials needed to be composited with conductive fillers to be used for electronic applications such as supercapacitor [88, 89].

2.1.5 Preparation of graphene by electrochemical exfoliation of graphite

The exfoliation of graphite by means of electrochemical reaction often involves the use of a sacrificial graphite electrode with the exfoliated materials being collected in the electrolytic solution. Among various electrolytes explored for the electrochemical exfoliation, aqueous H_2SO_4 acid or base like KOH, in combination of a suitable surfactant have been the most common [90]. A vital character of the surfactants is prevention of the reagglomeration of graphene layers. The hydrophobic groups of the surfactant bond to p-orbitals of graphene, whereas the hydrophilic groups help in stabilizing the dispersion of graphene sheets in water [91]. Still, as earlier explained, the removal of surfactants from the graphene after the exfoliation process is difficult [90], and significantly affect the properties of graphene [92]. H_2SO_4 is an excellent electrolyte for the exfoliation of graphite due to its ability to intercalate the graphite layers with $[SO_4]^{-2}$ ions [93]. Often KOH is added to suppress the oxidation brought about by the H_2SO_4. Both the cases yield a mixture of graphite flakes of different thicknesses and the few-layer graphene can be easily isolated after centrifugation.

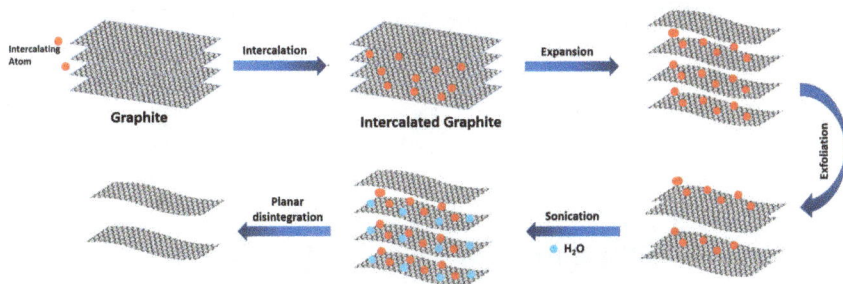

Figure 4. A schematic of the formation of graphene through the expansion of graphite.
[65]

The *in-situ* expansion of the graphite electrode is another strategy utilized for the electrochemical preparation of graphene. This expansion is followed by the sonication of electrode to obtain the exfoliated product [65, 94]. Lithium salts are often used to achieve this type of exfoliation by forming lithium salt GIC. The ultrasonic cavitation caused due to sonication induce thermal shock which separates the graphene layers [95]. The Li salt intercalated graphite when sonicated in a binary mixture of DMF and propylene carbonate yield >70% graphene [94]. Sonication in water even reports the yield in excess of 80% graphene, [65]. This is because Li reacts readily with water to liberate hydrogen gas resulting in added extent of exfoliation. Figure 4 demonstrates a typical mechanism of the formation of graphene by exfoliation of graphite using suitable chemical species.

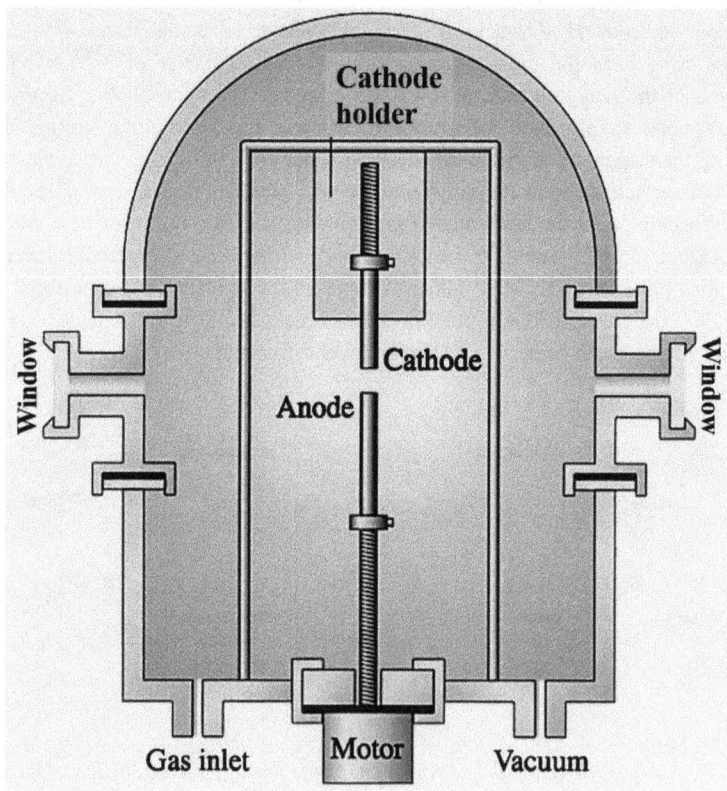

Figure 5. Schematic diagram of arc discharge method.

2.1.6 Arc discharge

Arc discharge is a highly specialized technique of passing a high intensity direct current through highly pure graphite electrodes separated by small distance in vacuum (Fig. 5). The preparation of carbon nanomaterials such as carbon nanotubes and fullerene has been commonly done using this method [96]. Arc discharge methods have also been used to synthesize graphene in presence of suitable buffer gases [97-99]. Among all the gases being employed, H_2 gas has been the main component of buffer gas as it terminates the dangling carbon bonds further inhibiting the self-rolling of graphitic sheets into nanotubes [97, 98]. Binary mixture of helium and hydrogen gas creates highly crystalline graphene sheets [99]. However, this method has been rarely used to synthesize graphene for supercapacitors.

2.1.7 Unzipping carbon nanotubes

Unzipping CNT is another way to produce graphene. The nature of graphene is often determined by the type of CNT being used. Single walled CNT lead to a mono-layer graphene whereas multi-walled CNT yield few-layer graphene. The unzipping can be carried out using wet chemistry methods with the help of strong oxidizing agents [100], or physical routes like laser irradiation [101] and plasma etching [102, 103] may be employed. The unzipped CNT are also referred to 'graphene nanoribbons' (GNR). The width of the GNR is dictated by the diameter of the CNT used for unzipping. However, GNR are a quasi-one-dimensional material with properties dissimilar to graphene. The properties of GNR depend on their width and the edge they possess (armchair/zigzag). Such dependence of properties of materials over the latter is termed as 'edge effect' [104, 105]. The defect sites present on the CNT are often preferred for the initiation of unzipping of the nanotubes leading to the irregular cleavage of C–C bond [106]. The unzipping of flattened CNT leads to the formation of a well-regulated GNR [107]. The cleavage of CNT in this method is preferentially along the bent edge. Synthesis of graphene from unzipping CNT has opened interesting opportunities as graphene is an attractive alternative to nanotubes.

2.2 Bottom-up approach

While the top-down approach focuses on breaking precursor like graphite, graphitic oxide or CNT into atomic layers from a stack, bottom up approach implements carbon molecules as building blocks. These carbon molecules are obtained from alternative sources. Although the bottom-up approach is not suitable to produce graphene sheets of very high purity with large surface area and conductivity as may be possible through the top down approach. Bottom-up approach offers the possibility of manufacturing graphene, GNR and graphene quantum dots in very large quantities[108].

2.2.1 Chemical vapor deposition (CVD)

Among various methods constituting bottom-up approach, CVD is possibly the most popularly used method to produce carbon nanostructures like CNT and graphene in bulk [109, 110].

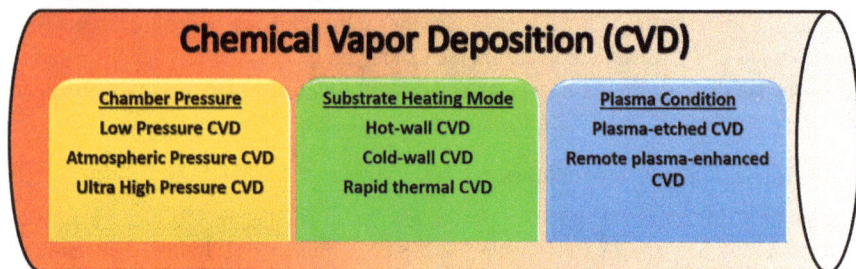

Figure 6. Various types of Chemical Vapor Deposition.

Generally speaking, CVD is essentially a chemical route to synthesize both high-quality and high-quantity, nanosized particles and thin-films. A typical CVD process involves a catalytic substrate being exposed to volatile gaseous precursors. The gaseous precursor reacts on or with the surface of substrate return desired product. The gaseous side-products are removed with the flow of carrier gas. Commonly known CVD types are classified as illustrated in Figure 6. CVD method may have two reaction routes:

- Homogenous gas-phase reactions followed by the physical deposition of the product on the substrate
- Heterogeneous chemical reactions on the surface substrate which acts as a catalyst.

The synthesis of graphene using CVD may be classified in four major classes:

- Growth on metal catalyst
- Substrate free growth
- Thermal CVD
- Plasma etched CVD

A schematic representation of graphene grown on CVD is given as Figure 7.

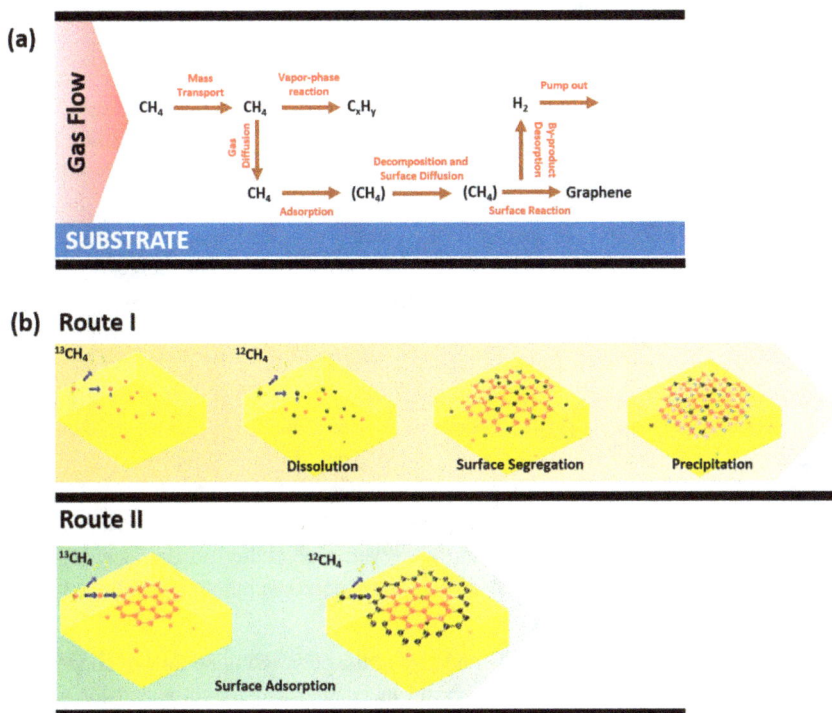

Figure 7. The mechanism of growth of graphene synthesized by CVD on a metal substrates [110]. a) Basic steps involved in growing graphene layers over a catalytic substrate. b) A Schematic of the dissolution-surface aggregation and surface adsorption routes of the growth of graphene on metal catalysts having high and low carbon solubility, respectively [111].

2.2.1.1 Growth on metal catalyst

Transition metals as substrate for obtaining CVD grown graphene have been widely popular. The growth of graphene via CVD can be either surface catalyzed or achieved by segregation methods. In a surface catalyzed reaction, the carbon containing species forms graphene at the metal surface. This is a "self-limiting" process where the metal catalyst is rendered ineffective once the monolayer graphene covers the surface. In segregation method, graphene is formed when the carbon atoms dissolved in the bulk of the metal substrate diffuse towards the surface. The low solubility of carbon in metals at lower temperatures means that there is no back diffusion once the substrate is cooled. The dissolution of carbon and the cooling rate decide the thickness of the graphene layers [112, 113]. Various metals, including Fe [114, 115], Ru [116-118], Co [119-122], Rh [123, 124], Ir [125, 126], Ni [127-130], Pd [131, 132], Pt [133], Cu [111, 134-138], Au [139], and alloys (Co–Ni [140-142], Au–Ni[143], Ni–Mo [144-146],188, Co-Mo[145], and stainless steel [147-149] have been reported to synthesize graphene. The conditions required for growing graphene depend on the metal used, pressure, temperature, and flow rate of carbon precursor. In addition, the number of ripples and defects formed in graphene are direct implication of the graphene–metal interaction. The strength of this interaction also determines the ease of the transfer of graphene to a different substrate. Among various metals and alloys studied. Copper and nickel have remained the most used catalysts for the growth of graphene with diagonal lengths of ~ 30 inches in case of polycrystalline copper foils [150]. The mechanism of growing graphene onto copper is a surface catalyzed one. This means that the monolayers of graphene may be grown on the copper surface in a wide range of reaction conditions [134]. On the other hand, nickel facilitates the graphene synthesis via segregation [151]. Therefore, it is difficult to control the thickness of graphene on nickel. However, the ability to form graphene in atmospheric condition means that it can be used more commonly for the synthesis of graphene as compared to copper.

Graphene growth starts at the grain boundaries of the polycrystalline foils of copper [152] and nickel [153]. Therefore, a single crystal metal precursor is not suited for growing large scale graphene films using CVD. Sputtering and electron beam evaporation are the most popular methods for preparing the polycrystalline metal thin films or foils which are further used as catalyst for growing large area graphene films. It has been reported earlier that the often seemingly continuous graphene films grown on the grain boundaries acquire defects like multi-layer graphene nucleation affecting the properties of graphene [128].

A major disadvantage of growing graphene onto metals is the requirement of harsh growth conditions, especially where UHV are needed. These conditions induce defects in the continuous monolayer graphene. However, modifying these parameters to achieve more mild conditions might lead to non-uniform films of graphene. Growth of graphene on copper at atmospheric pressure results in creation of smaller areas of multi-layer graphene unlike the mono-layer graphene observed in UHV conditions [154]. Selection of carbon sources which may lower the deposition temperatures is another way to optimize the growth of graphene [155, 156]. A recent summary of growth of graphene on Cu catalysts is provided in Table 2 [110].

Table 2*. Few of the selected CVD conditions from the literature.*

Technique	Substrate	Gas	Flow [sccm]	Pressure [Torr]	Temp. [°C]	Time	Size	μ [cm^2 V^{-1} s^{-1}]	$R_\square$ [$\Omega_\square^{-1}$]
Low Pressure CVD[150]	25 μm Cu foil	H$_2$:CH$_4$	24:8	0.09	1000	1 h	0.25 m^2	5100 @ Room Temperature	125
Atmospheric Pressure CVD[157]	25 μm Cu foil	Ar:H$_2$:CH$_4$	5000:125:0.15	760	1000	4 h	0.39 m^2	–	1–3 k
Low Pressure CVD[157]	25 μm Cu foil	H$_2$:CH$_4$	200:10	8	1000	6 h	0.3 m^2	3349 @ Room Temperature	460
Atmospheric Pressure CVD[158]	25 μm Cu foil	Ar:H$_2$:CH$_4$	3000:150:1	760	900–1040	3–5 min	0.8 m^2	6944 @ Room Temperature	500
Atmospheric Pressure CVD[159]	25 μm Cu foil	Ar:H$_2$:CH$_4$	1000:50:25	760	1000	1–40 cm min^{-1}	1 m	–	–
Low Pressure CVD[160]	18 μm Cu foil	H$_2$:CH$_4$	5:36	Low Pressure	1000	0–50 cm min^{-1}	5 m×5 cm	–	662
CVD[161]	50 μm Cu foil	H$_2$:CH$_4$	315:10	4	1010	2.5–50 cm min^{-1}	–	–	–
CVD[162]	40 μm Cu foil	N$_2$:H$_2$:CH$_4$	8000:160:2	760	1010	0.75 cm min^{-1}	20 m × 12 mm	5270–6040	–
CVD[163]	125 μm Cu foil	Ar:CH$_4$	95%:5%	150	1035	20 min	4.5 cm^2	14 000	1300
CVD[164]	36 μm Cu foil	H$_2$:CH$_4$	40:450	7.5	1000	10 cm min^{-1}	100 m×23 cm	–	500
CVD[165]	35 μm Cu foil	CH$_4$	50	0.55	970	5 min	0.12 m^2	5200 @ Room Temperature	–
CVD[166]	25 μm Cu foil	H$_2$:CH$_4$	0.4:1.4	0.01	1000	20 min	≈cm	3300 @ 1.4 K	–

Graphene films grown on metal catalysts cannot be used as such for various applications. In order to make them useful, the graphene films need to be transferred to an insulating substrate from the conductive metal films. There are variety of etchants in use to remove graphene sheets from the metal substrate [167]. However, this method of removal damages the graphene films. To prevent the films from cracking, a polymeric support is used [168-170]. A simple way to achieve this transfer is to coat the graphene side of the metal film with suitable polymer followed by etching the underlying metals. The polymer can later be dissolved in an organic solvent to yield free standing graphene films. Additionally, the transfer of the graphene on a flexible substrate may be achieved with hot press lamination [171] and roll-to-roll transfer [172]. Therefore, transfer of large area graphene films is possible. Transfer of graphene via electrochemical delamination is often advantageous requiring a slight etching of the metal, while letting the rest of the substrate reusable [173]. In various cases, solid carbon sources such as amorphous carbons [174], nanodiamond [175], polymers [176], and self-assembled monolayers (SAM) [177] have been used instead of the gaseous precursor for growing graphene films. As a fun fact, the thermal decomposition of unconventional wastes including cookies, chocolate, legs of cockroach, and grass have been shown to grow monolayer graphene onto copper foils [178].

2.2.1.2 Substrate free

Substrate free CVD produces graphene sheets instead of substrate deposited films. This method has advantage of being continuous (with collection of products being outside the furnace), hence allowing preparation of graphene in bulk. Another advantage is the avoidance of removal of graphene from the substrate. Among the pioneering works on the substrate free growth of graphene, Dato et al. produced monolayer and bilayer graphenes by microwave enhanced CVD of ethanol under atmospheric pressure [179, 180]. Thermal decomposition of sodium ethoxide in ethanol via substrate free CVD synthesis of few-layer graphene has been another such example [181]. However, the quality of the graphene obtained by substrate free route is often inferior to those obtain by the growth on metal substrate.

2.2.1.3 Epitaxial growth of graphene

Epitaxial growth or thermal decomposition of SiC involves the growth of graphene on the (0001) plane of a single crystal 6H-SiC [182, 183]. Graphene is formed on the surface of SiC on heating the H_2-etched surface of 6H-SiC to 1250-1450 °C for 1-20 minutes. The as-prepared graphene typically is 1 to 3 atomic layers thick. The number of layers in as-synthesized graphene is directly related to decomposition temperature. Epitaxial growth has been successfully used to produce mono-layer graphene [184]. This method is advantageous for producing very high purity defect free graphene [185]. Furthermore, this method is very lucrative for the semiconductor industries [186-188]. Epitaxial growth has also led to the fabrication of mm scale continuous films of graphene on a Ni thin film coated SiC substrate [189]. This was achieved at a low temperature of 750 °C which is significantly less than that used in CVD process. The ability to produce wafer-size graphene films with large area makes it favorable for industrial application [190]. Epitaxial growth is found to be selective in producing different structures of graphene on different planes of SiC crystal lattice. This is due to the difference in the arrangement of carbon atoms in both the lattice planes.

The epitaxial growth of graphene monolayers has been also observed on the (0001) plane of the single crystal Ru under ultra-high vacuum (4 × 10^{-11} Torr) [191]. To achieve this, Ru crystal was pretreated with the multiple cycles of Ar^+ sputtering, oxygen exposure, and heating to high temperature, respectively, followed by the adsorption of ethylene. Graphene film was formed due to the thermal decomposition of ethylene at 1000K. The controlled segregation of carbon diffusing outward of the bulk was the mechanism behind the formation of graphene films. However, the mono-layer graphene obtained was of exceptional purity and surface area with periodic ripples. Thermal decomposition of absorbed carbon atoms (although source not clear) on Ru(0001) surface, led to the formation of macroscopic single-crystalline domains of mono-layer to few-layer graphene [192]. Other single crystal transition metals used in synthesis of epitaxially grown graphene were Ir, Ni, Co, Pt, etc. [112].

The issues such as lack of absolute control over the attained thickness of graphene layer and obtaining the same quality graphene in repeated trials need to be addressed in order to make this process industrially viable. A thorough research is needed to be undertaken to realize and control the mechanisms of graphene growth by this method.

Conclusions and outlook

Graphene-based materials are proven candidates for forming electrode of electrochemical energy storage systems, such as electrochemical cells, fuel cells, and supercapacitors. Since the rediscovery of graphene (or should we say mono-layer graphene), considerable efforts have been put on perfecting various routes to produce graphene or their derivatives. These materials are aimed to be consistent in their design, production, performance, and their ability to exhibit the same electrochemical phenomena on repeated basis. To realize the full potential of graphene for supercapacitors, the stringent control over the obtaining reproducible quality and quantity of the materials needs further improvement. This improvement should be visible at the desired nano-, micro-, meso- and macroscales. This chapter summarizes the various routes to synthesize graphene which looked upon as being able to become commercial in the near future. Few key points on synthetic approach are:

- Chemical or electrochemical exfoliation of GIC is low cost and effective to produce graphene-based materials such as GO and rGO. However, before being able to apply this method for large-scale application, the challenges like stabilization of monolayer or few-layer graphene sheets in solvents and with consistent properties and performance need to be overcome.

- CVD grown graphene have shown potential to be rolled out both in batch and continuous process. However, the control over the induced defects still remains a major challenge.

- Epitaxial growth of graphene provides greater control over its thickness, defects, and properties among all the available methods. However, the technique still needs to be perfected for commercialization.

- Micromechanical exfoliation remains the most consistent method to produce research grade graphene of highest purity so far. Although the method is not suitable for commercial application.

The development of graphene-based materials for the electrodes of supercapacitors and other energy storage devices is possible via various routes such as chemical and electrochemical exfoliation, and CVD. These methods seem most consistent in their ability to produce graphene nanosheets in bulk with acceptable properties. The composites formed using these materials have shown to produce good results and these synthetic methods will be perfected in the near future to achieve consistently performing devices.

References

[1] Poonam, K. Sharma, A. Arora, S.K. Tripathi, Review of supercapacitors: Materials and devices, J. Energy Storage 21 (2019) 801-825. https://doi.org/10.1016/j.est.2019.01.010

[2] M. Winter, R.J. Brodd, What Are Batteries, Fuel Cells, and Supercapacitors?, Chem. Rev. 104 (2004) 4245-4270. https://doi.org/10.1021/cr020730k

[3] L.L. Zhang, X.S. Zhao, Carbon-based materials as supercapacitor electrodes, Chem. Soc. Rev. 38 (2009) 2520-2531. https://doi.org/10.1039/b813846j

[4] A. Balducci, R. Dugas, P.L. Taberna, P. Simon, D. Plée, M. Mastragostino, S. Passerini, High temperature carbon–carbon supercapacitor using ionic liquid as electrolyte, J. Power Sources 165 (2007) 922-927. https://doi.org/10.1016/j.jpowsour.2006.12.048

[5] R.R. Rajagopal, L.S. Aravinda, R. Rajarao, B.R. Bhat, V. Sahajwalla, Activated carbon derived from non-metallic printed circuit board waste for supercapacitor application, Electrochim. Acta 211 (2016) 488-498. https://doi.org/10.1016/j.electacta.2016.06.077

[6] C. Meng, O.Z. Gall, P.P. Irazoqui, A flexible super-capacitive solid-state power supply for miniature implantable medical devices, Biomedical Microdevices 15 (2013) 973-983. https://doi.org/10.1007/s10544-013-9789-1

[7] Q. Ke, J. Wang, Graphene-based materials for supercapacitor electrodes – A review, Journal of Materiomics 2 (2016) 37-54. https://doi.org/10.1016/j.jmat.2016.01.001

[8] M. Kaempgen, C.K. Chan, J. Ma, Y. Cui, G. Gruner, Printable thin film supercapacitors using single-walled carbon nanotubes, Nano Lett. 9 (2009) 1872-1876. https://doi.org/10.1021/nl8038579

[9] S. Ruben, Electric condenser, in: U.S.P. Office (Ed.) USA, 1925.

[10] H. Becker, Low voltage electrolytic capacitor: US, 2800616, 1957.

[11] P. Simon, Y. Gogotsi, Materials for electrochemical capacitors, Nanoscience And Technology: A Collection of Reviews from Nature Journals, World Scientific2010, pp. 320-329. https://doi.org/10.1142/9789814287005_0033

[12] M. Inagaki, H. Konno, O. Tanaike, Carbon materials for electrochemical capacitors, J. Power Sources 195 (2010) 7880-7903. https://doi.org/10.1016/j.jpowsour.2010.06.036

[13] S. Sarangapani, B.V. Tilak, C.P. Chen, Materials for electrochemical capacitors: Theoretical and experimental constraints, J. Electrochem. Soc. 143 (1996) 3791-3799. https://doi.org/10.1149/1.1837291

[14] K.S. Novoselov, A.K. Geim, S.V. Morozov, D. Jiang, Y. Zhang, S.V. Dubonos, I.V. Grigorieva, A.A. Firsov, Electric Field Effect in Atomically Thin Carbon Films, Science 306 (2004) 666. https://doi.org/10.1126/science.1102896

[15] A.K. Geim, K.S. Novoselov, The rise of graphene, Nature Mater. 6 (2007) 183. https://doi.org/10.1038/nmat1849

[16] T.J. Booth, P. Blake, R.R. Nair, D. Jiang, E.W. Hill, U. Bangert, A. Bleloch, M. Gass, K.S. Novoselov, M.I. Katsnelson, A.K. Geim, Macroscopic graphene membranes and their extraordinary stiffness, Nano Lett. 8 (2008) 2442-2446. https://doi.org/10.1021/nl801412y

[17] C. Lee, X. Wei, J.W. Kysar, J. Hone, Measurement of the elastic properties and intrinsic strength of monolayer graphene, Science 321 (2008) 385. https://doi.org/10.1126/science.1157996

[18] J. Xia, F. Chen, J. Li, N. Tao, Measurement of the quantum capacitance of graphene, Nat. Nanotechnol. 4 (2009) 505. https://doi.org/10.1038/nnano.2009.177

[19] Q. He, H.G. Sudibya, Z. Yin, S. Wu, H. Li, F. Boey, W. Huang, P. Chen, H. Zhang, Centimeter-Long and Large-scale micropatterns of reduced graphene oxide films: fabrication and sensing applications, ACS Nano 4 (2010) 3201-3208. https://doi.org/10.1021/nn100780v

[20] D.A.C. Brownson, D.K. Kampouris, C.E. Banks, An overview of graphene in energy production and storage applications, J. Power Sources 196 (2011) 4873-4885. https://doi.org/10.1016/j.jpowsour.2011.02.022

[21] M. Pumera, Graphene-based nanomaterials and their electrochemistry, Chem. Soc. Rev. 39 (2010) 4146-4157. https://doi.org/10.1039/c002690p

[22] X. Huang, X. Qi, F. Boey, H. Zhang, Graphene-based composites, Chem. Soc. Rev. 41 (2012) 666-686. https://doi.org/10.1039/C1CS15078B

[23] N. Savage, Materials science: Super carbon, Nature 483 (2012) S30. https://doi.org/10.1038/483S30a

[24] K.S. Kim, Y. Zhao, H. Jang, S.Y. Lee, J.M. Kim, K.S. Kim, J.H. Ahn, P. Kim, J.Y. Choi, B.H. Hong, Large-scale pattern growth of graphene films for stretchable transparent electrodes, Nature 457 (2009) 706. https://doi.org/10.1038/nature07719

[25] S. Park, J. An, J.W. Suk, R.S. Ruoff, Graphene-based actuators, Small 6 (2010) 210-212. https://doi.org/10.1002/smll.200901877

[26] A.C.H. Tsang, H.Y.H. Kwok, D.Y.C. Leung, The use of graphene based materials for fuel cell, photovoltaics, and supercapacitor electrode materials, Solid State Sci. 67 (2017) A1-A14. https://doi.org/10.1016/j.solidstatesciences.2017.03.015

[27] X. Cai, L. Lai, Z. Shen, J. Lin, Graphene and graphene-based composites as Li-ion battery electrode materials and their application in full cells, J. Mater. Chem. A 5 (2017) 15423-15446. https://doi.org/10.1039/C7TA04354F

[28] N. Kheirabadi, A. Shafiekhani, Graphene/Li-ion battery, J. Appl. Phys. 112 (2012) 124323. https://doi.org/10.1063/1.4771923

[29] R.S. Dey, H.A. Hjuler, Q. Chi, Approaching the theoretical capacitance of graphene through copper foam integrated three-dimensional graphene networks, J. Mater. Chem. A 3 (2015) 6324-6329. https://doi.org/10.1039/C5TA01112D

[30] H. Huang, L. Xu, Y. Tang, S. Tang, Y. Du, Facile synthesis of nickel network supported three-dimensional graphene gel as a lightweight and binder-free electrode for high rate performance supercapacitor application, Nanoscale 6 (2014) 2426-2433. https://doi.org/10.1039/C3NR05952A

[31] T. Kuila, A.K. Mishra, P. Khanra, N.H. Kim, J.H. Lee, Recent advances in the efficient reduction of graphene oxide and its application as energy storage electrode materials, Nanoscale 5 (2013) 52-71. https://doi.org/10.1039/C2NR32703A

[32] S. Stankovich, D.A. Dikin, R.D. Piner, K.A. Kohlhaas, A. Kleinhammes, Y. Jia, Y. Wu, S.T. Nguyen, R.S. Ruoff, Synthesis of graphene-based nanosheets via chemical reduction of exfoliated graphite oxide, Carbon 45 (2007) 1558-1565. https://doi.org/10.1016/j.carbon.2007.02.034

[33] M. Beidaghi, C. Wang, Micro-Supercapacitors Based on Interdigital Electrodes of Reduced Graphene Oxide and Carbon Nanotube Composites with Ultrahigh Power Handling Performance, Adv. Funct. Mater. 22 (2012) 4501-4510. https://doi.org/10.1002/adfm.201201292

[34] Y. Zhao, J. Liu, B. Wang, J. Sha, Y. Li, D. Zheng, M. Amjadipour, J. MacLeod, N. Motta, Supercapacitor Electrodes with Remarkable Specific Capacitance Converted from Hybrid Graphene Oxide/NaCl/Urea Films, ACS Appl. Mater. Interfaces 9 (2017) 22588-22596. https://doi.org/10.1021/acsami.7b05965

[35] K. Kakaei, M.D. Esrafili, A. Ehsani, Chapter 9 - Graphene-Based Electrochemical Supercapacitors, in: K. Kakaei, M.D. Esrafili, A. Ehsani (Eds.) Interface Science and Technology, Elsevier 2019, pp. 339-386. https://doi.org/10.1016/B978-0-12-814523-4.00009-5

[36] P.R. Wallace, The Band Theory of Graphite, Phys. Rev. 71 (1947) 622-634. https://doi.org/10.1103/PhysRev.71.622

[37] G. Ruess, F.J.M.f.C.u.v.T.a.W. Vogt, Höchstlamellarer Kohlenstoff aus Graphitoxyhydroxyd, 78 (1948) 222-242. https://doi.org/10.1007/BF01141527

[38] W.S. Hummers, R.E. Offeman, Preparation of Graphitic Oxide, J. Am. Chem. Soc. 80 (1958) 1339-1339. https://doi.org/10.1021/ja01539a017

[39] H. Boehm, A. Clauss, G. Fischer, U.J.Z.F.N.B. Hofmann, Dünnste kohlenstoff-folien, 17 (1962) 150-153. https://doi.org/10.1515/znb-1962-0302

[40] H.-P. Boehm, A. Clauss, G. Fischer, U.J.Z.f.a.u.a.C. Hofmann, Das adsorptions verhalten sehr dünner kohlenstoff-folien, 316 (1962) 119-127. https://doi.org/10.1002/zaac.19623160303

[41] S. Mouras, A. Hamm, D. Djurado, J.C. Cousseins, Synthesis of first stage graphite intercalation compounds with fluorides, J. Fluorine Chem. 24 (1987) 572-582.

[42] M. Bacon, S.J. Bradley, T. Nann, Graphene quantum dots, Part. Part. Syst. Char. 31 (2014) 415-428. https://doi.org/10.1002/ppsc.201300252

[43] T. Mahmoudi, Y. Wang, Y.-B. Hahn, Graphene and its derivatives for solar cells application, Nano Energy 47 (2018) 51-65. https://doi.org/10.1016/j.nanoen.2018.02.047

[44] R.S. Edwards, K.S. Coleman, Graphene synthesis: Relationship to applications, Nanoscale 5 (2013) 38-51. https://doi.org/10.1039/C2NR32629A

[45] K.S. Novoselov, D. Jiang, F. Schedin, T.J. Booth, V.V. Khotkevich, S.V. Morozov, A.K. Geim, Two-dimensional atomic crystals, PNAS 102 (2005) 10451. https://doi.org/10.1073/pnas.0502848102

[46] R. Verdejo, M.M. Bernal, L.J. Romasanta, M.A. Lopez-Manchado, Graphene filled polymer nanocomposites, J. Mater. Chem. 21 (2011) 3301-3310. https://doi.org/10.1039/C0JM02708A

[47] P. Blake, P.D. Brimicombe, R.R. Nair, T.J. Booth, D. Jiang, F. Schedin, L.A. Ponomarenko, S.V. Morozov, H.F. Gleeson, E.W. Hill, A.K. Geim, K.S. Novoselov, Graphene-based liquid crystal device, Nano Lett. 8 (2008) 1704-1708. https://doi.org/10.1021/nl080649i

[48] Y.J.N.N. Hernandez, Y. Hernandez, V. Nicolosi, M. Lotya, F.M. Blighe, Z. Sun, S. De, IT McGovern, B. Holland, M. Byrne, YK Gun'Ko, JJ Boland, P. Niraj, G. Duesberg, S. Krishnamurthy, R. Goodhue, J. Hutchison, V. Scardaci, A.C. Ferrari, J.N. Coleman, Nat. Nanotechnol. 3 (2008) 563. https://doi.org/10.1038/nnano.2008.215

[49] Y. Hernandez, M. Lotya, D. Rickard, S.D. Bergin, J.N. Coleman, Measurement of multicomponent solubility parameters for graphene facilitates solvent discovery, Langmuir 26 (2010) 3208-3213. https://doi.org/10.1021/la903188a

[50] U. Khan, A. O'Neill, M. Lotya, S. De, J.N. Coleman, High-concentration solvent exfoliation of graphene, Small 6 (2010) 864-871. https://doi.org/10.1002/smll.200902066

[51] U. Khan, H. Porwal, A. O'Neill, K. Nawaz, P. May, J.N. Coleman, Solvent-exfoliated graphene at extremely high concentration, Langmuir 27 (2011) 9077-9082. https://doi.org/10.1021/la201797h

[52] A. O'Neill, U. Khan, P.N. Nirmalraj, J. Boland, J.N. Coleman, Graphene dispersion and exfoliation in low boiling point solvents, J. Phys. Chem. C 115 (2011) 5422-5428. https://doi.org/10.1021/jp110942e

[53] M. Lotya, Y. Hernandez, P.J. King, R.J. Smith, V. Nicolosi, L.S. Karlsson, F.M. Blighe, S. De, Z. Wang, I.T. McGovern, G.S. Duesberg, J.N. Coleman, Liquid Phase

production of graphene by exfoliation of graphite in surfactant/water solutions, J. Am. Chem. Soc. 131 (2009) 3611-3620. https://doi.org/10.1021/ja807449u

[54] M. Lotya, P.J. King, U. Khan, S. De, J.N. Coleman, High-concentration, surfactant-stabilized graphene dispersions, ACS Nano 4 (2010) 3155-3162. https://doi.org/10.1021/nn1005304

[55] R.J. Smith, M. Lotya, J.N. Coleman, The importance of repulsive potential barriers for the dispersion of graphene using surfactants, New J. Phys. 12 (2010) 125008. https://doi.org/10.1088/1367-2630/12/12/125008

[56] D.C. Marcano, D.V. Kosynkin, J.M. Berlin, A. Sinitskii, Z. Sun, A. Slesarev, L.B. Alemany, W. Lu, J.M. Tour, Improved synthesis of graphene oxide, ACS Nano 4 (2010) 4806-4814. https://doi.org/10.1021/nn1006368

[57] J. Chen, B. Yao, C. Li, G. Shi, An improved Hummers method for eco-friendly synthesis of graphene oxide, Carbon 64 (2013) 225-229. https://doi.org/10.1016/j.carbon.2013.07.055

[58] S. Pei, H.-M. Cheng, The reduction of graphene oxide, Carbon 50 (2012) 3210-3228. https://doi.org/10.1016/j.carbon.2011.11.010

[59] C. Ogata, R. Kurogi, K. Awaya, K. Hatakeyama, T. Taniguchi, M. Koinuma, Y. Matsumoto, All-graphene oxide flexible solid-state supercapacitors with enhanced electrochemical performance, ACS Appl. Mater. Interfaces 9 (2017) 26151-26160. https://doi.org/10.1021/acsami.7b04180

[60] A.K. Das, S. Sahoo, P. Arunachalam, S. Zhang, J.-J. Shim, Facile synthesis of Fe3O4 nanorod decorated reduced graphene oxide (RGO) for supercapacitor application, RSC Advances 6 (2016) 107057-107064. https://doi.org/10.1039/C6RA23665K

[61] T. Purkait, G. Singh, D. Kumar, M. Singh, R.S. Dey, High-performance flexible supercapacitors based on electrochemically tailored three-dimensional reduced graphene oxide networks, Sci. Rep. 8 (2018) 640. https://doi.org/10.1038/s41598-017-18593-3

[62] L.M. Viculis, J.J. Mack, R.B. Kaner, A Chemical route to carbon nanoscrolls, Science 299 (2003) 1361. https://doi.org/10.1126/science.1078842

[63] C. Vallés, C. Drummond, H. Saadaoui, C.A. Furtado, M. He, O. Roubeau, L. Ortolani, M. Monthioux, A. Pénicaud, Solutions of negatively charged graphene sheets and ribbons, J. Am. Chem. Soc. 130 (2008) 15802-15804. https://doi.org/10.1021/ja808001a

Materials Research Forum LLC
https://doi.org/10.21741/9781644900550-2

[64] X. Li, G. Zhang, X. Bai, X. Sun, X. Wang, E. Wang, H. Dai, Highly conducting graphene sheets and Langmuir–Blodgett films, Nat. Nanotechnol.3 (2008) 538. https://doi.org/10.1038/nnano.2008.210

[65] H. Huang, Y. Xia, X. Tao, J. Du, J. Fang, Y. Gan, W. Zhang, Highly efficient electrolytic exfoliation of graphite into graphene sheets based on Li ions intercalation–expansion–microexplosion mechanism, J. Mater. Chem. 22 (2012) 10452-10456. https://doi.org/10.1039/c2jm00092j

[66] L.M. Viculis, J.J. Mack, O.M. Mayer, H.T. Hahn, R.B. Kaner, Intercalation and exfoliation routes to graphite nanoplatelets, J. Mater. Chem. 15 (2005) 974-978. https://doi.org/10.1039/b413029d

[67] D. Chung, Exfoliation of graphite, J. Mater. Sci. 22 (1987) 4190-4198. https://doi.org/10.1007/BF01132008

[68] T. Wei, Z. Fan, G. Luo, C. Zheng, D. Xie, A rapid and efficient method to prepare exfoliated graphite by microwave irradiation, Carbon 47 (2009) 337-339. https://doi.org/10.1016/j.carbon.2008.10.013

[69] S. Malik, A. Vijayaraghavan, R. Erni, K. Ariga, I. Khalakhan, J.P. Hill, High purity graphenes prepared by a chemical intercalation method, Nanoscale 2 (2010) 2139-2143. https://doi.org/10.1039/c0nr00248h

[70] W. Gu, W. Zhang, X. Li, H. Zhu, J. Wei, Z. Li, Q. Shu, C. Wang, K. Wang, W. Shen, F. Kang, D. Wu, Graphene sheets from worm-like exfoliated graphite, J. Mater. Chem. 19 (2009) 3367-3369. https://doi.org/10.1039/b904093p

[71] S.R. Dhakate, N. Chauhan, S. Sharma, J. Tawale, S. Singh, P.D. Sahare, R.B. Mathur, An approach to produce single and double layer graphene from re-exfoliation of expanded graphite, Carbon 49 (2011) 1946-1954. https://doi.org/10.1016/j.carbon.2010.12.068

[72] W. Fu, J. Kiggans, S.H. Overbury, V. Schwartz, C. Liang, Low-temperature exfoliation of multilayer-graphene material from FeCl3 and CH3NO2 co-intercalated graphite compound, Chem. Comm. 47 (2011) 5265-5267. https://doi.org/10.1039/c1cc10508f

[73] A. Safavi, M. Tohidi, F.A. Mahyari, H. Shahbaazi, One-pot synthesis of large scale graphene nanosheets from graphite–liquid crystal composite via thermal treatment, J. Mater. Chem. 22 (2012) 3825-3831. https://doi.org/10.1039/c2jm13929d

Materials Research Forum LLC
https://doi.org/10.21741/9781644900550-2

[74] N.-W. Pu, C.-A. Wang, Y. Sung, Y.-M. Liu, M.-D. Ger, Production of few-layer graphene by supercritical CO_2 exfoliation of graphite, Mater. Lett. 63 (2009) 1987-1989. https://doi.org/10.1016/j.matlet.2009.06.031

[75] L. Staudenmaier, Verfahren zur Darstellung der Graphitsäure, Berichte der deutschen chemischen Gesellschaft 31 (1898) 1481-1487. https://doi.org/10.1002/cber.18980310237

[76] Brodie, XIII. On the atomic weight of graphite, Trans. Royal Soc. London 149 (1859) 249-259. https://doi.org/10.1098/rstl.1859.0013

[77] D.R. Dreyer, S. Park, C.W. Bielawski, R.S. Ruoff, The chemistry of graphene oxide, Chem. Soc. Rev. 39 (2010) 228-240. https://doi.org/10.1039/B917103G

[78] W. Chen, L. Yan, P.R. Bangal, Preparation of graphene by the rapid and mild thermal reduction of graphene oxide induced by microwaves, Carbon 48 (2010) 1146-1152. https://doi.org/10.1016/j.carbon.2009.11.037

[79] X. Gao, J. Jang, S. Nagase, Hydrazine and Thermal Reduction of Graphene Oxide: Reaction Mechanisms, Product Structures, and Reaction Design, J. Phys. Chem. C 114 (2010) 832-842. https://doi.org/10.1021/jp909284g

[80] D. Li, M.B. Müller, S. Gilje, R.B. Kaner, G.G. Wallace, Processable aqueous dispersions of graphene nanosheets, Nat. Nanotechnol. 3 (2008) 101. https://doi.org/10.1038/nnano.2007.451

[81] H.-J. Shin, K.K. Kim, A. Benayad, S.M. Yoon, H.K. Park, I.S. Jung, M.H. Jin, H.K. Jeong, J.M. Kim, J.Y. Choi, Y.H. Lee, Efficient reduction of graphite oxide by sodium borohydride and its effect on electrical conductance, Adv. Funct. Mater. 19 (2009) 1987-1992. https://doi.org/10.1002/adfm.200900167

[82] D. Yang, A. Velamakanni, G. Bozoklu, S. Park, M. Stoller, R.D. Piner, S. Stankovich, I. Jung, D.A. Field, C.A. Ventrice, R.S. Ruoff, Chemical analysis of graphene oxide films after heat and chemical treatments by X-ray photoelectron and Micro-Raman spectroscopy, Carbon 47 (2009) 145-152. https://doi.org/10.1016/j.carbon.2008.09.045

[83] H.A. Becerril, J. Mao, Z. Liu, R.M. Stoltenberg, Z. Bao, Y. Chen, Evaluation of solution-processed reduced graphene oxide films as transparent conductors, ACS Nano 2 (2008) 463-470. https://doi.org/10.1021/nn700375n

[84] X. Wang, L. Zhi, K. Müllen, Transparent, conductive graphene electrodes for dye-sensitized solar cells, Nano Lett. 8 (2008) 323-327. https://doi.org/10.1021/nl072838r

[85] W. Gao, L.B. Alemany, L. Ci, P.M. Ajayan, New insights into the structure and reduction of graphite oxide, Nat. Chem. 1 (2009) 403. https://doi.org/10.1038/nchem.281

[86] M.J. Fernández-Merino, L. Guardia, J.I. Paredes, S. Villar-Rodil, P. Solís-Fernández, A. Martínez-Alonso, J.M.D. Tascón, Vitamin C is an ideal substitute for hydrazine in the reduction of graphene oxide suspensions, J. Phys. Chem. C 114 (2010) 6426-6432. https://doi.org/10.1021/jp100603h

[87] S. Pei, J. Zhao, J. Du, W. Ren, H.-M. Cheng, Direct reduction of graphene oxide films into highly conductive and flexible graphene films by hydrohalic acids, Carbon 48 (2010) 4466-4474. https://doi.org/10.1016/j.carbon.2010.08.006

[88] M. Faraji, A. Abedini, Fabrication of electrochemically interconnected MoO_3/GO/MWCNTs/graphite sheets for high performance all-solid-state symmetric supercapacitor, Int. J. Hydrogen Energy 44 (2019) 2741-2751. https://doi.org/10.1016/j.ijhydene.2018.12.015

[89] Y. Cheng, Y. Zhang, Q. Wang, C. Meng, Synthesis of amorphous MnSiO3/graphene oxide with excellent electrochemical performance as supercapacitor electrode, Colloids Surf. A: Physicochem. Eng. Aspects 562 (2019) 93-100. https://doi.org/10.1016/j.colsurfa.2018.11.011

[90] G. Wang, B. Wang, J. Park, Y. Wang, B. Sun, J. Yao, Highly efficient and large-scale synthesis of graphene by electrolytic exfoliation, Carbon 47 (2009) 3242-3246. https://doi.org/10.1016/j.carbon.2009.07.040

[91] J.M. Englert, J. Röhrl, C.D. Schmidt, R. Graupner, M. Hundhausen, F. Hauke, A. Hirsch, Soluble graphene: generation of aqueous graphene solutions aided by a perylenebisimide-based bolaamphiphile, Adv. Mater. 21 (2009) 4265-4269. https://doi.org/10.1002/adma.200901578

[92] D.A.C. Brownson, J.P. Metters, D.K. Kampouris, C.E. Banks, Graphene electrochemistry: Surfactants inherent to graphene can dramatically effect electrochemical processes, Electroanal. 23 (2011) 894-899. https://doi.org/10.1002/elan.201000708

[93] C.-Y. Su, A.-Y. Lu, Y. Xu, F.-R. Chen, A.N. Khlobystov, L.-J. Li, High-quality thin graphene films from fast electrochemical exfoliation, ACS Nano 5 (2011) 2332-2339. https://doi.org/10.1021/nn200025p

[94] J. Wang, K.K. Manga, Q. Bao, K.P. Loh, High-yield synthesis of few-layer graphene flakes through electrochemical expansion of graphite in propylene carbonate

electrolyte, J. Am. Chem. Soc. 133 (2011) 8888-8891.
https://doi.org/10.1021/ja203725d

[95] K. Suslick, G.J.A.R.M.S. Price, Interaction of acoustic waves and matter at a molecular or atomic level, J. Phys. D Appl. Phys. 29 (1999) 295-326. https://doi.org/10.1146/annurev.matsci.29.1.295

[96] S. Farhat, C.D. Scott, Review of the arc process modeling for fullerene and nanotube production, J. Nanosci. Nanotechnol. 6 (2006) 1189-1210. https://doi.org/10.1166/jnn.2006.331

[97] K.S. Subrahmanyam, L.S. Panchakarla, A. Govindaraj, C.N.R. Rao, Simple method of preparing graphene flakes by an arc-discharge Method, J. Phys. Chem. C 113 (2009) 4257-4259. https://doi.org/10.1021/jp900791y

[98] Y. Chen, H. Zhao, L. Sheng, L. Yu, K. An, J. Xu, Y. Ando, X. Zhao, Mass-production of highly-crystalline few-layer graphene sheets by arc discharge in various H_2–inert gas mixtures, Chem. Phys. Lett. 538 (2012) 72-76. https://doi.org/10.1016/j.cplett.2012.04.020

[99] B. Shen, J. Ding, X. Yan, W. Feng, J. Li, Q. Xue, Influence of different buffer gases on synthesis of few-layered graphene by arc discharge method, Appl. Surf. Sci. 258 (2012) 4523-4531. https://doi.org/10.1016/j.apsusc.2012.01.019

[100] D.V. Kosynkin, A.L. Higginbotham, A. Sinitskii, J.R. Lomeda, A. Dimiev, B.K. Price, J.M. Tour, Longitudinal unzipping of carbon nanotubes to form graphene nanoribbons, Nature 458 (2009) 872. https://doi.org/10.1038/nature07872

[101] P. Kumar, L.S. Panchakarla, C.N.R. Rao, Laser-induced unzipping of carbon nanotubes to yield graphene nanoribbons, Nanoscale 3 (2011) 2127-2129. https://doi.org/10.1039/c1nr10137d

[102] L. Jiao, L. Zhang, X. Wang, G. Diankov, H. Dai, Narrow graphene nanoribbons from carbon nanotubes, Nature 458 (2009) 877. https://doi.org/10.1038/nature07919

[103] L. Valentini, Formation of unzipped carbon nanotubes by CF_4 plasma treatment, Diamond Rel. Mater. 20 (2011) 445-448. https://doi.org/10.1016/j.diamond.2011.01.038

[104] K. Nakada, M. Fujita, G. Dresselhaus, M.S. Dresselhaus, Edge state in graphene ribbons: Nanometer size effect and edge shape dependence, Phys. Rev. B 54 (1996) 17954-17961. https://doi.org/10.1103/PhysRevB.54.17954

[105] L. Xie, H. Wang, C. Jin, X. Wang, L. Jiao, K. Suenaga, H. Dai, Graphene nanoribbons from unzipped carbon nanotubes: Atomic structures, Raman

spectroscopy, and electrical Properties, J. Am. Chem. Soc. 133 (2011) 10394-10397. https://doi.org/10.1021/ja203860a

[106] S. Cho, K. Kikuchi, A. Kawasaki, Radial followed by longitudinal unzipping of multiwalled carbon nanotubes, Carbon 49 (2011) 3865-3872. https://doi.org/10.1016/j.carbon.2011.05.023

[107] Y.-R. Kang, Y.-L. Li, M.-Y. Deng, Precise unzipping of flattened carbon nanotubes to regular graphene nanoribbons by acid cutting along the folded edges, J. Mater. Chem. 22 (2012) 16283-16287. https://doi.org/10.1039/c2jm33385f

[108] J.H. Warner, F. Schaffel, M. Rummeli, A. Bachmatiuk, Graphene: Fundamentals and emergent applications, Newnes2012.

[109] E. Baddour Carole, C. Briens, Carbon nanotube synthesis: A review, Int. J. Chem. Reactor Eng. 3 (2005). https://doi.org/10.2202/1542-6580.1279

[110] B. Deng, Z. Liu, H. Peng, Toward mass production of CVD graphene films, Adv. Mater. 0 (2018) 1800996. https://doi.org/10.1002/adma.201800996

[111] X. Li, W. Cai, L. Colombo, R.S. Ruoff, Evolution of graphene growth on ni and cu by carbon isotope labeling, Nano Lett. 9 (2009) 4268-4272. https://doi.org/10.1021/nl902515k

[112] J. Wintterlin, M.L. Bocquet, Graphene on metal surfaces, Surf. Sci. 603 (2009) 1841-1852. https://doi.org/10.1016/j.susc.2008.08.037

[113] M. Batzill, The surface science of graphene: Metal interfaces, CVD synthesis, nanoribbons, chemical modifications, and defects, Surf. Sci. Rep. 67 (2012) 83-115. https://doi.org/10.1016/j.surfrep.2011.12.001

[114] H. An, W.-J. Lee, J. Jung, Graphene synthesis on Fe foil using thermal CVD, Current Appl. Phys. 11 (2011) S81-S85. https://doi.org/10.1016/j.cap.2011.03.077

[115] D. Kondo, K. Yagi, M. Sato, M. Nihei, Y. Awano, S. Sato, N. Yokoyama, Selective synthesis of carbon nanotubes and multi-layer graphene by controlling catalyst thickness, Chem. Phys. Lett. 514 (2011) 294-300. https://doi.org/10.1016/j.cplett.2011.08.042

[116] E. Sutter, P. Albrecht, F.E. Camino, P. Sutter, Monolayer graphene as ultimate chemical passivation layer for arbitrarily shaped metal surfaces, Carbon 48 (2010) 4414-4420. https://doi.org/10.1016/j.carbon.2010.07.058

[117] E. Sutter, P. Albrecht, P.J.A.P.L. Sutter, Graphene growth on polycrystalline Ru thin films, Appl. Phys. Lett. 95 (2009) 133109. https://doi.org/10.1063/1.3224913

[118] P.W. Sutter, P.M. Albrecht, E.A. Sutter, Graphene growth on epitaxial Ru thin films on sapphire, Applied Physics Letters 97 (2010) 213101. https://doi.org/10.1063/1.3518490

[119] S. Wang, Y. Pei, X. Wang, H. Wang, Q. Meng, H. Tian, X. Zheng, W. Zheng, Y.J.J.o.P.D.A.P. Liu, Synthesis of graphene on a polycrystalline Co film by radio-frequency plasma-enhanced chemical vapour deposition, J. Phys. D Appl. Phys. 43 (2010) 455402. https://doi.org/10.1088/0022-3727/43/45/455402

[120] M.E. Ramón, A. Gupta, C. Corbet, D.A. Ferrer, H.C. Movva, G. Carpenter, L. Colombo, G. Bourianoff, M. Doczy, D.J.A.N. Akinwande, CMOS-compatible synthesis of large-area, high-mobility graphene by chemical vapor deposition of acetylene on cobalt thin films, ACS Nano 5 (2011) 7198-7204. https://doi.org/10.1021/nn202012m

[121] N. Zhan, G. Wang, J.J.A.P.A. Liu, Cobalt-assisted large-area epitaxial graphene growth in thermal cracker enhanced gas source molecular beam epitaxy, Appl. Phys. A 105 (2011) 341-345. https://doi.org/10.1007/s00339-011-6612-9

[122] H. Ago, Y. Ito, N. Mizuta, K. Yoshida, B. Hu, C.M. Orofeo, M. Tsuji, K.-i. Ikeda, S. Mizuno, Epitaxial chemical vapor deposition growth of single-layer graphene over cobalt film crystallized on sapphire, ACS Nano 4 (2010) 7407-7414. https://doi.org/10.1021/nn102519b

[123] E. Rut'kov, A.J.P.o.t.S.S. Kuz'michev, Carbon interaction with rhodium surface: Adsorption, dissolution, segregation, growth of graphene layers, Phys. Solid State 53 (2011) 1092-1098. https://doi.org/10.1134/S1063783411050246

[124] S. Roth, J. Osterwalder, T.J.S.S. Greber, Synthesis of epitaxial graphene on rhodium from 3-pentanone, Surf. Sci. 605 (2011) L17-L19. https://doi.org/10.1016/j.susc.2011.02.007

[125] F. Müller, S. Grandthyll, C. Zeitz, K. Jacobs, S. Hüfner, S. Gsell, M.J.P.R.B. Schreck, Epitaxial growth of graphene on Ir (111) by liquid precursor deposition, Phys. Rev. B 84 (2011) 075472. https://doi.org/10.1103/PhysRevB.84.075472

[126] C. Vo-Van, A. Kimouche, A. Reserbat-Plantey, O. Fruchart, P. Bayle-Guillemaud, N. Bendiab, J.J.A.p.l. Coraux, Epitaxial graphene prepared by chemical vapor deposition on single crystal thin iridium films on sapphire, Appl. Phys. Lett. 98 (2011) 181903. https://doi.org/10.1063/1.3585126

[127] A. Reina, X. Jia, J. Ho, D. Nezich, H. Son, V. Bulovic, M.S. Dresselhaus, J.J.N.l. Kong, Large area, few-layer graphene films on arbitrary substrates by chemical vapor deposition, Nano Lett. 9 (2008) 30-35. https://doi.org/10.1021/nl801827v

[128] Y.Z. LG De Arco, CW Schlenker, K. Ryu, ME Thompson, C. Zhou Continuous, highly flexible, and transparent graphene films by chemical vapor deposition for organic photovoltaics, ACS Nano 4 (2010) 2865. https://doi.org/10.1021/nn901587x

[129] S.J. Chae, F. Güneş, K.K. Kim, E.S. Kim, G.H. Han, S.M. Kim, H.J. Shin, S.M. Yoon, J.Y. Choi, M.H.J.A.M. Park, Synthesis of large-area graphene layers on poly-nickel substrate by chemical vapor deposition: Wrinkle formation, Adv. Mater. 21 (2009) 2328-2333. https://doi.org/10.1002/adma.200803016

[130] K.S. Kim, Y. Zhao, H. Jang, S.Y. Lee, J.M. Kim, K.S. Kim, J.-H. Ahn, P. Kim, J.-Y. Choi, B.H.J.n. Hong, Large-scale pattern growth of graphene films for stretchable transparent electrodes, Nature 457 (2009) 706. https://doi.org/10.1038/nature07719

[131] S.-Y. Kwon, C.V. Ciobanu, V. Petrova, V.B. Shenoy, J. Bareno, V. Gambin, I. Petrov, S.J.N.l. Kodambaka, Growth of semiconducting graphene on palladium, Nano Lett. 9 (2009) 3985-3990. https://doi.org/10.1021/nl902140j

[132] Y. Murata, S. Nie, A. Ebnonnasir, E. Starodub, B. Kappes, K. McCarty, C. Ciobanu, S.J.P.R.B. Kodambaka, Growth structure and work function of bilayer graphene on Pd (111), Phys. Rev. B 85 (2012) 205443. https://doi.org/10.1103/PhysRevB.85.205443

[133] B.J. Kang, J.H. Mun, C.Y. Hwang, B.J.J.J.o.A.P. Cho, Monolayer graphene growth on sputtered thin film platinum, J. Appl. Phys. 106 (2009) 104309. https://doi.org/10.1063/1.3254193

[134] X. Li, W. Cai, J. An, S. Kim, J. Nah, D. Yang, R. Piner, A. Velamakanni, I. Jung, E.J.s. Tutuc, Large-area synthesis of high-quality and uniform graphene films on copper foils, Science 324 (2009) 1312-1314. https://doi.org/10.1126/science.1171245

[135] Y. Hao, M. Bharathi, L. Wang, Y. Liu, H. Chen, S. Nie, X. Wang, H. Chou, C. Tan, B.J.S. Fallahazad, The role of surface oxygen in the growth of large single-crystal graphene on copper, Science (2013) 1243879. https://doi.org/10.1126/science.1243879

[136] Q. Li, H. Chou, J.-H. Zhong, J.-Y. Liu, A. Dolocan, J. Zhang, Y. Zhou, R.S. Ruoff, S. Chen, W. Cai, Growth of adlayer graphene on Cu studied by carbon isotope labeling, Nano Lett.13 (2013) 486-490. https://doi.org/10.1021/nl303879k

[137] X. Li, Y. Zhu, W. Cai, M. Borysiak, B. Han, D. Chen, R.D. Piner, L. Colombo, R.S. Ruoff, Transfer of large-area graphene films for high-performance transparent

conductive electrodes, Nano Lett. 9 (2009) 4359-4363.
https://doi.org/10.1021/nl902623y

[138] Z. Luo, Y. Lu, D.W. Singer, M.E. Berck, L.A. Somers, B.R. Goldsmith, A.C.J.C.o.M. Johnson, Effect of substrate roughness and feedstock concentration on growth of wafer-scale graphene at atmospheric pressure, Chem. Mater. 23 (2011) 1441-1447. https://doi.org/10.1021/cm1028854

[139] T. Oznuluer, E. Pince, E.O. Polat, O. Balci, O. Salihoglu, C.J.A.P.L. Kocabas, Synthesis of graphene on gold, Applied Physics Letters 98 (2011) 183101. https://doi.org/10.1063/1.3584006

[140] T. Wu, X. Zhang, Q. Yuan, J. Xue, G. Lu, Z. Liu, H. Wang, H. Wang, F. Ding, Q.J. Yu, Fast growth of inch-sized single-crystalline graphene from a controlled single nucleus on Cu–Ni alloys, Nat. Mater. 15 (2016) 43. https://doi.org/10.1038/nmat4477

[141] Y. Wu, H. Chou, H. Ji, Q. Wu, S. Chen, W. Jiang, Y. Hao, J. Kang, Y. Ren, R.D.J.A.N. Piner, Growth mechanism and controlled synthesis of AB-stacked bilayer graphene on Cu–Ni alloy foils, ACS Nano 6 (2012) 7731-7738. https://doi.org/10.1021/nn301689m

[142] S. Chen, W. Cai, R.D. Piner, J.W. Suk, Y. Wu, Y. Ren, J. Kang, R.S.J.N.l. Ruoff, Synthesis and characterization of large-area graphene and graphite films on commercial Cu–Ni alloy foils, Nano letters 11 (2011) 3519-3525. https://doi.org/10.1021/nl201699j

[143] R.S. Weatherup, B.C. Bayer, R. Blume, C. Ducati, C. Baehtz, R. Schlogl, S.J.N.l. Hofmann, In situ characterization of alloy catalysts for low-temperature graphene growth, Nano Lett. 11 (2011) 4154-4160. https://doi.org/10.1021/nl202036y

[144] M. Zeng, L. Tan, J. Wang, L. Chen, M.H. Rümmeli, L.J.C.o.M. Fu, Liquid metal: An innovative solution to uniform graphene films, Chem. Mater. 26 (2014) 3637-3643. https://doi.org/10.1021/cm501571h

[145] E.V. Lobiak, E.V. Shlyakhova, L.G. Bulusheva, P.E. Plyusnin, Y.V. Shubin, A.V. Okotrub, Ni–Mo and Co–Mo alloy nanoparticles for catalytic chemical vapor deposition synthesis of carbon nanotubes, J. Alloys Compd. 621 (2015) 351-356. https://doi.org/10.1016/j.jallcom.2014.09.220

[146] B. Dai, L. Fu, Z. Zou, M. Wang, H. Xu, S. Wang, Z. Liu, Rational design of a binary metal alloy for chemical vapour deposition growth of uniform single-layer graphene, Nat. Commun. 2 (2011) 522. https://doi.org/10.1038/ncomms1539

[147] R. John, A. Ashokreddy, C. Vijayan, T.J.N. Pradeep, Single-and few-layer graphene growth on stainless steel substrates by direct thermal chemical vapor deposition, Nanotechnology 22 (2011) 165701. https://doi.org/10.1088/0957-4484/22/16/165701

[148] H. Gullapalli, A.L. Mohana Reddy, S. Kilpatrick, M. Dubey, P.M.J.S. Ajayan, Graphene growth via carburization of stainless steel and application in energy storage, Small 7 (2011) 1697-1700. https://doi.org/10.1002/smll.201100111

[149] L.F. Dumée, L. He, Z. Wang, P. Sheath, J. Xiong, C. Feng, M.Y. Tan, F. She, M. Duke, S. Gray, A. Pacheco, P. Hodgson, M. Majumder, L. Kong, Growth of nano-textured graphene coatings across highly porous stainless steel supports towards corrosion resistant coatings, Carbon 87 (2015) 395-408. https://doi.org/10.1016/j.carbon.2015.02.042

[150] S. Bae, H. Kim, Y. Lee, X. Xu, J.S. Park, Y. Zheng, J. Balakrishnan, T. Lei, H. Ri Kim, Y.I. Song, Y.J. Kim, K.S. Kim, B. Özyilmaz, J.-H. Ahn, B.H. Hong, S. Iijima, Roll-to-roll production of 30-inch graphene films for transparent electrodes, Nat. Nanotechnol. 5 (2010) 574. https://doi.org/10.1038/nnano.2010.132

[151] L. Baraton, Z. He, C.S. Lee, J.L. Maurice, C.S. Cojocaru, Y.H. Lee, D. Pribat, Study of Graphene Growth Mechanism on Nickel Thin Films, in: L. Ottaviano, V. Morandi (Eds.) GraphITA 2011, Springer Berlin Heidelberg, Berlin, Heidelberg, 2012, pp. 1-7. https://doi.org/10.1007/978-3-642-20644-3_1

[152] Q. Yu, L.A. Jauregui, W. Wu, R. Colby, J. Tian, Z. Su, H. Cao, Z. Liu, D. Pandey, D.J. Wei, Control and characterization of individual grains and grain boundaries in graphene grown by chemical vapour deposition, Nat. Mater. 10 (2011) 443. https://doi.org/10.1038/nmat3010

[153] Y. Zhang, L. Gomez, F.N. Ishikawa, A. Madaria, K. Ryu, C. Wang, A. Badmaev, C.W. Zhou, Comparison of graphene growth on single-crystalline and polycrystalline Ni by chemical vapor deposition, J. Phys. Chem. Lett. 1 (2010) 3101-3107. https://doi.org/10.1021/jz1011466

[154] S. Bhaviripudi, X. Jia, M.S. Dresselhaus, J. Kong, Role of kinetic factors in chemical vapor deposition synthesis of uniform large area graphene using copper catalyst, Nano Lett. 10 (2010) 4128-4133. https://doi.org/10.1021/nl102355e

[155] S. Kumar, N. McEvoy, T. Lutz, G.P. Keeley, V. Nicolosi, C.P. Murray, W.J. Blau, G.S. Duesberg, Gas phase controlled deposition of high quality large-area graphene films, Chem. Comm. 46 (2010) 1422-1424. https://doi.org/10.1039/b919725g

[156] B. Zhang, W.H. Lee, R. Piner, I. Kholmanov, Y. Wu, H. Li, H. Ji, R.S. Ruoff, Low-temperature chemical vapor deposition growth of graphene from toluene on electropolished copper foils, ACS Nano 6 (2012) 2471-2476. https://doi.org/10.1021/nn204827h

[157] I. Vlassiouk, P. Fulvio, H. Meyer, N. Lavrik, S. Dai, P. Datskos, S. Smirnov, Large scale atmospheric pressure chemical vapor deposition of graphene, Carbon 54 (2013) 58-67. https://doi.org/10.1016/j.carbon.2012.11.003

[158] J. Xu, J. Hu, Q. Li, R. Wang, W. Li, Y. Guo, Y. Zhu, F. Liu, Z. Ullah, G. Dong, Z. Zeng, L. Liu, Fast batch production of high-quality graphene films in a sealed thermal molecular movement system, Small 13 (2017) 1700651. https://doi.org/10.1002/smll.201700651

[159] T. Hesjedal, Continuous roll-to-roll growth of graphene films by chemical vapor deposition, Appl. Phys. Lett. 98 (2011) 133106. https://doi.org/10.1063/1.3573866

[160] B. Deng, P.-C. Hsu, G. Chen, B.N. Chandrashekar, L. Liao, Z. Ayitimuda, J. Wu, Y. Guo, L. Lin, Y. Zhou, M. Aisijiang, Q. Xie, Y. Cui, Z. Liu, H. Peng, Roll-to-Roll encapsulation of metal nanowires between graphene and plastic substrate for high-performance flexible transparent electrodes, Nano Lett. 15 (2015) 4206-4213. https://doi.org/10.1021/acs.nanolett.5b01531

[161] E.S. Polsen, D.Q. McNerny, B. Viswanath, S.W. Pattinson, A. John Hart, High-speed roll-to-roll manufacturing of graphene using a concentric tube CVD reactor, Sci. Rep. 5 (2015) 10257. https://doi.org/10.1038/srep10257

[162] G. Zhong, X. Wu, L. D'Arsie, K.B.K. Teo, N.L. Rupesinghe, A. Jouvray, J. Robertson, Growth of continuous graphene by open roll-to-roll chemical vapor deposition, Appl. Phys. Lett. 109 (2016) 193103. https://doi.org/10.1063/1.4967010

[163] R. Piner, H. Li, X. Kong, L. Tao, I.N. Kholmanov, H. Ji, W.H. Lee, J.W. Suk, J. Ye, Y. Hao, S. Chen, C.W. Magnuson, A.F. Ismach, D. Akinwande, R.S. Ruoff, Graphene synthesis via magnetic inductive heating of copper substrates, ACS Nano 7 (2013) 7495-7499. https://doi.org/10.1021/nn4031564

[164] T. Kobayashi, M. Bando, N. Kimura, K. Shimizu, K. Kadono, N. Umezu, K. Miyahara, S. Hayazaki, S. Nagai, Y. Mizuguchi, Y. Murakami, D. Hobara, Production of a 100-m-long high-quality graphene transparent conductive film by roll-to-roll chemical vapor deposition and transfer process, Appl. Phys. Lett. 102 (2013) 023112. https://doi.org/10.1063/1.4776707

[165] J. Ryu, Y. Kim, D. Won, N. Kim, J.S. Park, E.K. Lee, D. Cho, S.P. Cho, S.J. Kim, G.H. Ryu, H.A.S. Shin, Z. Lee, B.H. Hong, S. Cho, Fast synthesis of high-performance graphene films by hydrogen-free rapid thermal chemical vapor deposition, ACS Nano 8 (2014) 950-956. https://doi.org/10.1021/nn405754d

[166] T.H. Bointon, M.D. Barnes, S. Russo, M.F. Craciun, High quality monolayer graphene synthesized by resistive heating cold wall chemical vapor deposition, Adv. Mater. 27 (2015) 4200-4206. https://doi.org/10.1002/adma.201501600

[167] R.S. Edwards, K.S. Coleman, Graphene film growth on polycrystalline metals, Acc. Chem. Res. 46 (2013) 23-30. https://doi.org/10.1021/ar3001266

[168] E. Auchter, J. Marquez, S.L. Yarbro, E. Dervishi, A facile alternative technique for large-area graphene transfer via sacrificial polymer, AIP Adv. 7 (2017) 125306. https://doi.org/10.1063/1.4986780

[169] X. Liang, B.A. Sperling, I. Calizo, G. Cheng, C.A. Hacker, Q. Zhang, Y. Obeng, K. Yan, H. Peng, Q. Li, X. Zhu, H. Yuan, A.R. Hight Walker, Z. Liu, L.M. Peng, C.A. Richter, Toward clean and crackless transfer of graphene, ACS Nano 5 (2011) 9144-9153. https://doi.org/10.1021/nn203377t

[170] N. Liu, Z. Pan, L. Fu, C. Zhang, B. Dai, Z. Liu, The origin of wrinkles on transferred graphene, Nano Res. 4 (2011) 996. https://doi.org/10.1007/s12274-011-0156-3

[171] V.P. Verma, S. Das, I. Lahiri, W. Choi, Large-area graphene on polymer film for flexible and transparent anode in field emission device, Appl. Phys. Lett. 96 (2010) 203108. https://doi.org/10.1063/1.3431630

[172] Z.Y. Juang, C.Y. Wu, A.Y. Lu, C.Y. Su, K.C. Leou, F.R. Chen, C.H. Tsai, Graphene synthesis by chemical vapor deposition and transfer by a roll-to-roll process, Carbon 48 (2010) 3169-3174. https://doi.org/10.1016/j.carbon.2010.05.001

[173] Y. Wang, Y. Zheng, X. Xu, E. Dubuisson, Q. Bao, J. Lu, K.P. Loh, Electrochemical delamination of CVD-Grown graphene film: Toward the recyclable use of copper catalyst, ACS Nano 5 (2011) 9927-9933. https://doi.org/10.1021/nn203700w

[174] M. Zheng, K. Takei, B. Hsia, H. Fang, X. Zhang, N. Ferralis, H. Ko, Y.-L. Chueh, Y. Zhang, R. Maboudian, A. Javey, Metal-catalyzed crystallization of amorphous carbon to graphene, Appl. Phys. Lett. 96 (2010) 063110. https://doi.org/10.1063/1.3318263

[175] J.M. García, R. He, M.P. Jiang, P. Kim, L.N. Pfeiffer, A. Pinczuk, Multilayer graphene grown by precipitation upon cooling of nickel on diamond, Carbon 49 (2011) 1006-1012. https://doi.org/10.1016/j.carbon.2010.11.008

[176] Z. Sun, Z. Yan, J. Yao, E. Beitler, Y. Zhu, J.M. Tour, Growth of graphene from solid carbon sources, Nature 468 (2010) 549. https://doi.org/10.1038/nature09579

[177] H.-J. Shin, W.M. Choi, S.M. Yoon, G.H. Han, Y.S. Woo, E.S. Kim, S.J. Chae, X.S. Li, A. Benayad, D.D. Loc, F. Gunes, Y.H. Lee, J.Y. Choi, Transfer-free growth of few-layer graphene by self-assembled monolayers, Advanced Materials 23 (2011) 4392-4397. https://doi.org/10.1002/adma.201102526

[178] G. Ruan, Z. Sun, Z. Peng, J.M. Tour, Growth of graphene from food, insects, and waste, ACS Nano 5 (2011) 7601-7607. https://doi.org/10.1021/nn202625c

[179] A. Dato, V. Radmilovic, Z. Lee, J. Phillips, M. Frenklach, Substrate-free gas-phase synthesis of graphene sheets, Nano Lett. 8 (2008) 2012-2016. https://doi.org/10.1021/nl8011566

[180] A. Dato, M. Frenklach, Substrate-free microwave synthesis of graphene: experimental conditions and hydrocarbon precursors, New J. Phys. 12 (2010) 125013. https://doi.org/10.1088/1367-2630/12/12/125013

[181] C.R. Herron, K.S. Coleman, R.S. Edwards, B.G. Mendis, Simple and scalable route for the 'bottom-up' synthesis of few-layer graphene platelets and thin films, J. Mater. Chem. 21 (2011) 3378-3383. https://doi.org/10.1039/c0jm03437a

[182] Z.S. Wu, W. Ren, L. Gao, B. Liu, C. Jiang, H.M. Cheng, Synthesis of high-quality graphene with a pre-determined number of layers, Carbon 47 (2009) 493-499. https://doi.org/10.1016/j.carbon.2008.10.031

[183] C. Berger, Z. Song, X. Li, X. Wu, N. Brown, C. Naud, D. Mayou, T. Li, J. Hass, A.N. Marchenkov, E.H. Conrad, P.N. First, W.A. de Heer, Electronic confinement and coherence in patterned epitaxial graphene, Science 312 (2006) 1191. https://doi.org/10.1126/science.1125925

[184] E. Rollings, G.H. Gweon, S.Y. Zhou, B.S. Mun, J.L. McChesney, B.S. Hussain, A.V. Fedorov, P.N. First, W.A. de Heer, A. Lanzara, Synthesis and characterization of atomically thin graphite films on a silicon carbide substrate, J. Phys. Chem. Solids 67 (2006) 2172-2177. https://doi.org/10.1016/j.jpcs.2006.05.010

[185] Y. Qi, S.H. Rhim, G.F. Sun, M. Weinert, L. Li, Epitaxial graphene on SiC(0001): More than just honeycombs, Phys. Rev. Lett. 105 (2010) 085502. https://doi.org/10.1103/PhysRevLett.105.085502

[186] T. Ohta, F. El Gabaly, A. Bostwick, J.L. McChesney, K.V. Emtsev, A.K. Schmid, T. Seyller, K. Horn, E. Rotenberg, Morphology of graphene thin film growth on SiC (0001), New J. Phys. 10 (2008) 023034. https://doi.org/10.1088/1367-2630/10/2/023034

[187] J. Hass, W.A. de Heer, E.H. Conrad, The growth and morphology of epitaxial multilayer graphene, J. Phys. Condensed Matter 20 (2008) 323202. https://doi.org/10.1088/0953-8984/20/32/323202

[188] Z.G. Cambaz, G. Yushin, S. Osswald, V. Mochalin, Y. Gogotsi, Noncatalytic synthesis of carbon nanotubes, graphene and graphite on SiC, Carbon 46 (2008) 841-849. https://doi.org/10.1016/j.carbon.2008.02.013

[189] Z.Y. Juang, C.Y. Wu, C.W. Lo, W.Y. Chen, C.F. Huang, J.C. Hwang, F.R. Chen, K.C. Leou, C.H. Tsai, Synthesis of graphene on silicon carbide substrates at low temperature, Carbon 47 (2009) 2026-2031. https://doi.org/10.1016/j.carbon.2009.03.051

[190] K.V. Emtsev, A. Bostwick, K. Horn, J. Jobst, G.L. Kellogg, L. Ley, J.L. McChesney, T. Ohta, S.A. Reshanov, J. Röhrl, E. Rotenberg, A.K. Schmid, D. Waldmann, H.B. Weber, T. Seyller, Towards wafer-size graphene layers by atmospheric pressure graphitization of silicon carbide, Nat. Mater. 8 (2009) 203. https://doi.org/10.1038/nmat2382

[191] A.L. Vázquez de Parga, F. Calleja, B. Borca, M.C.G. Passeggi, J.J. Hinarejos, F. Guinea, R. Miranda, Periodically rippled graphene: Growth and spatially resolved electronic structure, Phys. Rev. Lett.100 (2008) 056807. https://doi.org/10.1103/PhysRevLett.100.056807

[192] P.W. Sutter, J.-I. Flege, E.A. Sutter, Epitaxial graphene on ruthenium, Nat. Mater. 7 (2008) 406. https://doi.org/10.1038/nmat2166

Materials Research Forum LLC
https://doi.org/10.21741/9781644900550-3

Chapter 3

Two–Dimensional Graphene Materials for Supercapacitors

S. Hariganesh[1], S. Vadivel[1*], Bappi Paul [2], B. Saravanakumar[3], N. Balasubramanian[4], Vikas Gupta[5]

[1]Department of Chemistry, PSG College of Technology, Coimbatore 641004, India

[2]Department of Chemistry, National Institute of Technology Silchar, Silchar-788010, Assam, India

[3]Department of Physics, Dr. Mahalingam College of Engineering and Technology, Pollachi 642003, Tamil Nadu, India

[4]Department of Chemical Engineering, A.C Tech Campus, Anna University, Chennai 600025, India

Department of Chemistry, IFTM University, Moradabad, India

*vlvelu7@gmail.com

Abstract

Application of two-dimensional graphene-based materials as electrode materials for supercapacitors has been discussed. The properties such as high specific surface area, excellent electrical conductivity and strength and flexibility etc., have made graphene and graphene composites as suitable candidates for fabricating the electrodes of energy storage devices. Even though graphene has been extensively researched in the past as an electrode material still there are some limitations to utilise it in practical applications due to high cost, standard procedure for capacitance measurement and the charge storage mechanism etc. We also aimed to discuss the improvement of the electrochemical and surface properties due to various synthesis methods, heteroatom doping, and composite formation with metal oxides/hydroxides, metal sulphides and conducting polymers. We have briefly accounted the 2D graphene-based materials, their charge storage mechanism such as EDLC or pseudocapacitance and novel designs in device fabrications that have been explored in the recent past.

Keywords

2D-Graphene, Supercapacitor, Graphene Composites, Exfoliation

Contents

1.   Introduction

As the modern world has become more and more digitalised as it progresses, the usage of smartphones, electric vehicles and electronic gadgets have increased the requirement of electric energy to higher level. Hence, there is an urgent need to increase the electricity production to meet the dependency on electric energy. As the fossil fuels have already depleted, this raised alarm about the extensive use of non-renewable sources for energy production [1-4]. In the recent past the increased awareness on global environmental concern had led to usage of renewable energy sources such as solar energy, wind and water energies for energy production. But the main problems associated with renewable energy sources have been proper energy transportation and efficient energy storage technology. The extensive research on the energy storing devices in the past decades has led to the development of supercapacitors [5,6].

Supercapacitors (also called as ultracapacitors) as electrochemical energy storing device fills the void between the high energy density batteries and the traditional capacitors possessing high power density [7,8]. A supercapacitor consists of two electrodes separated by a separator (such as papers, polymers or fibres) which is permeable in nature. The electrodes can be either symmetrical or asymmetrical using electrolytes like aqueous and organic electrolytes [9,10]. The SC's can be categorized into two different types based on their energy storage mechanism such as (i) electrical double layer capacitors (EDLC's) and (ii) pseudocapacitors. In EDLC electrical energy is stored at the electrode-electrolyte interface through adsorption & desorption of ions in the electrolyte, so it has high power density on the other hand in pseudocapacitors, the energy is stored

Graphene as Energy Storage Material for Supercapacitors   Materials Research Forum LLC
Materials Research Foundations **64** (2020) 63-76   https://doi.org/10.21741/9781644900550-3

by fast Faradaic redox reactions that are reversible at the electrode surface and lead to higher capacitance and energy density compared to EDLC. Recently, a special category of SC's called hybrid capacitors has been developed, where the advantages of EDLC and pseudocapacitors have been coupled to give high power and energy densities as in the hybrid capacitors one electrode stores charge in EDL and other electrode stores charge in pseudocapacitance way [11-13].

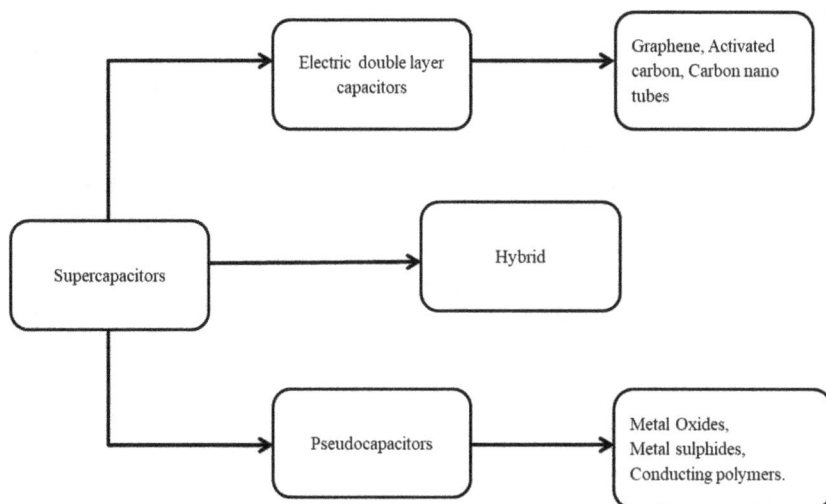

Fig.1. Classification of supercapacitors based on charge storage mechanism

2.   Graphene – as an electrode

Graphene is a two dimensional material with large surface area of about 2630 m^2/g, it has high strength and excellent electrical conducting and electron mobility properties [14]. These properties make graphene as a suitable candidate for EDLC electrode. It provides EDL capacitance of about 550 F/g when its surface area is completely used [15,16]. The limitation with graphene is that due to van der Waals interaction the graphene layers get stacked upon each other and hence the diffusion of ions in the electrolytes is affected. The agglomeration of graphene during synthesis also affects its properties [17,18]. Hence, till now researchers are unable to unlock the fullest of potential of graphene as an electrode material for supercapacitor.

Since pure graphene is expensive and difficult to synthesise in large quantities, reduced graphene oxide has been the most commonly used graphene alternative for extensive applications. Normally, graphene oxide is prepared form graphite via Hummers' method [or modified Hummers' method which includes an oxidation process [19,20]. During oxidation reaction, hydroxyl (–OH) and epoxy (C–O–C) functional groups are attached to the basal planes, whereas carbonyl (–C=O) and carboxyl (–COOH) functional groups are attached to the sheet edges. Due to these defects, the properties of graphene get altered such as the interlayer distance increases due to weakening of van der Waals interaction, hydrophilicity increases due to anionic character of the intercalated functional groups but finally GO has an insulating character whereas reduced GO has good electronic conductivity [21,22]. Hence, reduced GO has been practically suitable for the application as electrode for supercapacitors.

Apart from the functionalization, to enhance the capacitance property of graphene and to overcome its other limitations, various steps have been taken starting from different synthesis methods and precursors, heteroatom doping, coupling with metal oxides/hydroxides, polymers and metal sulphides [23]. In this chapter, we will discuss about synthesis, roles of graphene, synergistic effects of graphene composites and mainly the performance of 2D graphene-based materials as electrodes in supercapacitors.

3.   Modified graphene/heteroatom doped graphene

The main advantage of 2D graphene has been its specific surface area and hence to achieve larger surface area, various synthesis methods have been followed to exfoliate graphene sheets. Yang et al. reported a facile one-step process of rapid annealing of GO in Argon environment to exfoliate the stacked sheets of GO and to remove oxide groups. In this process GO was annealed at 500 °C at a heating rate of 70-80 °C/s. The effective reduction and exfoliation of graphene were indicated by interlayer distance of 3.53 Å (measured by XRD) and high C/O ratio of 7.9 (revealed by XPS). Thus, the symmetrical supercapacitor assembled using rapid annealed graphene electrodes provided enhanced specific capacitance of about 279 F/g and energy density of 135 W/kg [24]. Likewise, Senthilkumar et al. assembled a symmetric supercapacitor using nanoporous graphene synthesized via a new route where Mg/Zn metal strips were used to produce graphene oxide directly from graphene. Thus, obtained nanoporous graphene-GR(Mg/Zn) had high surface area as well as higher conductivity compared to one produced using sodium borohydride-GR(SBH). BET surface area of GR(Mg/Zn) was 1294.6 m^2g^{-1} where GR(SBH) had 798.9 m^2g^{-1}. The GR(Mg/Zn) showed a maximum specific capacitance of 204 F/g at a current density of 10 A/g compared to specific capacitance of 176 F/g showed by the GR(SBH) in 6M KOH electrolyte solution. The GR(Mg/ZN) electrode also retained 95% of its initial capacitance even after 20000 cycles at the current density of 10 A/g [25]. Silva et al. produced graphene sheets by unzipping carbon nanotubes (CNT) and studied their electrochemical properties. In this synthesis process, they have unzipped MWCNT based on hydrogen plasma by microwave chemical vapour deposition and also created open edges on CNT walls to improve the surface area. The unzipped CNT producing graphene exhibited maximum specific capacitance of 217.1 F/g compared to 148.1 F/g for MWCNT at current density of 4 A/g [26]. Edge carboxylated graphene nanoplatelets (ECG) have been prepared by Deb Nath et al. via ball milling method and studied their electrochemical properties comparing with nitrogen doped carboxylic graphene (NGOOH). The edge carboxylation increased the surface area, hydrophilicity and wettability of graphene. The maximum specific capacitance obtained by ECG was as high as 365.72 F/g compared to 175.05 F/g for NGOOH at current density of 1 A/g where the improved property was due to the enhanced ion adsorption rate and ion diffusion rate of ECG than NGOOH. The electrode showed excellent stability of retaining 93% of its initial capacitance after 10,000 cycles at 2 A/g. The structural defect level was about 16.2% for ECG prepared by ball milling method whereas it was 48.9 % for NGOOH prepared by solution-exfoliation method indicating the importance of synthesis methodology influencing the properties of graphene [27]. Thus, such modifications in synthesis methods showed to have improvement in the electrochemical properties. Doping of hetero atoms in graphene causes change in sp^2 hybridisation as the heteroatom replaces some carbon in the lattice resulting in alteration of its properties [28, 29]. Recently, Zeng et al. reported synthesis of stretchable N doped graphene films stuck on poly acrylate as an electrode for supercapacitor which exhibited excellent capacitance of 300 F/g at current density 1 A/g and also retained 88.7% of its initial capacitance after 10,000 cycles at 1 A/g [30]. Similarly, Yue et al. also synthesized flexible nanoporous nitrogen – doped graphene film through facile hydrothermal ammonia reaction and applied as an electrode for supercapacitor. The maximum specific capacitance obtained was 468 F/g at a scan rate of 2 mV/s. Symmetric supercapacitor with two electrode system and 6M KOH as an electrolyte was fabricated which delivered maximum power density of 5000W/kg at energy density of 4.17 W h/kg and maximum energy density of 6.43 W h/kg at power density of 150 W Kg. The doping of nitrogen increased the space between the layers of graphene sheets which thereby increases the surface area for enhanced ion transport and electrolyte access leading to an improved capacitance property [31]. Likewise restacking of graphene layers were prohibited by doping functional groups like thiol with graphene and its analogous, Kanappan et al. reported synthesis of thiolated graphene using NaSH which showed excellent specific capacitance of 196 F/g at 2.5A/g [32]. Also composites of graphene with carbon compounds were reported for supercapacitor application, Ansaldo et al. proposed a hybrid composite of graphene/SWCNT as an electrode for high-performance supercapacitor. The electrode exhibited electrode gravimetric capacitance of 104 F/g at scan rate $2mV^{-1}$ and maximum specific power of 92.3 kW/kg in an aqueous electrolyte of 3M LiNO$_3$ [33].

4.   Graphene composites/metal oxide

Qin et al. reported simple facile one-step chemical bath method to synthesize AgVO$_3$/graphene composite where the AgVO$_3$ was synthesized in-situ to allow it to anchor strongly onto the graphene surface. The composite with 30$wt\%$ of AgVO$_3$ showed a maximum capacitance of 250F/g at 10A/g with retention capacity of 83.3% after 10000 cycles in 6M KOH solution electrolyte. The assembled symmetric capacitor exhibited energy density of 10 W h/kg at power density of 25 W/kg and power density of 2045 W/kg at energy density of 6 Wh/kg where in all aspects the AgVO$_3$/graphene composite was better than the graphene composite with 30$wt\%$ of activated carbon [34]. Deng et al. proposed binder free BiVO$_4$/graphene monolith composite electrode where graphene sheets played an important role as the backbone of the electrode which when used in a three electrode system in 2M NaOH solution as an electrolyte showed specific capacitance of 479 F/g which was higher than the 224 F/g for BiVO$_4$ and 40 F/g for graphene electrode at current density of 5 A/g. The charge transfer resistance value of BiVO$_4$ was about 541.1 Ω but on adding graphene it decreased to 84.8 Ω which indicated the increase in the electrical conductivity and higher capacitance property due to the addition of graphene [35]. In the recent past much work has been done on MnO$_2$/graphene composite as an electrode material for supercapacitors, Singu et al. have synthesized rGO-MnO$_x$ and rGO-Mn$_3$O$_4$ composites via one-pot approach in which the rGO was highly exfoliated. On comparing the capacitance property of the above two composites, the rGO-MnO$_x$ showed maximum capacitance of 398.8 F/g at 5 mV/s whereas rGO-Mn$_3$O$_4$ showed 248.1 F/g in the same condition. The difference in activity was due to the synergistic effect of rGO with MnO$_x$ and the spherical and cubical morphology of the MnO$_x$. The rGO-MnO$_x$ showed good stability even after 5000 cycles which retained about 80% of its initial value at 2 A/g [36]. Similarly Arguello prepared MnO$_2$ and graphene nanoplatelets(GNP) coated graphite paper through aqueous electrophoretic deposition method. The electrochemical study was carried out on two electrode cells, with cellulose separator in 1M Na$_2$SO$_4$ electrolyte. Different composites with various combinations and coating sequence of MnO$_2$ and GNP were prepared, amongst them the specific capacitance value was as high as 422 F/g for GNP+ MnO$_2$ (composites prepared by coating GNP first and then MnO$_2$) which was higher compared to 222 F/g for MnO$_2$ and 35 F/g for GNP at current density of 0.1 A/g [37]. Also recently another MnO_2 composite with 1,3-dicarbonyl- functionalized reduced graphene oxide was reported by Xing et al. which exhibited excellent capacitance of 267.4 F/g at 0.5 A/g in 1M sulphuric acid. The mass of MnO_2 of the composite that showed high specific capacitance was about 32.1% [38]. Apart from metal oxide – graphene composites, binary metal oxide – graphene composites were also studied for their electrochemical pseudocapacitive properties. Cheng et al. synthesized $MnSiO_3$/graphene oxide delivering specific capacitance value of 262.5 F/g which was higher compared to 168.8 F/g for $MnSiO_3$ and 4.6 F/g for graphene oxide at 0.5 A/g in 1M Na_2SO_4 electrolyte. The enlarged surface due to the incorporation of graphene oxide sheets increase the electrical conductivity thereby increased the capacitance value [39]. $NiCoO_2$/rGO composites were synthesized via simple template-assisted method by Guan et al. The $NiCoO_2$/rGO composite containing 1wt% graphene showed remarkable specific capacitance value of 1375 F/g at 1 A/g which decreased to 742 F/g at 10A/g and also showed good stability where the capacitance retention was about 778 F/g at 1 A/g after 3000 cycles [40].

5.   Graphene/metal sulphide composites

Vikraman et al. reported a facile one-pot chemical method to prepare MoS_2/graphene composite which exhibited specific higher capacitance value than MoS_2. The cyclic stability was tested for 10000 cycles where it retained 88% of its initial value. The MoS_2/graphene composite showed maximum energy density of about 26.6 Wh/kg at power density of 125 W/kg [41]. Xiong et al. reported a two-step hydrothermal synthesis method to obtain Co_9S_8 nanotube arrays grown on graphene papers. It served as a binder-free flexible electrode for supercapacitor application. It showed excellent capacitance value of about 653 F/g at 0.5 A/g which decreased to 469 F/g at 10A/g, good rate capability and also 91.8% of cyclic stability after 2500 cycles[42]. Likewise Xie et al. also reported Co_9S_8/graphene electrode that achieved maximum specific capacitance of 1140F/g at 4A/g which decreased to 849F/g at 30A/g, the capacity retention was 93.9% after 1000 cycles. The notable feature of this electrode was excellent charge capacity of 540C/g within 1 min and after discharge time of 14s it retained 74.5% of capacitance. Hybrid supercapacitor assembled with the Co_9S_8/graphene electrode gave utmost energy density of 37Wh/kg at 170W/kg and power density of 12kW/kg at 15.3Wh/kg [43]. Binary metal sulphides such as $CoNi_2S_4$ and $NiCo_2S_4$ were coupled with graphene to improve their electrochemical properties. Lin et al. applied $CoNi_2S_4$/graphene nanocomposite synthesized via solvothermal method as supercapacitor electrodes, It performed as an excellent pseudocapacitive material yielding value of 1621 F/g at 0.5A/g with remarkable cyclic stability of 100% even after 2500 cycles at 5A/g. The amount of graphene loading was 30mg for the electrode with maximum capacitance [44]. Yang et al. prepared $NiCo_2S_4$/graphene nanosheet hybrid electrode which delivered specific capacitance of 1063 F g-1 at 2 A g^{-1}, good rate capability, better cycling life and 82% retention of its capacitance value after 10000 cycles [45]. Even ternary composites were developed by Lin et al. where MoS_2/WS_2/graphene heterostructure was synthesized and applied as an electrode for supercapacitor which showed higher specific capacitance of 365 F/g than 247 F/g, 253 F/g and 161 F/g for rGO-WS_2, rGO-MoS_2 and rGO respectively at current density of 1 A/g. It was also found that the optimum loading of WS_2-MoS_2 to achieve maximum specific capacitance was 53 wt% [46]. The enhancement in the capacitance value of the composite was due to the high surface area provided by rGO, various oxidation states of the transition metal dichalcogenides and the reduced charge transfer resistance of the formed heterostructure.

6.   Graphene/conducting polymer composite

Graphene sheets coupled with conducting polymers such as polyaniline (PANI), Polypyrrole (Ppy) and PEDOT have attracted researchers in the recent past, because of the fact that conducting polymers were excellent electrodes for supercapacitors with their remarkable electrochemical properties, but they suffer from certain limitations arisen due to structural instability, volume expansion and contraction while charge-discharge process [47-49]. Hence, hybridising the polymers with graphene and its analogous was found to be an excellent way to increase its potential as an electrode. Khalid et al. synthesized highly corrugated graphene/PANI composite films on PVC plastic sheets by thermal method using HI(hydroiodic acid) as reducing agent. The prepared composite electrode displayed specific gravimetric capacitance of 1041F/g which was over 6 times greater than graphene film on PVC electrode which showed 170 F/g at 0.125A/g. Symmetrical solid-state capacitor fabricated using gel electrolyte exhibited volumetric energy density of 3.1mW h cm^{-3} at 73.9 mW cm^{-3}[50]. Similarly Tang et al. reported rGO/PANI composite synthesized by *in-situ* polymerisation of aniline followed by GO reduction and re-oxidation of PANI. The maximum specific capacitance achieved by the composite was 596F/g which was higher than 256 F/g for PANI at 1 A/g, at the same current density the cyclic life was studied which showed 85% capacitance retention after 1000 cycles. This enhancement in capacitance was mainly attributed to the increased electrical conductivity of rGO/PANI (125.6 S cm^{-1}) compared to PANI (1.3 S cm^{-1}) [51]. Obeidat et al. assembled asymmetric PEDOT and graphene electrode based supercapacitor using ionic liquid gel electrolyte ([BMIM][BF_4]+PVDF). In this process, PEDOT/graphene electrode was synthesized by pulsed current electro polymerisation method. The device exhibited capacitance of 71.4 F/g at 2.7 V operating voltage which was significant for the solid-state supercapacitors, energy density of 14.9 Wh/kg at 9.8 kW/kg [52].

7. Summary

In this chapter, recent progress of 2D graphene-based electrodes for supercapacitor application has been discussed. Remarkable progress has been made in the field of device fabrication by these graphene electrodes, even binder-free, flexible, solid state capacitors having superior specific capacitance value were assembled. In recent times, researchers have started to develop new technologies and methodologies to synthesize practically applicable reduced graphene oxide in bulk quantity which will ease the progress of commercialization of more supercapacitors in the near future. Also novel hybrid graphene composites were synthesized and their electrochemical performances were studied. To further the research of 2D graphene electrodes in the future following aspects such as (i) synthesis of high-quality 2D graphene-based materials by simple methods in bulk quantities, (ii) detailed study about the charge-discharge mechanism at the electrode-electrolyte interface (iii), focus on preparing flexible and wearable supercapacitors for simple, portable, compact & lightweight energy storage systems and, (iv) the use of environment-friendly electrode materials, electrolytes, separators, current collectors and substrate films need to be considered.

References

[1] M. Chen, W. Li, X. Shen, G. Diao, Fabrication of Core-Shell α-Fe_2O_3@$Li_4Ti_5O_{12}$ composite and its application in the lithium-ion batteries, ACS Appl. Mater. Interfaces 6 (2014) 4514–4523. https://doi.org/10.1021/am500294m

[2] D. Lindley, Smart grids: The energy storage problem, Nature 463 (2010) 18–20. https://doi.org/10.1038/463018a

[3] S. Zheng, H. Xue, H. Pang, Supercapacitors based on metal coordination materials, Coord. Chem. Rev. 373 (2018) 2–21. https://doi.org/10.1016/j.ccr.2017.07.002

[4] D. Larcher, J. M. Tarascon, Towards greener and more sustainable batteries for electrical energy storage, Nat. Chem. 7 (2014) 19. https://doi.org/10.1038/nchem.2085

[5] P. Simon, Y. Gogotsi, Capacitive Energy storage in nanostructured carbon–electrolyte systems, Acc. Chem. Res. 46 (2013) 1094–1103. https://doi.org/10.1021/ar200306b

[6] Chen, George Z., Supercapacitor and supercapattery as emerging electrochemical energy stores, Int. Mater. Rev. 62 (2017) 173–202. https://doi.org/10.1080/09506608.2016.1240914

[7] A. González, E. Goikolea, J.A. Barrena, R. Mysyk, Review on supercapacitors: Technologies and materials, Renew. Sustain. Energy Rev. 58 (2016) 1189–1206. https://doi.org/10.1016/j.rser.2015.12.249

[8] J.R. Miller, P. Simon, Electrochemical capacitors for energy management, Science 321 (2008) 651-652. https://doi.org/10.1126/science.1158736

[9] A.G. Pandolfo, A.F. Hollenkamp, Carbon properties and their role in supercapacitors, J. Power Sources 157 (2006) 11–27. https://doi.org/10.1016/j.jpowsour.2006.02.065

[10] P. Sharma, T.S. Bhatti, A review on electrochemical double-layer capacitors, Energy Convers. Manag. 51 (2010) 2901–2912. https://doi.org/10.1016/j.enconman.2010.06.031

[11] Y. Zhai, Y. Dou, D. Zhao, P.F. Fulvio, R.T. Mayes, S. Dai, Carbon materials for chemical capacitive energy storage, Adv. Mater. 23 (2011) 4828–4850. https://doi.org/10.1002/adma.201100984

[12] Winter M, Brodd R. What are batteries, fuel cells, and supercapacitors?, Chem. Rev. 104 (2004) 4245-4269. https://doi.org/10.1021/cr020730k

[13] A.S. Aricò, P. Bruce, B. Scrosati, J.-M. Tarascon, W. van Schalkwijk, Nanostructured materials for advanced energy conversion and storage devices, Nat. Mater. 4 (2005) 366-377. https://doi.org/10.1038/nmat1368

[14] A. Peigney, C. Laurent, E. Flahaut, R.R. Bacsa, A. Rousset, Specific surface area of carbon nanotubes and bundles of carbon nanotubes, Carbon 39 (2001) 507–514. https://doi.org/10.1016/S0008-6223(00)00155-X

[15] J. Xia, F. Chen, J. Li, N. Tao, Measurement of the quantum capacitance of graphene, Nat. Nanotechnol. 4 (2009) 505–509. https://doi.org/10.1038/nnano.2009.177

[16] Q. Ke, J. Wang, Graphene-based materials for supercapacitor electrodes-A review, J. Materiomics 2 (2016) 37–54. https://doi.org/10.1016/j.jmat.2016.01.001

[17] X. Zhang, H. Zhang, C. Li, K. Wang, X. Sun, Y. Ma, Recent advances in porous graphene materials for supercapacitor applications, RSC Adv. 4 (2014) 45862–45884. https://doi.org/10.1039/C4RA07869A

[18] W. Yang, M. Ni, X. Ren, Y. Tian, N. Li, Y. Su, X. Zhang, Graphene in supercapacitor applications, Curr. Opin. Colloid Interface Sci. 20 (2015) 416–428. https://doi.org/10.1016/j.cocis.2015.10.009

[19] W.S. Hummers, R.E. Offeman, Preparation of graphitic oxide, J. Am. Chem. Soc. 80 (1958) 1339. https://doi.org/10.1021/ja01539a017

[20] D. Maruthamani, S. Vadivel, M. Kumaravel, B. Saravanakumar, B. Paul, S. Sankar, A. Habibi-yangjeh, A. Manikandan, G. Ramadoss, Fine cutting edge shaped Bi_2O_3 rods/reduced graphene oxide (RGO) composite for supercapacitor and visible-light photocatalytic applications, J. Colloid Interface Sci. 498 (2017) 449–459. https://doi.org/10.1016/j.jcis.2017.03.086

[21] S. Stankovich, D.A. Dikin, R.D. Piner, K.A. Kohlhaas, A. Kleinhammes, Y. Jia, Y. Wu, S.T. Nguyen, R.S. Ruoff, Synthesis of graphene-based nanosheets via chemical reduction of exfoliated graphite oxide, Carbon 45 (2007) 1558–1565. https://doi.org/10.1016/j.carbon.2007.02.034

[22] S.I. Wong, J. Sunarso, B.T. Wong, H. Lin, A. Yu, B. Jia, Towards enhanced energy density of graphene-based supercapacitors: Current status, approaches, and future directions, J. Power Sources. 396 (2018) 182–206. https://doi.org/10.1016/j.jpowsour.2018.06.004

[23] M. Kim, H. Min, G.H. Park, H. Lee, Graphene-based composite electrodes for electrochemical energy storage devices: Recent progress and challenges, FlatChem 6 (2017) 48–76. https://doi.org/10.1016/j.flatc.2017.08.002

[24] H. Yang, S. Kannappan, A.S. Pandian, J.-H. Jang, Y.S. Lee, W. Lu, Rapidly annealed nanoporous graphene materials for electrochemical energy storage, J. Mater. Chem. A 5 (2017) 23720–23726. https://doi.org/10.1039/C7TA07733E

[25] E. Senthilkumar, V. Sivasankar, B.R. Kohakade, K. Thileepkumar, M. Ramya, G. Sivagaami Sundari, S. Raghu, R.A. Kalaivani, Synthesis of nanoporous graphene and their electrochemical performance in a symmetric supercapacitor, Appl. Surf. Sci. 460 (2018) 17–24. https://doi.org/10.1016/j.apsusc.2017.10.221

[26] A.A. Silva, R.A. Pinheiro, A.C. Rodrigues, M.R. Baldan, V.J. Trava-Airoldi, E.J. Corat, Graphene sheets produced by carbon nanotubes unzipping and their performance as supercapacitor, Appl. Surf. Sci. 446 (2018) 201–208. https://doi.org/10.1016/j.apsusc.2018.01.214

[27] N.C. Deb Nath, I.-Y. Jeon, M.J. Ju, S.A. Ansari, J.-B. Baek, J.-J. Lee, Edge-carboxylated graphene nanoplatelets as efficient electrode materials for electrochemical supercapacitors, Carbon. 142 (2019) 89–98. https://doi.org/10.1016/j.carbon.2018.10.011

[28] Y. Liu, Y. Li, F. Su, L. Xie, Q. Kong, Easy one-step synthesis of N-doped graphene for supercapacitors, Energy Storage Mater. 2 (2016) 69–75. https://doi.org/10.1016/j.ensm.2015.09.006

[29] M. Li, Z. Wu, W. Ren, H. Cheng, N. Tang, The doping of reduced graphene oxide with nitrogen and its effect on the quenching of the material's photoluminescence, Carbon 50 (2012) 5286–5291. https://doi.org/10.1016/j.carbon.2012.07.015

[30] Q. Zeng, Z. Ullah, M. Chen, H. Zhang, R. Wang, L. Gao, L. Liu, G. Tao, Q. Li, Assembly of highly stable aqueous dispersions and flexible films of nitrogen-doped graphene for high- performance stretchable supercapacitors, J. Mater. Sci. 52 (2017) 12751-12760. https://doi.org/10.1007/s10853-017-1336-7

[31] S. Yue, H. Tong, Z. Gao, W. Bai, L. Lu, J. Wang, X. Zhang, Fabrication of flexible nanoporous nitrogen-doped graphene film for high-performance supercapacitors, J. Solid State Electrochem. 21 (2017) 1653–1663. https://doi.org/10.1007/s10008-017-3538-y

[32] S. Kannappan, H. Yang, K. Kaliyappan, R.K. Manian, A. Samuthira Pandian, Y.S. Lee, J.H. Jang, W. Lu, Thiolated-graphene-based supercapacitors with high energy density and stable cycling performance, Carbon 134 (2018) 326–333. https://doi.org/10.1016/j.carbon.2018.02.036

[33] A. Ansaldo, P. Bondavalli, S. Bellani, A.E. Del Rio Castillo, M. Prato, V. Pellegrini, G. Pognon, F. Bonaccorso, High-power graphene–carbon nanotube hybrid supercapacitors, ChemNanoMat 3 (2017) 436–446. https://doi.org/10.1002/cnma.201700093

[34] J. Qin, M. Zhang, S. Rajendran, X. Zhang, R. Liu, Facile synthesis of graphene-AgVO$_3$ nanocomposite with excellent supercapacitor performance, Mater. Chem. Phys. 212 (2018) 30–34. https://doi.org/10.1016/j.matchemphys.2018.01.040

[35] L. Deng, J. Liu, Z. Ma, G. Fan, Z. Liu, Free-standing graphene/bismuth vanadate monolith composite as a binder-free electrode for symmetrical supercapacitors, RSC Adv. 8 (2018) 24796–24804. https://doi.org/10.1039/C8RA04200D

[36] B.S. Singu, K.R. Yoon, Exfoliated graphene-manganese oxide nanocomposite electrode materials for supercapacitor, J. Alloys Compd. 770 (2019) 1189–1199. https://doi.org/10.1016/j.jallcom.2018.08.145

[37] J.A. Argüello, J.M. Rojo, R. Moreno, Electrophoretic deposition of manganese oxide and graphene nanoplatelets on graphite paper for the manufacture of supercapacitor electrodes, Electrochim. Acta 294 (2019) 102–109. https://doi.org/10.1016/j.electacta.2018.10.091

[38] R. Xing, R. Li, X. Ge, Q. Zhang, B. Zhang, C. Bulin, H. Sun, Y. Li, Synthesis of 1,3-dicarbonyl-functionalized reduced graphene oxide/MnO$_2$ composites and their

electrochemical properties as supercapacitors, RSC Adv. 8 (2018) 11338–11343. https://doi.org/10.1039/C7RA13394D

[39] Y. Cheng, Y. Zhang, Q. Wang, C. Meng, Synthesis of amorphous $MnSiO_3$/ graphene oxide with excellent electrochemical performance as supercapacitor electrode, Colloids Surf. A 562 (2019) 93–100. https://doi.org/10.1016/j.colsurfa.2018.11.011

[40] X.H. Guan, M. Li, H.Z. Zhang, L. Yang, G.S. Wang, Template-assisted synthesis of $NiCoO_2$ nanocages/reduced graphene oxide composites as high-performance electrodes for supercapacitors, RSC Adv. 8 (2018) 16902–16909. https://doi.org/10.1039/C8RA02267D

[41] D. Vikraman, K. Karuppasamy, S. Hussain, A. Kathalingam, A. Sanmugam, J. Jung, H.-S.Kim, One-pot facile methodology to synthesize MoS_2-graphene hybrid nanocomposites for supercapacitors with improved electrochemical capacitance, Compos. Part B Eng. 161 (2019) 555–563. https://doi.org/10.1016/j.compositesb.2018.12.143

[42] D. Xiong, X. Li, Z. Bai, J. Li, Y. Han, D. Li, Vertically aligned Co_9S_8 nanotube arrays onto graphene papers as high-performance flexible electrodes for supercapacitors, Chem. Eur. J. 24 (2018) 2339–2343. https://doi.org/10.1002/chem.201704239

[43] B. Xie, M. Yu, L. Lu, H. Feng, Y. Yang, Y. Chen, H. Cui, R. Xiao, J. Liu, Pseudocapacitive Co_9S_8/graphene electrode for high-rate hybrid supercapacitors, Carbon 141 (2019) 134–142. https://doi.org/10.1016/j.carbon.2018.09.044

[44] J. Lin, S. Yan, P. Liu, X. Chang, L. Yao, Facile synthesis of $CoNi_2S_4$/graphene nanocomposites as a high-performance electrode for supercapacitors, Res. Chem. Intermed. 44 (2018) 4503-4518. https://doi.org/10.1007/s11164-018-3400-6

[45] X. Yang, H. Niu, H. Jiang, Z. Sun, Q. Wang, F. Qu, One-step synthesis of $NiCo_2S_4$/graphene composite for asymmetric supercapacitors with superior performances, ChemElectroChem 5 (2018) 1576–1585. https://doi.org/10.1002/celc.201800302

[46] T.W. Lin, T. Sadhasivam, A.-Y. Wang, T.Y. Chen, J.Y. Lin, L. Shao, Ternary Composite Nanosheets with MoS_2/WS_2/graphene heterostructures as high-performance cathode materials for supercapacitors, ChemElectroChem 5 (2018) 1024–1031. https://doi.org/10.1002/celc.201800043

[47] N. Pal, S. Chauhan, M. Mozafari, N. Singh, K. Meghwal, R. Ameta, S.C. Ameta, High-performance supercapacitors based on polyaniline–graphene nanocomposites:

Some approaches, challenges and opportunities, J. Ind. Eng. Chem. 36 (2016) 13–29.
https://doi.org/10.1016/j.jiec.2016.03.003

[48] K. Pal, V. Panwar, S. Bag, J. Manuel, J.H. Ahn, J.K. Kim, Graphene oxide–polyaniline–polypyrrole nanocomposite for a supercapacitor electrode, RSC Adv. 5 (2015) 3005–3010. https://doi.org/10.1039/C4RA14614J

[49] A.V. Murugan, T. Muraliganth, A. Manthiram, Rapid, Facile Microwave-solvothermal synthesis of graphene nanosheets and their polyaniline nanocomposites for energy strorage, Chem. Mater. 21 (2009) 5004–5006. https://doi.org/10.1021/cm902413c

[50] M. Khalid, L.T. Quispe, C.C. Pla Cid, A. Mello, M.A. Tumelero, A.A. Pasa, The synthesis of highly corrugated graphene and its polyaniline composite for supercapacitors, New J. Chem. 41 (2017) 4629–4636. https://doi.org/10.1039/C7NJ00024C

[51] L. Tang, Z. Yang, F. Duan, M. Chen, Fabrication of graphene sheets/polyaniline nanofibers composite for enhanced supercapacitor properties, Colloids Surf. A. 520 (2017) 184–192. https://doi.org/10.1016/j.colsurfa.2017.01.083

[52] A.M. Obeidat, A.C. Rastogi, Electrochemical energy storage performance of asymmetric PEDOT and graphene electrode-based supercapacitors using ionic liquid gel electrolyte, J. Appl. Electrochem. 48 (2018) 747-764. https://doi.org/10.1007/s10800-018-1182-6

Chapter 4

Three-Dimensional Graphene Materials for Supercapacitors

Gurjinder Kaur[1#], Narasimha Vinod Pulagara[1#], Indranil Lahiri[1,2*]

[1]Nanomaterials and Applications Lab., Metallurgical and Materials Engineering Department, Indian Institute of Technology Roorkee, Roorkee, Uttrakhand-247667, India

[2]Centre of Excellence: Nanotechnology, Indian Institute of Technology Roorkee, Roorkee, Uttrakhand-247667, India

*indrafmt@iitr.ac.in

#authors with equal contributions

Abstract

Three-dimensional (3D) graphene architectures have allured remarkable attention for supercapacitor (SC) applications owing to their highly accessible surface area, low density, structural interconnectivity (micro-, meso- and macro-interconnected pores), excellent electrical conductivity and good mechanical strength. Overall supercapacitance performance of 3D graphene-based SCs has been due to enhanced accessibility of the electrode surface to electrolyte ions, which also provides conductive channels for electron transfer. In addition, 3D graphene structures provide an ideal template for active material decoration. In this book chapter, an intense review on types of three-dimensional graphene-related materials, and their synthesis methods as well as electrochemical performance for SC applications is presented.

Keywords

3D graphene, Foam, Gels, Spheres, Fibers, Supercapacitors

Contents

1.   Introduction

Electrical energy is the most widely used form of energy in the world and is an essential part of modern life. The robust world economic growth boosted the global demand for electric energy. Global energy demand was predicted to increase further by 30% till 2040 [1]. Fossil fuels, coal, oil and natural gases are mostly in demand as global energy sources. However, fossil fuel burning releases harmful gases, resulting in continuous deterioration of the global environment. Moreover, fossil fuel resources are expensive, continuously depleting, limited in supply and non-renewable. In recent years, renewable energy sources have become incredibly popular because these are sustainable, pollution-free, efficient and recyclable [1]. Despite of the vast potential and advantages of renewable energy resources, there are some shortcomings also associated with them. The renewable energy sources, particularly solar and wind can be capricious. The output energy produced from renewable energy sources is highly variable and has excessive fluctuations due to the stochastic nature of renewable energy sources [2]. Thus, it is inconvenient to adjust them in response to the demand requirements. Zhenguo "Gary" Yang examined the problem of renewable integration and found that crux of the issue was the lack of cost-effective, reliable technology to store the energy during peak production and release as demand picks up [3]. There are a myriad of advantages associated with electrical energy storage (EES) systems. EES systems reduce the electricity cost, improve the reliability, stability and flexibility of power supply. It also improves the power quality of renewable energy generation network by sustaining the frequency and voltage at the required levels [2,4]. Figure 1 shows that most of energy storage devices have either high energy or power.

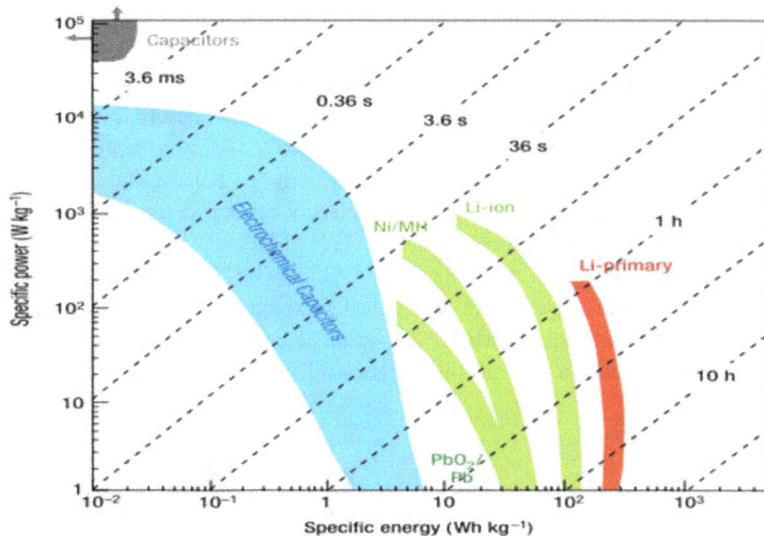

Figure 1: Power vs energy characteristics of ESS (Reproduced with permission Ref [5])

Both maximum power and energy capabilities are preferred for supercapacitors which make them ideal to use in high -energy and power storage devices [6,7]. Power and energy density ranges of supercapacitors are amid 1-10 $Whkg^{-1}$ and 1,000-5,000 Wkg^{-1}, respectively [8]. Supercapacitor follows the similar basic mathematical relations just as the traditional capacitors. These employ high specific surface area electrodes and thin dielectrics to attain much higher capacitances than conventional capacitors. Because of these attributes, Supercapacitors (SCs) can offer higher energy density than classic capacitors [9]. In addition, SCs exhibit higher power density and outstanding cycle performance, more than 100 times compared to batteries. Some other advantages of SCs over the conventional energy storage devices include their quick rate of charge and discharges, good life cycle, long operational life, robust temperature operating range, flexible packaging, lightweight, low maintenance, more capability and high current efficiency [6,7,10]. However, SCs possess lower energy density compared to available batteries, major obstacle to their widespread applicability. The energy density of economical supercapacitor cells is limited to 10 Wh/kg (which is satisfactory for several applications), while common lead-acid batteries' energy density can reach 35–40 Wh/kg. For Li-ion batteries, energy density can be higher than 100 Wh/kg [11]. Therefore, significant research efforts have been made to devise ways to enhance the energy density of SCs, to increase it to a level comparable to batteries and make them cost competitive with conventional batteries. The overall performance of SC depends upon the electrode material, electrolyte, operating voltage, internal resistance and voltage range. The capacity of SC is primarily determined by the type and structure of electrode material. Parameters, related to electrode of SC, such as specific surface area of the electrode material, distribution of pore size, electrical conductivity, pore shape, and structure along with surface functionality strongly affect the electrochemical performance of SC and hence its electrical energy storage capacity [12]. Electrode materials should have adequate surface area and high electrical conductivity in addition to be easily functionalized and possess excellent electrochemical properties [13]. Effective research on new electrode materials, which are important to improve the performance of the supercapacitor, has taken a big jump during the past few years (Figure 2).

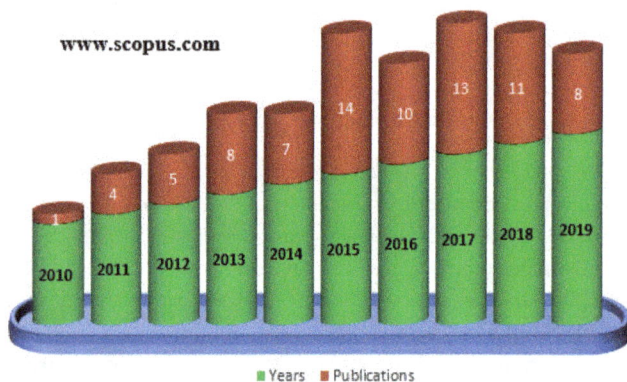

Figure 2: Year wise publication trend for graphene based supercapacitors (up to January 2019).

Depending upon the charge storage mechanisms, SCs could be classified into two distinct types: pseudocapacitors (PC) and electric double layer capacitors (EDLC) [5]. In EDLCs, the capacitance is achieved via electrostatic charge accumulation at the electrode, electrolyte interface, on both sides of the separator. Therefore, high storage capacity can be achieved by controlling the pore size, electrical conductivity and specific surface area of electrodes. For EDLCs, more specific-surface area (> 1000 m^2/g) can be obtained by using nanoporous carbon materials like carbon nanotubes (CNTs) as active electrode materials, leading to higher capacitance than traditional electrostatic capacitors. For PCs, the charge storage mechanism relies on the fast reversible multi-electron Faradaic process (redox reactions) at the surface of the electrodes. These possess much bigger specific capacitances and energy densities as compared to EDLCs. However, low electrical conductivity of PC electrodes results in unsatisfactory electrochemical performance and lower cycle life. Performance of SCs was improved by combining the properties of EDLC and PCs in one system, called hybrid supercapacitors [14]. There are three categories of electrode materials depending on their usage for EDLCs, PCs, and hybrid supercapacitors (Table 1).

Table 1. Types of SCs and their electrode materials

Electric Double Layer Capacitors	<ul><li>Carbon aerogels</li><li>Activated carbons</li><li>Carbon fibers</li><li>Carbon nanotubes</li></ul>
Pseudo – Capacitors	<ul><li>Metal oxides</li><li>Conducting polymers</li></ul>
Hybrid/Combination Capacitors ❖ Asymmetric ❖ Composite	<ul><li>Carbon materials, conducting polymers</li><li>Carbon materials, metal oxides</li></ul>

Though a large number of electrode materials have been reported for supercapacitors, carbon nanomaterials attracted great attention due to their physicochemical nature and electrochemical properties. In particular, graphitic carbon nanomaterials have an important role to increase overall performance of SCs. Demand for high-performance next-generation supercapacitors thrilled the scientists to search for carbon nanomaterials those have high specific surface area, are easily available, electrically conductive, clean and green in nature and low in cost. In EDLC supercapacitors, activated carbon materials have been mostly preferred due to more specific surface area (500 -2000 m^2/g) [15,16]. However, the whole area does not contribute to energy storage due to poor pore size distribution, small pore size (< 2nm [17] or up to 0.3 nm [18]) and poor electrical conductivity. The micropores of activated carbon are inaccessible to electrolyte ions [19], while the macropores lead to a low surface-to-volume ratio. Both these factors result in a low specific capacitance. CNTs exhibit moderate surface area and high electrical conductivity. However, these CNTs need to combine with pseudocapacitive material to achieve a good supercapacitive performance [20]. In addition, difficult manufacturing process and high cost of synthesis limit their actual application in supercapacitor devices.

To overcome the aforementioned limitations, graphene seems to be the most promising candidate for the EDL supercapacitors electrode material. Graphene, with its high electrical conductivity, unique lattice structure, specific surface area as well as immense chemical stability has received ample attraction for possible application as electrodes in supercpacitors. The intention of this book chapter is to mainly summarise contemporary progress in the development of three dimensional graphene-based supercapacitors followed by analytical approaches for engineering materials to enhance device efficiency. A very brief discussion on those properties of graphene which are important for its fulfillment in SCs and dimensionality based application of graphene for SCs is also presented in the following sections.

2.   Graphene-based materials for supercapacitor

Since the first demonstration of SC in 1957 by H.I. Becker using porous carbon electrodes [21], much progress has been made into this research area. As mentioned earlier that carbon nanomaterials have been basically used as electrode materials in SCs. Graphene has received special attention in this field because of marvelous properties like high carrier mobility, good thermal and electrical conductivities , lightweight, excessive specific surface area, high mechanical strength (~1 TPa), good flexibility, easy functionalization, good chemical stability, ecologically friendly and low cost of production. Among all the carbon-related materials as electrode to EDLCs, graphene has higher and immensely tunable surface area (up to 2700 m^2/g). In addition, electrochemical performance of EDLCs with graphene as electrode material does not rely on the spread of pore sizes and graphene has open pores which enhance the ion transport kinetics. Moreover, graphene is 2D atomically thick, lightweight material having both surfaces readily available for electrolyte. If the whole specific surface area of graphene could be utilized, the theoretical capacitance of graphene-based SCs could reach up to 21 μF/cm^2 or 550 F/g. With the aforementioned attributes, graphene takes a lead role in the recent research activities for the development of electrode material for the next-generation of high-performance SCs. Graphene has become the potential candidate for SC applications starting from portable electronic devices to electric vehicles.

Graphene dimensions can be tuned and assembled into various distinct structures; zero-dimensional (graphene quantum dot (GQDs) or particles), one- dimensional (graphene nanoscrolls, fibers, yarns etc), two-dimensional (graphene and graphene related 2D films), three-dimensional (graphene foams, sponges, aerogels, hydrogels and composites).

2.1   Zero-dimensional graphene based materials for supercapacitors

Owing to their novel chemical and physical properties associated with both graphene and QDs, GQDs have emerged as one of the promising materials for supercapacitors. These outstanding properties of GQDs include their nanometer size, high specific surface area, good electrical conductivity, sublime transporting ability, chemical inertness, abundant edge defects, excellent stability and good dispersion in several organic solvents [22,23]. Graphene dots and particles could be prepared through top-down or bottom-up synthetic routes. Top-down method, in contrast to bottom-up method, is a highly successful process to restraint the size distribution under the simplest synthesis conditions. GQD powder synthesis via top-down approach involves chemical exfoliation of graphite into graphene oxide (GO) followed thermal/chemical reduction of graphene oxide. Zhang et al. [23] prepared GQDs with uniform sizes (< 5nm) via innovative top-down method using nitric acid. GQDs were preferred as electrode materials for EDLC SC. In this work, specific capacitance of GQDs was observed to be twice that of graphene. This was because the GQDs with size less than 5 nm effectively decreased the degree of restacking of graphene sheets. Moreover, edge defects contributed to the ion diffusion between electrode material and electrolyte, thus enhancing the electrochemical performance. This ELDC SC with GQD as electrodes showed more specific capacitance of 307.6 F/g at rate of scan 5 mV/s, a satisfactory energy density of 41.2 W h/kg, superb stability of cycle conjointly a retention of capacitance over 97.6% after 5000 charge-discharge cycles.

GQDs based SCs were also prepared by combining GQDs with conducting polymers [22,24], 2D graphene [25], 3D graphene [26], CNTs [27] and halloysite nanotubes [28]. Very recently, GQDs were combined with transition metal oxide, MnO_2, to enhance the performance of the supercapacitor [29]. In that study, GQD/MnO_2 heterostructural material was used as an electrode. It is interesting to observe that the operating voltage window was increased from 0-1 to 0-1.3V. Besides, a very high specific capacitance (1170 F/g) at 5 mV/s was achieved between 0 and 1.3 V. It also showed excellent cycle performance with capacitance retention of 92.7% after 10000 cycles, good power and energy density of 12351 W/kg and 118 Wh/kg, respectively.

2.2   One-dimensional graphene based materials for supercapacitors

Graphene-based nanoscrolls (GNs) are prepared through rolling the graphene sheets into an uncommonly one-dimensional tubular geometry along with open ends. This new topological structure (helical nanostructure) of GNs has the characteristics of both graphene sheets and unique features of nanoscrolls such as scrolled conformation, porous structure, open ends and tunable inter-layer spacing. These are morphologically similar to multiwalled carbon nanotubes, but their properties enable GNs to inherit the excellent characteristics of both CNTs and graphene. Due to the open ends of GNs, the surface area of graphene sheets could be utilized more productively in GNs. Due to these attributes, GNs distinct from graphene sheets and CNTs to become promising candidates for SC applications [30]. Some reports describing the details of synthesis and characterization of GNs are available in the literature [31–33]. However, only a few reports dealing with the application of GNS as a promised electrode in SC have been published. Gao and his group [34] reported large production of GNs by enhanced spray freeze drying approach. The prepared GNs have specific surface area of 386.4 m^2/g with micro-pores of size ~0.7 nm and meso-pores of size ~2.7 nm. The synthesized GNs exhibited specific capacitance about 90–100 Fg^{-1} at 1 Ag^{-1}. Dhar and co-workers [30], in a recent article, have demonstrated GNs synthesis with adjustable lengths ranging from micro- to nano-scale using a renewable cellulose nanocrystal template. The GNs showed 223-357 F/g of specific capacitance (at 1A/g) along with attractive long-term cyclic stability (retention ~93.5–96.4% at ~10000 cycles). Very recently, the fast cooling of chemically reduced graphene through liquid nitrogen has been mentioned by Mohanpriya and co-workers [35]. The constructed GNs based SC showed a specific capacitance of 309.8 F/g at 0.5A/g along with long-term cyclic stability (retention ~90.6% after 4000 cycles). The power and energy densities achieved for GNS were 10800 W/kg and 27.5 Wh/kg, respectively. In order to enhance electrochemical efficiency of GNs based SCs, GNs were hybridised with metal oxides (MnO_2 and Co_3O_4) and conductive polymers. Yan et al. [36] reported MnO_2 nanowire templated GNs indicating the specific capacitance of 317 F/g at 1 A/g. Gho and co-authors [37] prepared GN/polyaniline composite which displayed specific capacitance of 320 F/g at 1 A/g, though it is not good enough for energy storage device applications.

Graphene-based yarns and fibers are of practical importance for next-generation supercapacitors due to their small volume, flexibility, less cost, lightweight, weave-ability, shape-ability and ease of functionalization [38,39]. This specific capacitance of graphene foams was evaluated to be ~100 F/g, and while specific capacitance of graphene based fibers, yarns based supercapacitors was reported to be 409 F/g [40]. Yu et al. reported the increased capacitance (300 F/cm^3) of GNFs by introduction of CNTs to create extra hierarchical pores in GNFs [41]. Other materials like transition metal oxides (MnO_2, Bi_2O_3) and conductive polymers were also added to GFs to improve the SC performance [33,42,43]. Liu et al. [44] have described the preparation of hierarchical composite/hybrid electrode composed of Ni coated cotton yarns and graphene sheets for SC. The device exhibited a high power density, volumetric energy density of 1400 mW/cm^3 and 6.1 mWh/cm^3, respectively. In order to get more details about the synthesis of graphene fibers or yarns and their SC applications, interested readers may refer to the relevant review papers [38,39,45].

2.3   Two-dimensional graphene based materials for supercapacitors

The one atomic layer thick 2D planar sheet structure of graphene makes a favorable candidate to SCs because of its characteristics. Stoller et al. [46] were the first who developed, in 2008, the chemically modified graphene (CMG) from a single layer graphene. The specific capacitance of CMG electrode based SC was found to be 135 F/g in a 5.5 molar KOH aqueous electrolyte. In organic electrolytes, the SC exhibited specific capacitance of 94 Fg^{-1} and 99 Fg^{-1} in tetraethylammonium tetrafluoroborate ($TEABF_4$)/acetonitrile (AN) and 1 molar $TEABF_4$)/propylene carbonate (PC) electrolytes, respectively. Since this first report on 2D graphene-based SCs, several research groups started to work on the direct utilization of graphene and its hybrids in SCs. Vivekchand et al. reported the use of graphene as an electrode material for SC applications. Graphene was synthesized by three different techniques to produce different porosities and surface areas, though other properties of graphene were also affected. These different techniques were thermal exfoliation of graphitic oxide (EG), heating of nanodiamond at 1650°C in a helium environment (NG) and decomposition of camphor over Ni nanoparticles (CG). The graphene produced by the first technique (EG) exhibited the highest surface area of 925 m^2g^{-1} and capacitance of 117 Fg^{-1}. DG, CG displayed specific capacitance of 35 Fg^{-1}, 6 Fg^{-1}, respectively. Graphene papers were also used in SCs because of their adjustable thickness, highly ordered layered structure, structural flexibility, light weight, useful mechanical as well as electrical properties [47–49]. Several synthesis methods such as spin coating, electrochemical reduction of GO, laser irradiation reduction of GO, reduction, self-assembly, vacumm filtration, Chemical Vapor Deposition (CVD), layer by layer stacking, have been developed to made high quality graphene-based 2D films and papers for SC applications [49–53].

However, aggregation or restacking at macroscopic scale of 2D graphene sheets restricts its application in supercapacitor devices in spite of the superb unique properties. This restracking into graphite results from strong van der Waals interactions amid two (side by side) graphene layers and interlayer $\pi–\pi$ stacking interactions causing decreased number of ions accessible through effective surface areas. To outrival this barrier, 3D graphene based nanostructure materials were introduced [46,54–57].

3.   Three-dimensional graphene based materials for supercapicitors

Three dimensional (3D) graphene architectures were designed to enhance the efficiency of graphene based SCs. 3D graphene based materials provide a functional structure having high accessible surface area, low density, structural interconnectivities (micro, meso and macro-interconnected pores), high electrical conductivity and good mechanical strength. This form of graphene possesses all the properties of graphene along with advanced functions of the 3D structure, which can enhance the power and energy densities, and hence, the supercapacitor performance [58,59]. The overall performance of 3D graphene based SCs was improved due to enhanced accessibility of electrode surface to electrolyte ions (provided by the porous nature of 3D materials to provide more active surface area) and the conductive channels (due to graphene properties) for electron transfer. In addition, 3D graphene structures provide an ideal template for active material decoration. 3D graphene structures include graphene foams, sponges, hydrogels, aerogels, graphene porous films, graphene spheres and graphene fibers (Figure 3). All these structures have been extensively investigated for supercapacitor applications. Following subsections discuss different morphologies of 3D graphene materials and their supercapacitor applications.

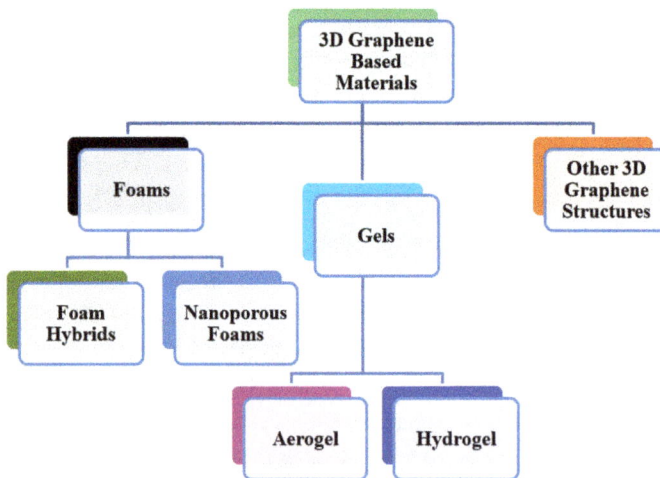

Figure 3: Classification of the 3D graphene dependent materials to SC applications

3.1   Graphene foam

Graphene foam (GF) is a 3-dimensional porous network composed of pores and pore walls made up of graphene. GF is a highly porous, lightweight, electrically conductive and flexible material. Due to these alluring inherent characteristics, 3D graphene foams became a keenly researched material for both researchers as well as related industries. In 2011, Z. Chen et al. reported a series of studies involving this material – template approach (CVD [60], powder metallurgy process[61]), without template approach, assembly of GO using hydrothermal process, three dimensional printing and sugar blowing approach and so forth [58,62–64]. Few other methods to synthesise 3D graphene foams and their supercapacitive performance are described below.

Because of non-polluting and economically favourable nature, the electrochemical exfoliation process is known to be one of the best processes to produce larger amounts of graphene. This is a top-down approach to synthesize graphene from graphite. Y. Ping and co-workers [65] prepared 3D graphene foam from graphite via the electrochemical exfoliation approach with two electrode system. Graphite (carbon source) foil, before its use as an anode, was immersed in liquid nitrogen for 20 s and then immediately shifted to the boiled deionized water. Thereafter, it was dried at 80 °C to achieve pre-processed expanded graphite. This pre-processed graphite was used as anode, Pt foil was considered as counter electrode, along with (1.0M) H_2SO_4 electrolyte solution. A distance of 3 cm was maintained between the anode (graphite) and counter electrode. DC bias (3V) was applied to the cell to carry out the electrochemical exfoliation process. After the completion of exfoliation process, the product of exfoliation was filtered with polytetrafluoroethylene (PTFE) membrane having 0.22 μm pore size and the product was cleaned multiple times with deionized water to remove the ionic impurities. After that, the final product of the graphite exfoliation was dispersed using sonication in DMF for 30 minutes. The dispersed solution was kept for 24 h for the impurities to settle down and the supernatant was dried for 12 h at 80 °C. In order to form graphene foam on a Ni foam structure, a mixture of proper mass ratio of exfoliated graphene, conductive agent (acetylene) and bonding agent PVDF (poly vinylidene fluoride) was prepared in a solvent of N-methyl pyrrolidone (NMP). This mixture was grinded for 2 h in an agate mortar to form homogeneous slurry (a semi liquid mixture). At the end, a semi liquid mixture was covered on nickel foam template and then baked at 120 °C (12 h) to produce three-dimensional graphene/nickel foam composite. The nickel foam template was later removed by an etching agent having 1:1 molar ratio ($FeCl_3$: HCl solution) for obtaining a final product of three-dimensional graphene foam. In this work, 2D graphene paper was also prepared via filtration of graphene dispersion in ethanol for comparison purpose.

Scanning electron microscope images of three dimensional graphene foam (Figure 4a) at low magnification shows that graphene foam exhibits a porous network, while the high magnification image (Figure 4b) shows the fine features, such as wrinkles, ripples at the edges, which result in lowering the free energy of the system and hence stabilize the graphene by avoiding agglomeration. The supercapacitor performances of the 3D graphene foam and 2D graphene paper were checked with cyclic voltammetry (CV), galvanostatic charge-discharge (GCD), including electrochemical impedance spectroscopy (EIS). These experiments were performed with 3 electrode system in 6M KOH electrolyte in the range of voltages 0-0.8 V, at a scan rate of 5-100 mV/s. CV curves for 3D graphene foam (Figure 4c) are almost rectangular and symmetrical, even at higher scan rates as compared to 2D graphene paper (Figure 4d). This shows that three dimensional graphene foam has outstanding rate performance. These curves also reveal the better capacitance and reversible cycle performance for three dimensional graphene foam, compared to two dimensional graphene paper electrodes. Absence of redox peaks during the scanning shows excellent stability and electrochemical performance for three dimensional graphene foam electrode material. The calculated capacitances at various rates of scans are displayed in Table 2.

Table 2. The capacitance values of three dimensional graphene foam than for two dimensional graphene paper

Scan rate (mVs^{-1})	5	10	20	30	50	100
Capacitance of three dimensional GF (Fg^{-1})	136.6	121.8	100.6	86.8	66.6	40.4
Capacitance of two dimensional graphene paper (Fg^{-1})	77.1	63.8	49.8	39.9	30.2	14.5

Figures 4(g,h) illustrate the electrochemical impedance spectra for 3D GF along with 2D graphene paper to determine AC impedance. The EIS curves were obtained in the frequency range 0.1Hz – 100Hz by applying AC voltage of 5mV amplitude. Typical Nyquist plots contain valuable information in three frequency regions, i.e. high, middle, and low frequency regions. At high frequency region, information about ionic resistance of the electrolyte, contact resistance (R_c) between current collector and electrode material, and intrinsic resistance of electrode are obtained. Another resistance term, charge transfer resistance, R_{ct} could be acquired from the semicircle diameter. Typically, when the diameter of semicircle is high, obviously then the charge transport resistance is more. The value of the R_c for 3D graphene foam is found to be 1.84 Ω, while its value for two dimensional graphene paper is (1.94 Ω). The R_{ct} value for three dimensional graphene foam was also less (1.28Ω) compared to that of two dimensional graphene paper (0.25Ω). In intermediate frequency range, Warburg resistance is obtained, which represents the spreading procedure of the electrolyte. In ideal conditions, it is a $45°$ oblique line. Here, $20°$ and $50°$ oblique lines for two dimensional graphene paper and three dimensional graphene foam, respectively, were observed, implying that three dimensional graphene performance was closer to the ideal conditions. Generally, in low frequency region, almost right angle line is observed, theoretically. Figures 4(g,h) show a more upright line (around $80°$) for 3D graphene as compared to that in 2D graphene paper (around $60°$). All these parameters obtained from Figure 4 indicate that 3D graphene foam exhibits good supercapactive performance than 2D graphene paper [65].

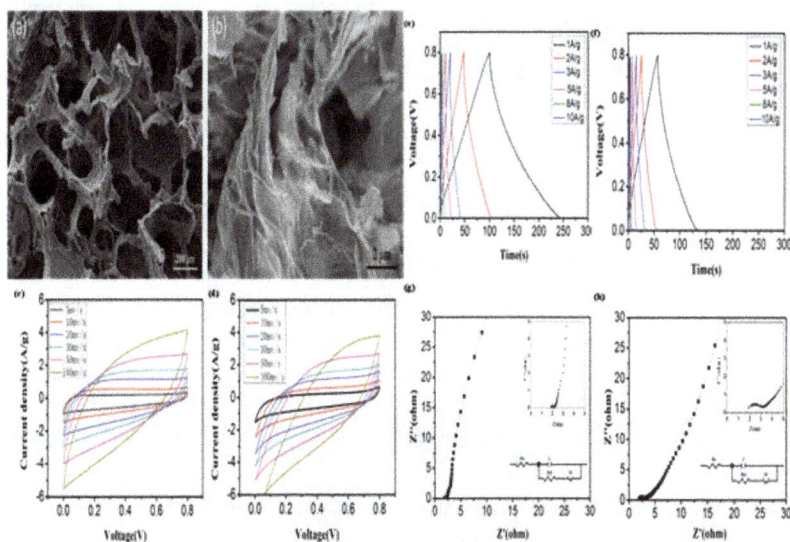

Figure 4: Scanning Electron Microscope images of (a) low, (b) high magnification of three dimenisonal graphene foam, CV and charge/discharge curves of (c,e) three dimensioal graphene foam; (d,f) two dimensional graphene paper electrode materials. Nyquist plots of (g) three dimensional graphene foam; (h) two dimensioal graphene paper electrode in 6 M KOH (Reproduced with permission Ref [65]).

3D porous graphene foams were also prepared by another top-down approach in which graphene oxide (GO) was prepared through Hummer's method, followed by chemical reduction to reduced graphene oxide (rGO) [66,67]. In this context, Yang et al. have demonstrated that GO, prepared via modified Hummers method, could be directly reduced on Ni foam, without any reducing agent. Ni foam was simply dipped in an acidic water suspension of GO sheets (pH=2) at room temperature. This was reduced to rGO in 15 days, without using any reducing agent (Figure 5). The composite structure of reduced graphene oxide with Ni foam was employed as electrode for supercapacitor applications. This composite exhibited electrochemical properties; areal capacitances were observed to be dependent on reduction time to various rGO loadings. The areal capacitance was found to be increased from 26.0 to 136.8 mF/ cm^2 at 0.5 mA/ cm^2, on the increase of the reaction time from 3 to 15 days. The electrochemical performance was further increased by extending the temperature to 70 °C. The authors observed that the rGO/Ni foam composite, synthesized at 70 °C with 5 h reduction, showed better electrochemical performance compared to the ones synthesized at room temperature with a 15 days reduction time. The 5 h reduced graphene oxide /nickel foam composite electrode demonstrated an areal capacitance of 206.7 mFcm^{-1} (at 0.5 mAcm^{-1}). It also showed an excellent cycling performance with areal capacitance retention of 97.4% after 10,000 cycles (3 mAcm^{-2}). The electrochemical properties were observed to improve further on increase of depletion time to 9 h (70 °C). The 5 h reduced graphene oxide/nickel foam composite delivered areal capacitance of 323 mFcm^{-2} (at 0.5 mAcm^{-2}) while maintaining a good rate performance and stability cycles [68].

In another interesting study, S Sivaprakash and co-workers prepared the chemically derived rGO foam on Ni template [69]. In synthesis process, GO was synthesised via improved Hummer's method and deposited on nickel foam via electrophoretic deposition method. After deposition, nickel foam was kept in the autoclave along with the remaining electrolyte at 180°C for 5 h to get rGO on Ni foam. By using etching agents (FeCl$_3$ / HCl), Ni foam was later removed to get free-standing three dimensional rGO foam.

The electrochemical properties of synthsized three dimensional rGO foam electrode and rGO on Ni foam electrodes were studied through CV analysis. CV tests were done at various scan rates (5mV/s – 200mV/s). The determined specific capacitance of reduced graphene oxide/nickel foam at various scan rates was found to be lesser compared to 3D rGO foam electrode, at corresponding rates of scans. However, this trend was reversed at highest rate of scan 200 mV/sec, depicting a high retention of capacitance at higher scan rate for rGO/Ni foam than the 3D-GF electrode. However, the electrochemical performance showed an opposite trend i.e. electrochemical performance of 3D rGO foam electrode was better compared to rGO/Ni foam due to high active surface area produced after Ni etching. GCD test was also carried out for both electrodes and the specific capacitance of reduced graphene oxide/nickel foam electrode was observed to be lower than that of the 3D graphene foam electrode (310 F/g at 20 A/g). Therefore, porous rGO was the perfect choice of electrode material for supercapacitors [69].

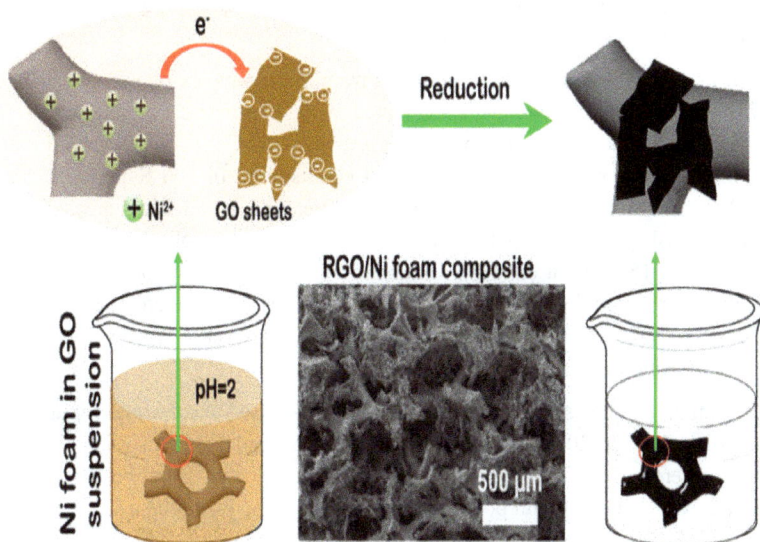

Figure 5: Procedure of 3D rGO/Ni foam composite (Reproduced with permission from Ref [68]).

Recently, Miao et al. [70] fabricated 3D graphene on nikel foam template through self assembly of graphene sheets 3D interconnected network using a hydrothermal approach. In this work, the Ni foam template was not removed and supercapacitor electrode was prepared by depositing Ni-Co-S nanosheets on an assembly of 3D graphene/Ni foam. The resultant electrode demonstrated maximum specific capacitance of $2526\,Fg^{-1}$ at $2\,Ag^{-1}$ and of $1916\,Fg^{-1}$ even at $10\,Ag^{-1}$. Further, the as-synthesized electrode showed cycle stability (capacitance retention of 77.0% after 2000 cycles at 20 A/g). Porous 3D foam like graphene architectures were also prepared by leavening process [71]. Authors were inspired by the ability of leavening the process to convert compact dough to light and porous bread with highly porous structure. GO films were prepared from GO suspension through vacuum filtration through a porous anodized aluminium oxide (AAO) membrane. Graphene films were taken-off from the AAO membrane and then kept in autoclave (heated to 90° C – 10 h) filled with hydrazine vapours. Hydrazine vapours were used as a leavening and reducing agent to convert GO film into porous 3D graphene. The gas evolved during conversion of GO to rGO by hydrazine caused a fifty-fold volume expansion and 30% mass loss to produce rGO foam with interconnected porous structures. A flexible rGO foam based symmetric solid-state supercapacitor was prepared using reduced graphene oxide foams as both current collectors, and working electrodes displayed a specific capacitance of 110 Fg^{-1}.

3D porous graphene could also be prepared from microwave exfoliated graphene oxides [72,73]. In this process, GO was exfoliated with microwave irradiation and further, chemically activated with KOH to produce activated MEGO (a-MEGO) with extremely high SSA of 3100 m^2/g and good electrical conductivity of 500 S/m. KOH activation created nanosize (1 to 10 nm) pores in graphene. The symmetric supercapacitor was constructed using a-MEGO both as agile electrode, current collector and has shown a capacitance of 150 Fg^{-1} at 0.8 Ag^{-1}. Further working on a-MEGO based supercapacitors [72], authors showed that the capability of supercapacitor to operates within wide range of temperature (-50 to 80 °C) along with specific capacitance of 100 F/g (20 mV/s), below room temperature.

Apart from Ni templates, polymer templates were also used to prepare 3D graphene foam like architectures. In 2012, 3D macroporous bubble graphene foam was prepared using PMMA spheres as hard templates. In this process, vacuum filtration of a mixture of graphene oxide hydrosol with polymer spheres suspension produced a sandwich structure of GO sheets and polymer spheres. The PMMA template was removed by calcination of the sandwich structure at 800°C. GO was converted to rGO simultaneously due to thermal reduction. The resultant 3D graphene foam, with controllable and uniform macro-pores and tunable microstructure, exhibited capacitance retention (67.9%) on raising the rate of scan rate from 3 to 1000 mV/sec. As the temperature of calcination was high, the graphene sheets aggregated together to reduce the specific surface area.

Choi and co-workers [74] synthesized similar kinds of three dimensional macro-porous GF with polystyrene colloidal particles as templates. After the synthesis of graphene, the polymer template was removed by acetone. This low temperature synthesis procedure prevented restacking of graphene sheets. The resultant macro-porous graphene films displayed superior electrical conductivity (1204 S/m). This porous 3D graphene foam was used as a scaffold to incorporate MnO$_2$ film into it. Good surface area including more electrical conductivity of MnO$_2$/ chemically modified graphene facilitated fast ionic transport and the as-synthesized electrode exhibited good electrochemical performance; specific capacitance of 389 Fg^{-1} at 1Ag^{-1} and 97.7% retention of capacitance when rate of scan was veried from 1 to 35 A/g. The asymmetrical supercapacitor fabricated by MnO$_2$/chemically modified graphene (cathode) and chemically modified graphene (anode) showed an energy and power density of 44 Wh/kg, 11.2 kW/g, respectively, at 2.0 V.

The 3D graphene architectures prepared from chemical reduction of GO contain morphological defects and chemical infomegeneities, which were continued during exfoliation and reduction processes. Though the products were high in yield, but the final products were defective as well as had less electrical conductivity [75]. On the other hand, the chemical vapor depostion approach could make large-area including high-quality desired 3D graphene materials with high electrical conductivity. In chemical vapor deposition process, the carbon atoms are directly deposited on targeted network (metallic foam template). The template/support acts as the catalyst to decompose the hydrocarbon precursors at the process temperature to form graphene in the three dimensional construction.

Chen et al. [60] were the first who synthesized three dimensional GF via chemical vapor deposition method employing nickel metallic foam as scaffold. Methane (CH_4) was used as carbon source and graphene was formed at 1000°C at ambient pressure controlled atmosphere of Ar and H_2 gases. After the synthesis of graphene films on Ni foam, a thin film of poly(methyl methacrylate) (PMMA) was coated on the top of graphene to avoid the collapsing of the graphene network during Ni foam etching. Thereafter, Ni was etched using $FeCl_3$/HCl solution. In the next step to obtain free-standing graphene foam, PMMA film was removed by hot acetone. The resultant free-standing graphene foam was extremely light (ultralow density of ~5 mg/cm^3), highly porous (approx. 99.7% porosity), and possessed good specific surface area (approx. 850 m^2/g) as well as good electrical conductivity of ~10 S/cm. This paper reported only the preparation of three dimensional GF morpholgy, however supercapacitor properties of graphene foam were not studied. But all the properties possessed by prepared graphene foam were desirable features for synthesis of high-performance supercapacitors.

The electrochemical efficiency of free-standing three dimensional GF was explored for the first time by Down et al. [76]. GF was manufactured on nickel foam via CVD approach at 1000°C under ambient pressure using methane as carbon source. As described in the above-mentioned method, Ni was removed after the graphene synthesis on Ni foam. This 3D graphene foam was directly used in symmetrical supercapacitor (two electrode system). All measurements were performed in a two-electrode system using aqueous and ionic liquid electrolytes. The 3-dimensional GF based SC gave a specific capacitance of 266 μFg^{-1} at 16.6 μAg^{-1} in an aqueous electrolyte. A notable development was noticed with the use of ionic liquids electrolytes. In ionic liquids, the device exhibited a capacitance from 287 up to 636 $\mu F/g$ at 6.66 mA/g down to 16.6 $\mu A/g$.

The power and energy densities were measured to be 29.33 kW/kg and 40.94 Wh/kg, respectively.

3.1.1 Graphene foam hybrids to enhance SC performance

The 3D graphene architectures can act as ideal templates to host transition metal oxides, conductive polymers for enhancement of electrochemical performance of SCs. Thus 3D graphene networks possessing low density, ample surface area, immense electrical conductivity, as well as high porosity have been combined with metal oxides [77,78], metal hydroxides [79,80], and polymers [81,82] to fabricate more efficient 3D graphene foam based composites/hybrids [78]. High conductivity of 3D graphene foam can facilitate the rapid electron mobility in SCs. High surface area of graphene foam could be helpful in progressing electrolyte ions very fast to the metal oxide surface. Zhang and co-authors [78] fabricated a 3D nanocomposite, consisting of graphene foam and NiO, for supercapacitor electrodes. Ethanol, which is safe and less expensive than explosive CH_4 gas, was chosen as the carbon source to produce GF on nickel foam via chemical vapor deposition (CVD) technique. The 3D graphene foam was further used as template to build three dimensional graphene/metal oxide nanocomposites. NiO was electrochemically deposited on the three dimensional GF to produce NiO-graphene three dimensional nanocomposite. This hybrid electrode material presented a big specific capacitance of approximately 816 F/g at 5 mV/s along with good rate capability. This supercapacitor exhibited a stable cycling performance with an improved specific capacitance after 2000 cycles.

Cobalt oxide can also be incorporated with graphene foam to produce Co_3O_4-graphene 3D nanocomposite materials for supercapacitor applications [77,83,84]. In a typical example, Dong et al. [84] constructed Co_3O_4/graphene foam nanocomposite. Graphene foam was synthesized by CVD method on Ni foam using ethanol as carbon source. Cobalt oxide nanowires were grown on graphene foam to construct 3D Co_3O_4/graphene nanocomposite material. The supercapacitor device with this electrode presented a good specific capacitance of 768 Fg^{-1} at 10 Ag^{-1}. Moreover, the specific capacitance increased to 1100 Fg^{-1} (at 10 Ag^{-1}) after 500 cycles and it remained stable afterwards. Another metallic oxide, MnO_2 was also combined with graphene foam to construct 3D graphene/MnO_2 nanocomposite networks [85]. Graphene foam was prepared by CVD of CH_4 source onto the Ni foam. The obtained flexible 3D graphene foam, after the Ni removal, possessed very low weight (0.70-0.75 mg/cm^2), ultra-small thickness (< 200 μm) and high conductivity (55 S/cm). The resultant 3D graphene foam being flexible, may prevent the collapsing and cracking during bending operation because of the hollow internal structure. Thus it can be easily applied as electrode in flexible SCs. After the withdrawal of Ni foam, MnO_2 was electrochemically deposited on 3D graphene foam. This hybrid electrode has shown specific capacitance of 465 Fg^{-1}, at a scan rate of 2 mVs^{-1}, for mass loading of 0.1 $mgcm^{-2}$.

Not only metal oxides, conductive polymers (CP) have also been used in pseudocapacitors because of their superior redox properties, low cost, great conductivity in doped state and simple synthesis [99]. However, the less cycling stability due to their expansion/contraction during charge and discharge process has been the problem with CP based SCs [100]. To overcome this problem related to life cycle stability, CPs (polypyrrole (PPY) and polyaniline (PANI)) were incorporated in 3D graphene foam [81,82,101]. Chabi et al. reported the electrochemical performance of 3D flexible graphene foam/PPY composite. The as synthesized graphene foam/PPY composite was directly employed as electrode in SC without any carbon additives. The SC presented a specific capacitance of 660 F/g, specific energy of 71 Wh/kg/, specific power of 2.4 kW/kg, superb life cycle with retention of capacitance almost 100% after 6000 charge/discharge cycles [101].

Furthermore, Xiong et al. [102] synthesized a 3D nonporous rGO-CNTs-PANI hybrid on Ni foam for SC applications. Graphene oxide was deposited on nickel foam through electrophoretic deposition, followed by thermal reduction in Ar/H_2 atmosphere. CNTs were grown on rGO-Ni foam via floating catalyst CVD. Polymer (PANI) was coated onto rGO-CNTs through *in-situ* anodic electrochemical polymerization of aniline monomer. SC prepared with rGO-CNTs-PANI hybrids as an electrode delivered specific capacitance of 741 F/g, better energy density of 92.4 Wh/kg, good power density of 6.3 kW/ kg, prolonged life cycle with retention of 95% capacitance after 5000 cycles.

Supercapacitor performance of 3D graphene foam based arcitechtectures was also improved by doping them with hetero-atoms like nitrogen, oxygen and boron [103,104]. It was observed that nitrogen doping into carbon enhanced the electrical conductivity and free charge-carrier density. The addition of hetero-atoms can increase pseudocapacitance through the introduction of N- or B-enriched functional groups. Huang et al. [105] synthesized N-doped rGO on Ni foam (NG/Ni) using urea as nitrogen containing precursor. The electrochemical efficiency was studied in two-electrode system using NG/ NF both as working electrode and collector electrode. This device displayed specific capacitance of 223 F/g at 0.29 A/g, but it decreased to 184 F/g at 35.7 A/g. The device showed a superior capability rate (83% retention) of capacitance for current density up to 35.7 A/g. It also offered cycling stability of 206 F/g after 3000 cycles.

Table 3. Supercapacitor properties of 3D graphene foam based hybrids

Graphene/Composite Foam	Specific capacitance	Energy density	Power density	Cyclic Performance	Electrolyte	Surface Area	Electrical conductivity	Ref
NiO/GF	~816 F/g at 5 mV/s			2000 (increased By 15%)	3 M KOH			[78
MnO_2/GF	465 F/g at 2 mV/s	6.8 Wh/Kg	62 W/Kg	5000 (29.8 F/g to 24.2)	0.5 M Na_2SO_4		55 S/m	[85
Co_3O_4/GF	768 F/g at 10 A/g.			500 (increased to 1100)	0.1 M NaOH			[84
$Co_3 O_4$/GN	1765 F/g at 1A/g	227.9 Wh/Kg	6000 W/Kg	5000 Cycles	0.1 M NaOH			[84
$Ni(OH)_2$/UGF	166 F/g at 0.5 A/g	13.4 Wh/Kg	85.0 Wh/Kg	10000 Cycles (5A/g)	6M KOH	20 Cm^2 Per Nominal Area		[86
$Ni_3S_2@Ni(OH)_2$/3D GF	1277 F/g at 2 mV/s	70.6 Wh/Kg	1.3 KW/Kg	2000 Cycles	3M KOH			[87
$Co(OH)_2$/GF	1139 F /g at 10 A/g	13.9 Wh/ Kg	18 KW/Kg	1000 Cycles At 10 A/g	1M KOH			[88
MnO_2/3D Nanoporous GF	310 F/g at 0.5 A/g			1000 Cycles	1 M Na_2SO_4			[89
$Ni(OH)_2$/GF	1450 F/g at 5 A/g			1000 Cycles	6 M KOH			[79
α-MnO_2 Nanofibers/3D GF	670 F/g at 10 A/g				1 M Na_2SO_4			[90
PANI/GF	346 F/g at 4 A/g			600 Cycles	1 M H_2SO_4			[81
CoO Nb/GF	352.75 F/g at 1 A/g			1000 Cycles	3 M NaOH			[91
Ag nanoparticles/GF	528 F /g at 1 A/g			3000 Cycles	4.0 M KOH			[92
RuO_2/ Graphene and CNT Hybrid Foam	502.78 F/g at 1 mA/cm^2	39.28 Wh/Kg	128.01 KW/Kg	8100 Cycles	2 M Li_2SO_4			[93
CO MoO_4/GF	2740 F/g at 1.43 A/g	21.1 Wh/Kg	6000 W/Kg	100000 Cycles At 400 A/g	2M KOH			[94
PANI/rGO film	438.8 F /g at 0.5 A/g			2000 Cycles	1 M H_2SO_4			[95
$CO_3(PO_4)_2.4H_2O$/GF	39.8 mAH/g	24 Wh/ Kg	468 W/Kg	10000 Cycles	6M KOH			[96
PPY/GF	258 mF/cm^2 at 1 mA/cm^2	22.9 μWh/ Cm^2	0.56 mW/cm^2	1000 Cycles	PVA/H_3PO_4 Gel		103 S/m	[82
N&S Co-doped/rGO foam	306.3 F/g at 1 A/g			10,000 Cycles	6 M KOH	505.7 m^2 /g		[97
V_2O_5/GF	73 mAH/g at 1 A/g	39 Wh/ Kg	947W/Kg	10,000 Cycles	6 M KOH	9.5 m^2/g		[98

N/O co-doped graphene flexible graphene foam on melamine foams was also reported to be used for SCs. Melamine foam worked as a sacrificial template to prevent the cluster of graphene. The supercapacitor manufactured through N/O graphene demonstrated huge areal capacitance of 375 mF/cm^2 at 1 A/g, ultra-high cycling stability of above 100% with capacitance retention of above 100% after 5000 cycles at 10 A/g, a maximal energy density of 16 Wh/kg and maximum power density of 17 kW /kg [104]. Many of these graphene foam hybrids, which were used for enhancing supercapacitor performance, were obtained from research reports. A summary of those materials along with their properties is presented in table 3.

3.1.2 Nanoporous graphene foam

Another approach to improve the supercapacitor performance of 3D GF is dependent on controlling the pore size of graphene foams. In much of the aforementioned reports, pore sizes and shapes, hole geometries of the three dimensional graphene foam were randomly arranged. In particular, pore morphology of CVD produced graphene is decided by the dimensions of metallic template foam. Thus graphene foam with macropores ($\sim$300 μm [106]) can only be formed on commercially available metallic foam templates. In order to counter this problem, many attempts were made to synthesize nanoporous graphene for supercapacitor applications [17,89,107–109].

Nanaji et al. [17] prepared nanoporous graphene (pore diameter of 3.4−3.8 nm) using green, recyclable bio-waste (jute stick) through carbon activation process at different activation temperatures. The fabrication process was facile, less cost, and eco-friendly. Jute sticks were baked at 80°C for 12 h and then pre-carbonized at 450°C in Ar environment to produce hydro char. This hydro char, having fine-grained powder, was then chemically agiled with KOH. Activated hydro-char was then annealed at various temperatures (800-1100°C) in an Ar environment. Thereafter, the pyrolyzed carbons were cleaned several times with 1 M HCl and deionized water to get pH neutrality. Finally, the products were baked at 100°C to prepare graphene nanoporous structure. The nanoporous graphene displayed a maximum specific surface area of 2396 m^2/g. The SC prepared with nanoporous graphene presented a specific capacitance of 282 F/g, rate capability (70% capacitance retention at high-rise current rates), and good energy density of 20.6 W h/kg with power density of 33 600 W/kg. In this report, a cylindrical supercapacitor device was constructed utilizing 1 meter length nanoporous graphene electrode covered on aluminum foil (Figure 6(C)).

Very recently, Senthilkumar et al. [107] have synthesized nanoporous graphene (2–5 nm) from GO dispersion through wet chemical method using Mg/Zn reducing agent.

The supercapacitor with nanoporous graphene electrodes demonstrated specific capacitance of 201 F/g even at 500 mV/s.

Similar to incorporation of metallic oxides with macroporous graphene, the nanoporous graphene was also combined with metallic oxides to enhance the supercapacitive performance [89,110]. For example, MnO_2/nanoporous 3D graphene composites were prepared for supercapacitor applications. The SC with MnO_2/nanoporous 3D graphene electrode has shown good specific capacitance of 310 F/g at 0.5 A/g including cyclic stability (retention of capacitance nearly 88.3% afterwards 1000 cycles) [89].

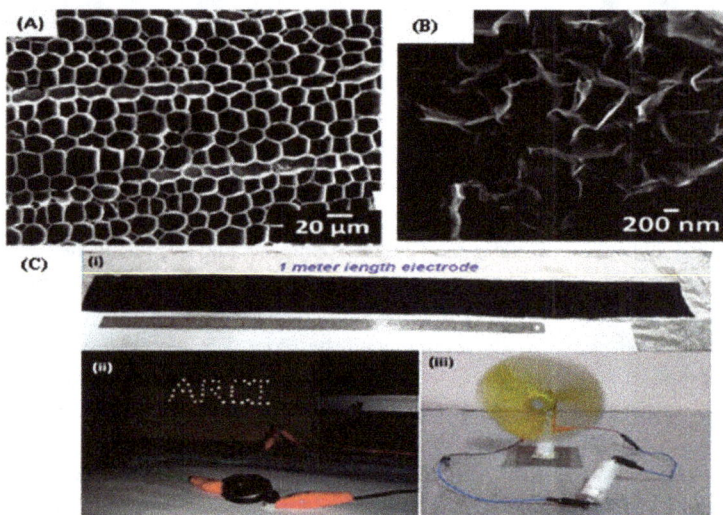

Figure 6: FESEM images of biowaste-acquire nanoporous carbon sample, (A) Cross-section of carbonized biowaste, (B) JC 1000 and (C) application of the biowaste extracted nanoporous carbon (Reproduced with permission Ref [17]).

3.2 Graphene gels (aerogels and hydrogels)

Graphene gels (aerogels and hydrogels) are also the members of 3D graphene family. These are beneficial for supercapacitor applications because of their unique 3D graphene macrostructures which avoid the restacking of graphene sheets and these also work as spacer to allow electrolyte ions to freely move to interior of the system [111].

Aerogel is the combination of two words 'aero' and 'gel'. 'Aero' is related to air and 'gel' means a semi-solid colloidal suspension of a solid dispersed in a liquid. Aerogel, also called frozen smoke or solid air or blue smoke or solid smoke, exhibits high porosity

(90-99.8%) and the nanosize pores, large surface area, less thermal conductivity (0.014 W/m/K) and less density (0.0011 g/cm^{-3}) [112]. Figure 7 shows different types of aerogels along with their period of invention.

Among aerogels was mentioned in Figure 7. graphene aerogels have attracted remarkable attention due their special characterstics like large surface area, huge electrical conductivity, well defined and controlled pore structure with tunable porosity, and reasonable thermal as well as mechanical stabilities [113,114]. Graphene aerogel is almost identical to graphene foam, but possesses extremely low density (ultralight, 0.16 mg/cm^3, lighter than air). These have immense surface-area-to-volume and strength-to-weight ratios [103]. Graphene foams are normally prepared by CVD technique on Ni foam which is removed after graphene growth, thus GFs are more conductive than graphene aerogels.

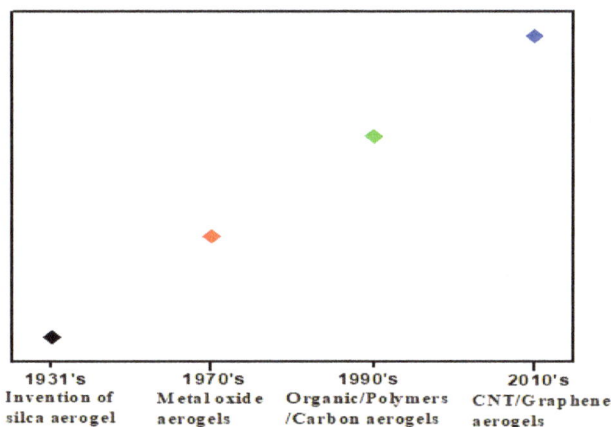

Figure 7: Types of aerogel with timeline of invention (Reproduced with permission from Ref [113]).

A typical synthesis procedure for gels is shown in Figure 8. In this method, firstly hydrogels were fabricated by the sol–gel method from molecular precursors followed by the removal of solvent from the pores of the hydrogel to produce aerogels. Carbon aerogels were synthesized in four steps; preparation of sol gels, aging of sol-gels, drying and carbonization. The third step involves the removal of water or organic molecules from the wet gels (hydrogels). This step may be performed by any of three different methods, such as super-critical drying, freeze drying and direct evaporation. Aerogels are usually prepared by freeze drying.

Graphene gels are usually produced with various methods such as sol-gel processing [115], self-assembly/ hydrothermal reduction [116], chemical reduction [117], cross-linking [118] and 3D printing process [119] etc. In this section, few of the preparation methods and their efficiency of the electrochemical performance for supercapacitor applications will be discussed.

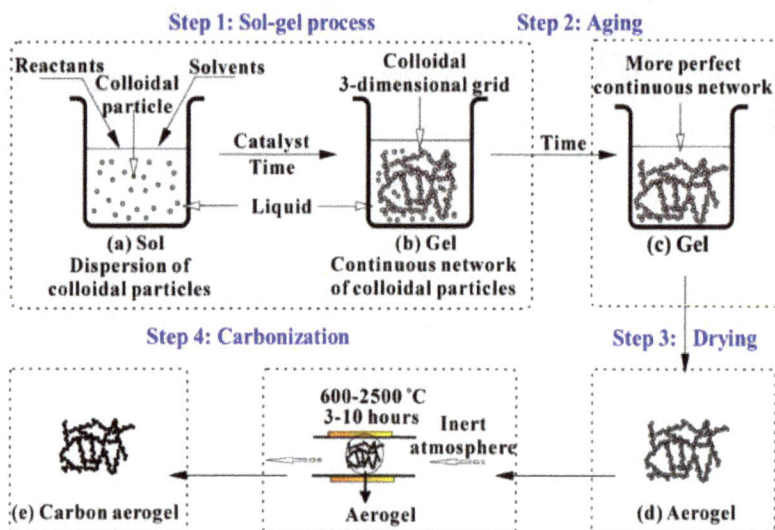

Figure: 8 A conventional procedure of aerogel fabrication(Reproduced with permission from Ref [113]).

GO is used as a precursor to prepare graphene hydrogels and aerogels as a result of its high-rise dispersion in aqueous media and its easy functionalization. GO gels were synthesized through gelation of GO dispersions. In a stable GO dispersion, the graphene oxide sheets remain well dispersed in an aqueous solvent due to force balance among the interplanar van der Waals attractions from the basal planes of graphene oxide sheets including the electrostatic repulsions from the functional groups of graphene oxide sheets. During gelation process of GO dispersion, this force balance is broken and GO sheets self-assemble to form GO hydrogels with pore sizes in the sub-micrometre to micrometre range. These 3D architectures of GO hydrogels can be changed to graphene hydrogels through chemical or hydrothermal reduction processes. Graphene hydrogels can be further converted to graphene aerogels by the removal of solvent from the pores of hydrogels [58,59]. The resulting graphene aerogels are porous, less dense material and exhibit great mechanical strength [114]. The outside disturbances to break the force balance and hence to induce the self assembly of GO sheets include (i) dynamics due to the pH value of graphene oxide dispersion [120], (ii) addition of cross-linking agents [121], and (iii) ultrasonication of the GO dispersion [122].

Alternatively, reduction of GO directly to graphene through chemical [123] or hydrothermal [124] reductions can also produce 3D rGO gel architectures. In these cases, the π–π stacking and hydrophobic interactions among the reduced graphene oxide (rGO) nanosheets are responsible for self assembly of graphene sheets. Figure 9 illustrates the self-assembly mechanism for graphene sheets. Graphene sheets may self-assemble through the interaction of hydrogen bonds in two ways. Ordered structures may form when two parallel GO sheets approach each other (Figure 9a), while amorphous structures may form when two GO sheets approach each other in non-parallel (Figure 9b)[125].

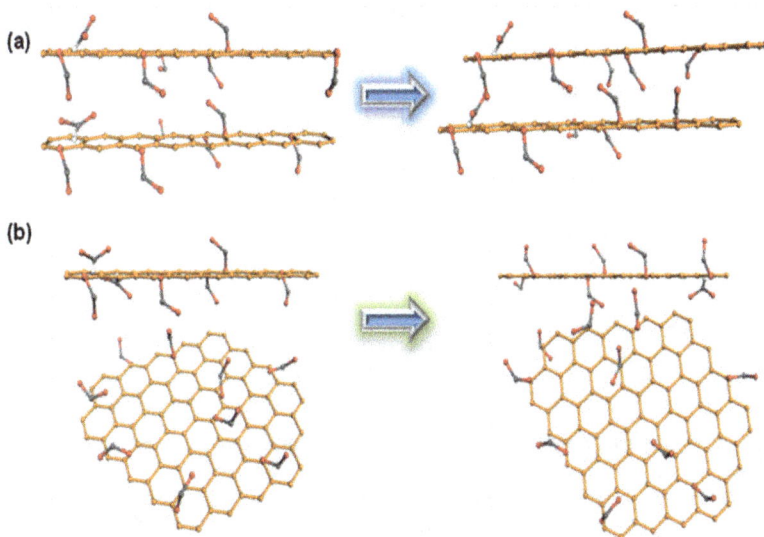

Figure 9: Self assembly mechanism of two (a) parallel and (b) non - parallel graphene oxide sheets (Reproduced with permission from Ref [125]).

Using sol-gel process, Zhang and co-workers [115] have developed conductive GAs for supercapacitor applications. The suspension of GO solution, as shown in Figure 10a, was mixed with L-ascorbic acid for reduction of GO to graphene (rGO). This mixture was heated to form a hydrogel of graphene through strong π–π interaction among these rGO sheets as shown in Figure 10b. After that, two different approaches, i.e super-critical CO_2 drying and freeze drying were used to produce GA (Figure 10c). In both cases, the morphology was almost similar and both materials showed of macro and mesoporous network structures (Figure 10d & e). Out of the two, the GA synthesized through super-critical CO_2 drying exhibited better properties like high surface area (512 m^2/g) as compared to GA prepared by freeze drying (surface area is 11.8 m^2/g). Due to this reason, the authors tested electrochemical performance for GA prepared through super-critical CO_2 drying, as high surface area is one of the crucial parameters in the energy storage applications.

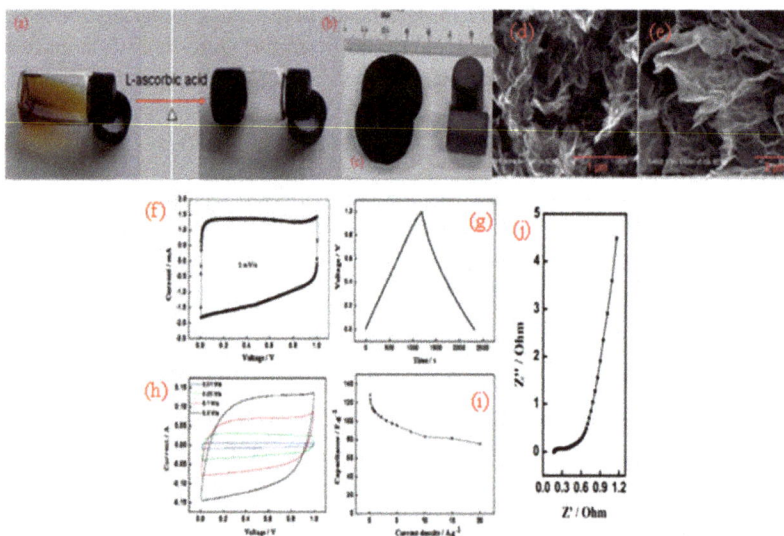

Figure 10 (a) Suspension of the Graphene Oxide. (b) graphene hydrogel (c) Combination of graphene oxide and L-ascorbic acid before stirring super-critical CO_2 dried (left) and freeze dried (right) of GAs (PTFE as a binder into pellets), SEM images of (d) super-critical of CO2 dried (e) freeze dried, (f) Cyclic Voltammetry at 2mV/s rate of scan (g) galvanostatic charge/discharge plot (h) CV curves with various scan rates (i) specific capacitance with current density (j) Nyquist curve of the graphene aerogel electrode (Reproduced with permission from Ref [115]).

GA prepared through super-critical CO_2 drying was used as active electrode and KOH as electrolyte to test the electrochemical efficiency of SC through cyclic voltammetry (CV), galvanostatic charge-discharge (GCD) and electrochemical impedance spectroscopy (EIS). The cyclic voltammetry plot (Figure 10f) of GA with $46mg/cm^3$ density shows the rectangular shape at 2mV/s scan rate. The charge-discharge of the galvanostatic test curve was conducted at 50 mA/g current density (Figure 10g) and specific capacitance determined from slope of discharge curve was found 128 F/g. The prepared SC showed a good rate capacity because the cyclic voltammetry curves maintained the rectangular shape at highest rate of scan 200 mV/s (Figure 11h). The performance rate of GA was even tested through GCD at high current densities (Figure 10i). At low current density, the specific capacitance decreases very slowly and after that it remained at 76 F/g as the current density rises to 20 A/g. The small diameter of the semicircle for GA in Nyquist curve (Figure 10j), was indicative to small internal resistance. These results showed that the graphene aerogel could be recommended as an electrode for electrochemical systems [115].

Similarly, Zhang et al. [95] prepared 3D self-assembled GH via chemical reduction of GO dispersion, with three different types of carbohydrates (glucose, sucrose and fructose) as reducing agents. The SC prepared by using GH as working electrode and KOH as electrolyte presented specific capacitances of 153.5, 145.0 and 150.3 F/g at 0.3 A/g for fructose reduced, glucose reduced and sucrose reduced GH respectively. All the GH electrodes showed good rate capability and good cycling durability.

Similar to graphene foam, graphene gels were also used as scaffolds to add metal oxides [126,127], conductive polymers [128,129] to improve the electrochemical performance of supercapacitor. As mentioned earlier also, most of the energy storage applications preferred PPY conductive polymer because of its properties such as higher conductivity, economical-synthesis, high hydrophobicity, good energy storage ability and environment stability [130]. Shibing Ye and co-workers [116] developed a 3D hybrid aerogel of graphene and PPY nanotube for supercapacitor applications. Hybrid synthesis process is shown in Figures 11(a & b). The PPY nanotubes were added and dissolved in GO solution, followed by their ultrasonication for 4 h. After the addition of ethylenediamine, the graphene oxide/PNTs suspensions were kept at 95°C for half hour. The mixture suspensions were transformed into hydrogels after another 4.5 hr. Hydrogel hybrid was converted to 3D hybrid aerogel of graphene and PPY nanotubes through freeze-drying. The authors prepared pure graphene aerogel (for comparisons) following the similar conditions, without the addition of PPY nanotubes. SEM image (Figure 11c) of the PPY nanotubes represents hollow structure with 5-10 μm length and 60-80 nm diameter.

GO/PNT aerogel exhibited a 3D network containing graphene oxide sheets and PNTs with endless macro-pores ranged from hundreds of nanometer to tens of μm (Figure11). The specific capacitances were determined from GCD curves (0.5 A/g) and found to be 144, 180, 253 F/ g for GA, PPY nanotubes, and hybrid aerogel of GO/PNTs, respectively. The prepared hybrid showed an excellent cyclic stability with ~ 95% capacitance retention even after 2000 cycles.

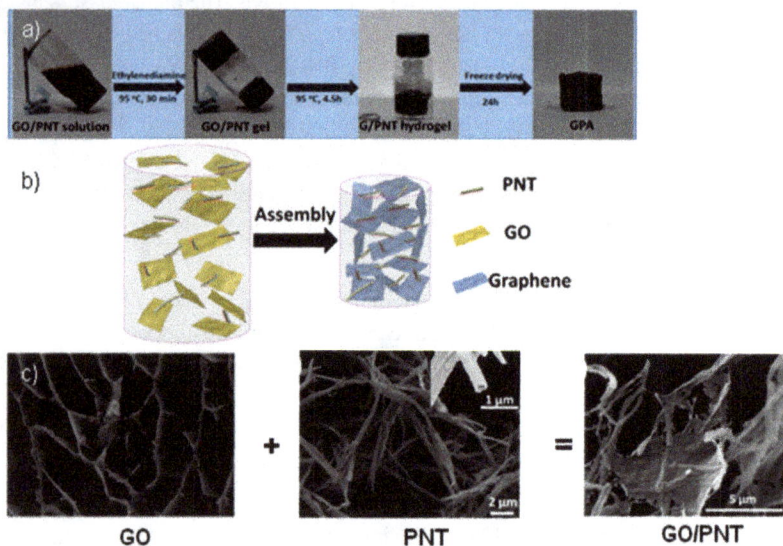

Figure 11: (a & b) Synthesis process of GPAs, (c) SEM images of GO, PNT and combination of GO/PNT (Reproduced with permission Ref [116]).

3D printing method:

Typical 3D printing technique has been used to prepare three-dimensional blocks through a layer by layer manner, with controlled computerized feedback and instructions. Material preparation is an important factor to make a 3D printing product – such materials are compulsory to be taken in the form of aqueous medium. Thus, this process was called by another name 'Direct Ink Writing'. Normally, this technique is used in rapid prototyping and additive manufacturing process, but newly research trends show its use in the fabrication of 3D graphene aerogel objects. Here, a big challenge to everyone is how to prepare the graphene based ink. Zhu et al. had been pioneer in accepting the challenge and prepared GO ink from both sol – gel chemistry and gelation methods, as seen in the Figure 12, for energy storage applications[131].

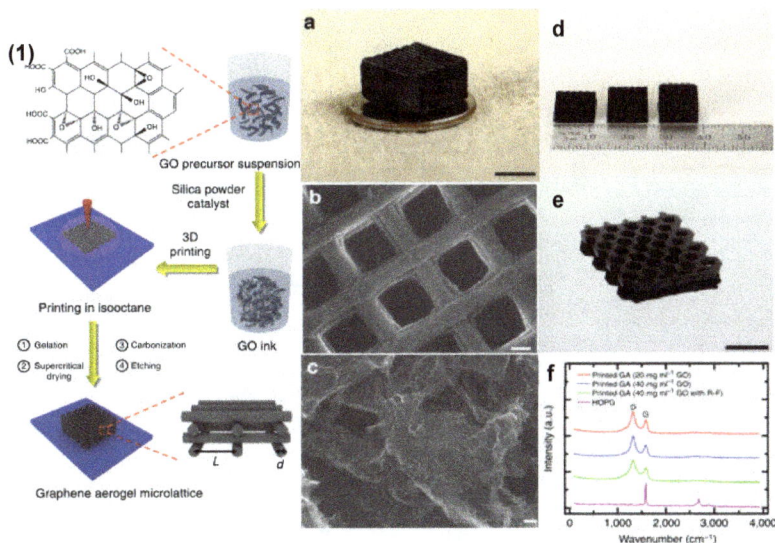

Figure: 12 (1)Illustrative of the synthesis process a) Optical picture, b) SEM image of three dimensional printed GA, c) GA without R-F after etching, d) different thickness of 3D GA microlattices, e) Honeycomb structure of 3D printed GA, f) Raman spectra of GA with different composition of GO (Reproduced with permission from Ref [131]).

According to the conventional approaches, inadequate bonding between the interfaces causes unsuitable voids to be formed. To overcome this issue, Zhang and co-workers fabricated the 3D-printed GA via a facile and easy approach. GO solution was deposited on the top of frozen material (ice support) continuously to fill the voids of ice support. This assembly was then immersed in liquid nitrogen for 30 minutes, then kept for 24h in the freezer at -80°C to self assemble of graphene sheets and afterwards, annealing process was done at 1000°C (30 minutes) in presence of Ar gas. Finally, the 3D graphene aerogel was prepared through the super critical CO_2 drying process [132].

The 3D carbon aerogel was preferred for electrochemical application due to its unique properties. However, the capacitance was reduced because of slow ion diffusion in bulk materials due to large thickness of the object or electrode. Recently, Santa Cruz and their team cracked this problem by developing a hybrid or composite of 3D printed GA/MnO_2 via electrodeposition process. It was noticed that the areal capacitance of the 3D

GA/MnO$_2$ hybrid electrode was around 25 times higher than that of 3D GA.The 3D GA/MnO$_2$ hybrid electrode showed 92.9% capacitance retention after 20,000 cycles [133]. Table 4 summarizes three dimensional graphene aerogel based hybrids with properties for supercapacitor applications.

Table 4. Supercapacitor properties of 3D graphene aerogel based hybrids

Graphene/com posite aerogel	Specific capacitance	Energy density	Power density	Cyclic performance	Electrolyte	Surface area	Electrical conductivity	Ref.
MnO$_2$/G aerogel	410 F/g			50000 cycles	0.5M Na$_2$SO$_4$	793 m^2/g	147 S/m	[127]
Graphene/Mn O$_2$ aerogel	320 F/g	11.1 Wh/kg	2.5 kW/kg	1000 cycles (90% strain)	PVA/H$_2$SO$_4$ gel	207 m^2/g		[134]
Graphene/ SWCNT/ MnO$_2$/ aerogels	626 F/g	~2.5 Wh/kg	~0.9kW /kg	10000 cycles (50% strain)	1M Na$_2$SO$_4$			[135]
MnO$_2$/ P-GA	175.8 F/g	13.9W h/kg	13.3 kW/kg	3000 cycles (89.6%)	0.1M Na$_2$SO$_4$	328 m^2/g		[136]
3D MoS2/CMG aerogel	268 F/g			1000 cycles (93%)	1M Na$_2$SO$_4$	149.3 m^2/g		[136]
NiO/Graphene aerogel	489.9 F/g				2M KOH			[137]
Melamine/Gra phene aerogel	106 F/g				6M KOH			[138]
PANI/ N-doped G aerogel	382 F/g	42 Wh/kg	2077 W/kg		1M H$_2$SO$_4$	288 m^2/g		[139]
MnO2/N doped G aerogel	456 F/g	39 Wh/kg	1953 W/kg		1M Na$_2$SO$_4$	58 m^2/g		[139]
PANI/ N doped G aerogel	489 F/g	43 Wh/kg	1981 W/kg		1M H$_2$SO$_4$			[139]
OMC/G aerogel	197 F/g	~6.2 Wh/kg	~3545 W/kg	~1000 cycles (96.2%)	6M KOH	254 m^2/g		[140]
Flower like FeS$_2$/G aerogel	313.6 F/g	22.86 Wh/kg	400 W/kg	2000 cycles (88.2%)	6M KOH	8.39 m^2/g		[141]
rGO-g-PANI aerogel	1600 F/g	545 Wh/kg	1538 W/kg	3000 cycles (91.3%)	1M H$_2$SO$_4$			[142]
rGO/S-PANI aerogel	480 F/g			10000 cycles (96.14%)	1M H$_2$SO$_4$	335 m^2/g		[143]
CoMoO$_4$/rGO hybrid aerogel	798.9 F/g			1000 cycles (84.7%)	3M KOH			[143]
CuO/GO aerogel	211 F/g				1M Na$_2$SO$_4$			[144]
G/Ni$_x$Co$_{1-x}$O aerogel	697.8 F/g	27.2 Wh/kg	725 W/kg	10000 cycles (86%)	6M KOH			[145]

4. Other 3D graphene structures for supercapacitors applications:

Apart from the graphene networks discussed in previous sections, other 3D architectures, such as graphene balls [3,146–149], graphene fibres and graphene pallets [106] have been used for supercapacitor applications. Lee et al. [3] reported, mass production of meso-porous graphene nanoballs (MGB) (Figure 13a) on gram scale through CVD technique. MGB exhibited average pore diameter of 4.27 nm, surface area of 508 m^2/g, and the electrical conductivity of 1.7 S/cm. The electrical conductivity of the MGB was raised to 6.5 S/cm by p-doping, achieved by dipping in H_2SO_4. The p-doped MGB electrode showed good electrochemical performance (Figure 13b) and it delivered specific capacitance of 206 F/g (5 mV/s). It showed an excellent rate capability with little decrease in capacitance to 191 F/g at high rate of scan 20 mV/s and then remained constant even when scan rate was increased to 100 mV/s (figure 13c). It display a good cyclic stability higher than 96% retention of capacitance after 10000 cycles at 20 A/ g (Figure 13d). Graphene nanoballs could also be combined with metal oxides to further improve their electrochemical performance.

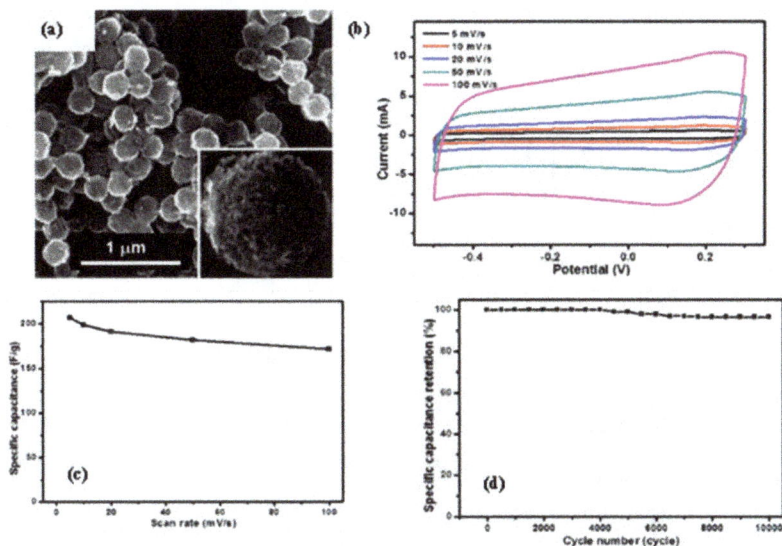

Figure 13: Figure: (a) SEM image of MGB obtained by CVD process. (b) CV plots, (c) specific capacitance with increasing rate of scan, and (d) Cyclic stability of p-doped MGB-based SC (20 A/g) (Reproduced with permission from Ref [3]).

He et al. [150] described the synthesis of three-dimensional hollow porous graphene balls (HPGBs) from coal tar pitch via a facile MgO scaffold approach coupled with KOH chemical activation. Graphene balls possessed higher specific surface area of 1871 m^2/g, capacitance of 321 F/g at 0.05 A/g, retention of capacitance about 94.5 % after 1000 cycles at 0.1 A/g. Mao et al. [146] synthesized graphene–Mn_3O_4 composite balls on indium tin oxide substrate using an aerosolization process. The resultant electrode showed specific capacitance of 1027 F/g at 5 A/g, cyclic stability with 78% capacitance retention afterward 1000 cycles at 20A/g. Zhang et al. [106] presented synthesis of three-dimensional graphene pallet, having nanosized pores ($\sim$ 2 nm) and high electrical conductivity (148 S/cm), through CVD method. In this method, Ni pallet was initially prepared from Ni powders by compression. Graphene was then grown on Ni pallet by CVD using methane as carbon source. After graphene growth, Ni pallet was then etched with HCl to get 3D nanoporous graphene in the form of pallet. This 3D graphene pallet was combined with MnO_2 and the obtained graphene pallet/MnO_2 hybrid electrode represented good capacitance of 395 F/g at 1A/g. This graphene pallet has shown specific capacitance of 7813 F/g at 10 A/g when combined with a hydroquinone and benzoquinone additive electrolyte.

Table 5. The electrochemical performance of graphene related fibres to SC applications

Graphene fibers	Specific capacitance	Energy density	Power density	Cyclic performance	Electrolyte	Surface area	Electrical conductivity	Ref.
G Fiber	279 F/g	5.76 Wh/kg	47.3 W/kg	1000 cycles (92%)	PVA/ H_2SO_4 gel		9.67$\pm$0.34 (10^2 S/m)	[159]
G Coaxial, Fiber	182 F/g	15.5 Wh/kg	8.19 x 10^{-4} W/cm^2	10000 cycles (100%)	H_2SO_4/PVA gel	2.25 mg/cm^2	6000 S/m	[160]
Activated Carbon/G fiber	43.8 F/g	3.96 mWh/g	5 mW/cm^3	10000 cycles (90.4%)	H_3PO_4/PVA gel	1476.5 m^2/g	185 S/m	[161]
CNT/rGO composite fibers		3.4 Wh/cm^3	0.7 W/cm^3	10000 cycles (93%)	H_3PO_4/PVA		210.7 S/cm	[162]
PVA/G hybrid fibers	216 F/g	5.32 mWh/g	23.9 mW/g	1000 cycles (85%)	1M H_2SO_4		13.9 S/cm	[158]
G/PPY fiber	107.2 mF/cm^2	9.7 $\mu Wh/cm^2$		1000 cycles	H_2SO_4/PVA gel		144 S/m	[155]
MnO_2/G fiber	42.02 mF/cm^2	1.46 x 10^{-3} mWh/cm^2	2.94 mW/cm^2	1000 cycles (92%)	PVA/H_3PO_4 gel		2.64 x 10^4 S/m	[152]
PANI/rGO/ C fiber	~257 mF/cm	19.6 Wh/kg	2.1 kW/kg	1000 cycles (~70%)	1M Na_2SO_4			[163]
GF/NiCo2S4	388 F/cm^3	12.3 mWh/cm^3	1600 mW/cm^3	2000 cycles (~92%)	PVA/KOH gel		39 S/cm	[41]
SGF	228 mF/cm^2	4.0 mWh/cm^3	1047.9 $\mu W/cm^2$	5000 cycles (98.7%)	PVA/H_2SO_4	839 m^2/g	567 S/cm	[164]

Flexible SCs require a lightweight, flexible and wearable electrode. Flexible fibers can fulfill these requirements. Compared to conventional carbon fibers, graphene fibers exhibit higher mechanical flexibility, higher surface area, better electrical conductivity, flexibility, lightweight and easy functionalization [151]. Various pseudocapacitive materials (metal oxides, metal sulfides, conductive polymers) have been employed to synthesize graphene fiber related SCs. Examples include the use of MnO_2 [42,152], Mn_3O_4 [153], Bi_2O_5 [43], WS_2 [48], MoS_2 [154], PPY [155], PANI [156], poly(3,4-ethylene dioxythiophene) (PEDOT) [157] and polyvinyl alcohol (PVA) [158]. Table 5 and 6 show the electrochemical performance of graphene related fibers and balls/spheres for supercapacitor applications.

Table 6. Electrochemical performance of graphene balls/spheres for supercapacitor applications.

3D Graphene balls/spheres	Specific capacitance	Energy density	Power density	Cyclic performance	Electrolyte	Surface area	Electrical conductivity	Ref.
MGB	206 F/g			10,000 cycles (96%)	1M H_2SO_4	508 m^2/g	6.5 S/cm	[3]
HPGBs	321 F/g	11.12 Wh/kg		1000 cycles (94.4%)	6M KOH	1947 m^2/g		[150]
CNT_n/CGB	162.3 F/g	4.9 Wh/kg	14.6 kW/kg	10,000 cycles (98.1%)	6M KOH	260 m^2/g		[165]
Crumpled rGO balls	396 F/g				6M KOH		4000 S/m	[166]
SMEGO	174 F/g	74 Wh/kg	338 kW/kg	1000 cycles (94%)	Ionic liquids	3290 m^2/g		[167]
PCG	20.7 µF/cm				6M KOH	753.5 m^2/g		[168]
SGR	182 F/g				5M KOH	216 m^2/g		[169]
GOS	306 F/g			10,000 cycles (93%)	1M KOH	695 m^2/g		[169]
NGHS	159 F/g			5000 cycles (99.24%)	6M KOH	402.9 m^2/g		[170]
Mn_3O_4/GO	225 F/g	34.1 Wh/kg	251 W/kg	6000 cycles (99.5%)	0.5M Na_2SO_4			[171]

Summary

It is worth to mention that graphene, in its different forms (0D, 1D, 2D), are the potential electrode materials for supercapacitor applications. However, the 3D graphene architectures have proven to be superior and excellent candidate for high-performance growth because these structures offer simultaneously the special characteristics of graphene and the advanced functions of three-dimensional morphology. 3D graphene architectures were further combined with nano-carbon structures (CNTs) and pseudocapacitive materials (metallic oxides/hydroxides and conductive polymers) to enhance the supercapacitor performance. Capacitance was also enhanced through doping of graphene with heteroatoms, such as N, P, B and S. This chapter included a detailed discussion regarding the preparation and application of graphene-based 3D materials (foams, hydrogels, aerogels, spheres, fibers) for supercapacitor applications. The 3-dimensional graphene building blocks have been prepared by different synthesis methods like hydrothermal self-assembly, template-assisted synthesis and chemical cross-linking. In spite of being in focus of the researcher community worldwide, some challenges still loom large on 3D graphene structures, such as controllable pore size and porosity of 3D graphene architectures. Most of the reports available in the open literature mentioned ample pore size distribution (few hundred nanometers to several micrometers). Thus, in spite of many existing construction approaches for 3D graphene buildings, it is still a challenge to find a controllable design which results in a 3D graphene architecture with enhanced specific area and controlled and optimized pore size. Another challenge is the electrical conductivity. Though electrical conductivity for 3D graphene architectures is high, but is still less than the theoretical capacitance of monolayer graphene. In addition, the mechanical properties of three-dimensional graphene architectures should be enhanced to their practical application in wearable and flexible supercapacitors. Therefore, more efforts and experiments (low cost and eco-friendly) are needed to obtain 3D graphene architectures with high surface area, controlled pore size, excellent electrical conductivity and high mechanical strength to enhance their practical application value.

List of abbreviations

CP	Conductive Polymers
CNTs	Carbon Nanotubes
CV	Cyclic Voltammetry
EIS	Electrochemcial Impedance Spectroscopy

EES	Electrical Energy Storage
EDLC	Electrical Double Layer Capacitor
GCD	Galvanostatic Charge Discharge
GO	Graphene Oxide
GQDs	Graphene Quantum Dots
GNs	Graphene-based nanoscrolls
GF	Graphene Foam
PC	Pseudocapacitor
rGO	Reduced Graphene Oxide
SC	Supercapacitor

References

[1] F. Mohamad, J. Teh, Impacts of energy storage system on power system reliability: A systematic review, Energies 11 (2018) 1-23. https://doi.org/10.3390/en11071749.

[2] A. Chatzivasileiadi, E. Ampatzi, I. Knight, Characteristics of electrical energy storage technologies and their applications in buildings, Renew. Sustain. Energy Rev. 25 (2013) 814–830. https://doi.org/10.1016/j.rser.2013.05.023.

[3] J.S. Lee, S.I. Kim, J.C. Yoon, J.H. Jang, Chemical Vapor Deposition of Mesoporous Graphene Nanoballs for Supercapacitor, ACS Nano 7 (2013) 6047–6055. https://doi.org/10.1021/nn401850z.

[4] K. Chatzivasileiadi, Large stationary batteries for deployment in grid-connected photovoltaic and other renewable energy power plants, 7th International Renewable Energy Storage Conference and Exhibition, November 2012.

[5] P. Simon, Y. Gogotsi, Materials for electrochmeical capacitors, Nat. Mater. (2008) 845–854. https://doi.org/10.1038/nmat2297.

[6] R.M.A.S. Rajakaruna, Small signal transfer functions of the classical boost converter supplied by ultracapacitor banks, Second IEEE conference on Ind. Electron. and applications (2007) 692–697.

[7] T. Wei, S. Wang, Z. Qi, A supercapacitor based ride-through system for industrial drive applications, Proc. 2007 IEEE Int. Conf. Mechatronics Autom. ICMA 2007. (2007) 3833–3837. https://doi.org/10.1109/ICMA.2007.4304186.

[8] S.C. Smith, P.K. Sen, Ultracapacitors and energy storage: Applications in electrical power system, 40th North Am. Power Symp. NAPS2008. (2008) 1–6. https://doi.org/10.1109/NAPS.2008.5307299.

[9] M. Vangari, T. Pryor, L. Jiang, Supercapacitors: Review of Materials and Fabrication Methods, J. Energy Eng. 139 (2013) 72–79. https://doi.org/10.1061/(ASCE)EY.1943-7897.0000102.

[10]   Y. Wang, Supercapacitor devices based on graphene materials, J. Phys. Chem. C. 113 (2009) 13103–13107. https://doi.org/10.1021/jp902214f.

[11]   V.V.N. Obreja, Supercapacitors specialities-Materials review, AIP Conf. Proc. 1597 (2014) 98–120. https://doi.org/10.1063/1.4878482.

[12]   Z.S. Iro, C. Subramani, S.S. Dash, A brief review on electrode materials for supercapacitor, Int. J. Electrochem. Sci. 11 (2016) 10628–10643. https://doi.org/10.20964/2016.12.50.

[13]   B. Francesco, C. Luigi, Y. Guihai, M. Stoller, V. Tozzini, C.F. Andrea, Graphene, related two-dimensional crystals, and hybrid systems for energy conversion and storage, Science 347 (2015) 27–43. https://doi.org/10.1126/science.1246501.

[14]   A. González, E. Goikolea, J.A. Barrena, R. Mysyk, Review on supercapacitors: Technologies and materials, Renew. Sustain. Energy Rev. 58 (2016) 1189–1206. https://doi.org/10.1016/j.rser.2015.12.249.

[15]   J.A. Fernández, T. Morishita, M. Toyoda, M. Inagaki, F. Stoeckli, T.A. Centeno, Performance of mesoporous carbons derived from poly(vinyl alcohol) in electrochemical capacitors, J. Power Sources 175 (2008) 675–679. https://doi.org/10.1016/j.jpowsour.2007.09.042.

[16]   D. Qu, Studies of the activated carbons used in double-layer supercapacitors, IFMBE Proc. 41 (2014) 1108–1110. https://doi.org/10.1007/978-3-319-00846-2_274.

[17]   K. Nanaji, V. Upadhyayula, T.N. Rao, S. Anandan, Robust, environmentally benign synthesis of nanoporous graphene sheets from biowaste for ultrafast supercapacitor application, ACS Sustain. Chem. Eng. 7 (2019) 2516-2529. https://doi.org/10.1021/acssuschemeng.8b05419.

[18]   Y. Wang, Y. Xia, Recent progress in supercapacitors: From materials design to system construction, Adv. Mater. 25 (2013) 5336–5342. https://doi.org/10.1002/adma.201301932.

[19]   C. Liu, Z. Yu, D. Neff, A. Zhamu, B.Z. Jang, Graphene-based supercapacitor with an ultrahigh energy density, Nano Lett. 10 (2010) 4863–4868. https://doi.org/10.1021/nl102661q.

[20]   H. Pan, J. Li, Y.P. Feng, Carbon nanotubes for supercapacitor, Nanoscale Res. Lett. 5 (2010) 654–668. https://doi.org/10.1007/s11671-009-9508-2.

[21]   H.I. Becker, United States Patent Office, (1940). https://doi.org/10.13189/cme.2016.040202.

[22]   M. Dinari, M.M. Momeni, M. Goudarzirad, Nanocomposite films of polyaniline/graphene quantum dots and its supercapacitor properties, Surf. Eng. 32 (2016) 535–540. https://doi.org/10.1080/02670844.2015.1108047.

[23]   S. Zhang, L. Sui, H. Dong, W. He, L. Dong, L. Yu, High-performance supercapacitor of graphene quantum dots with uniform sizes, ACS Appl. Mater. Interfaces 10 (2018) 12983–12991. https://doi.org/10.1021/acsami.8b00323.

[24]   S. Mondal, U. Rana, S. Malik, Graphene quantum dot-doped polyaniline nanofiber as high performance supercapacitor electrode materials, Chem. Commun. 51 (2015) 12365–12368. https://doi.org/10.1039/c5cc03981a.

[25]   K. Lee, H. Lee, Y. Shin, Y. Yoon, D. Kim, H. Lee, Highly transparent and flexible supercapacitors using graphene-graphene quantum dots chelate, Nano Energy 26 (2016) 746–754. https://doi.org/10.1016/j.nanoen.2016.06.030.

[26]   Q. Chen, Y. Hu, C. Hu, H. Cheng, Z. Zhang, H. Shao, L. Qu, Graphene quantum dots-three-dimensional graphene composites for high-performance supercapacitors, Phys. Chem. Chem. Phys. 16 (2014) 19307–19313. https://doi.org/10.1039/c4cp02761b.

[27]   Y. Hu, Y. Zhao, G. Lu, N. Chen, Z. Zhang, H. Li, H. Shao, L. Qu, Graphene quantum dots–carbon nanotube hybrid arrays for supercapacitors, Nanotechnology 24 (2013) 1-7. https://doi.org/10.1088/0957-4484/24/19/195401.

[28]   A.B. Ganganboina, A. Dutta Chowdhury, R.A. Doong, New avenue for appendage of graphene quantum dots on halloysite nanotubes as anode materials for high performance supercapacitors, ACS Sustain. Chem. Eng. 5 (2017) 4930–4940. https://doi.org/10.1021/acssuschemeng.7b00329.

[29]   H. Jia, Y. Cai, J. Lin, H. Liang, J. Qi, J. Cao, J. Feng, W.D. Fei, Heterostructural graphene quantum dot/MnO_2 nanosheets toward high-potential window electrodes for high-performance supercapacitors, Adv. Sci. 5 (2018) 1–10. https://doi.org/10.1002/advs.201700887.

[30]   P. Dhar, S.S. Gaur, A. Kumar, V. Katiyar, Cellulose nanocrystal templated graphene nanoscrolls for high performance supercapacitors and hydrogen storage: An experimental and molecular simulation study, Sci. Rep. 8 (2018) 1–15. https://doi.org/10.1038/s41598-018-22123-0.

[31]   Z. Xu, B. Zheng, J. Chen, C. Gao, Highly efficient synthesis of neat graphene nanoscrolls from graphene oxide by well-controlled lyophilization, Chem. Mater. 26 (2014) 6811–6818. https://doi.org/10.1021/cm503418h.

[32]   F. Zeng, Y. Kuang, Y. Wang, Z. Huang, C. Fu, H. Zhou, Facile preparation of high-quality graphene scrolls from graphite oxide by a microexplosion method, Adv. Mater. 23 (2011) 4929–4932. https://doi.org/10.1002/adma.201102798.

[33]   C.A. Amadei, I.Y. Stein, G.J. Silverberg, B.L. Wardle, C.D. Vecitis, Fabrication and morphology tuning of graphene oxide nanoscrolls, Nanoscale. 8 (2016) 6783–6791. https://doi.org/10.1039/c5nr07983g.

[34]   B. Zheng, Z. Xu, C. Gao, Mass production of graphene nanoscrolls and their application in high rate performance supercapacitors, Nanoscale. 8 (2016) 1413–1420. https://doi.org/10.1039/c5nr07067h.

[35]   K. Mohanapriya, N. Jha, Fabrication of one dimensional graphene nanoscrolls for high performance supercapacitor application, Appl. Surf. Sci. 449 (2018) 461–467. https://doi.org/10.1016/j.apsusc.2017.12.186.

[36]   M. Yan, F. Wang, C. Han, X. Ma, X. Xu, Q. An, L. Xu, C. Niu, Y. Zhao, X. Tian, P. Hu, H. Wu, L. Mai, Nanowire templated semihollow bicontinuous graphene scrolls: Designed construction, mechanism, and enhanced energy storage performance, J. Am. Chem. Soc. 135 (2013) 18176–18182. https://doi.org/10.1021/ja409027s.

[37]   B.N. Zheng, C. Gao, Preparation of graphene nanoscroll/polyaniline composites and their use in high performance supercapacitors, New Carbon Mater. 31 (2016) 315–320. https://doi.org/10.1016/S1872-5805(16)60015-X.

[38]   H. Cheng, C. Hu, Y. Zhao, L. Qu, Graphene fiber: A new material platform for unique applications, NPG Asia Mater. 6 (2014) e113-13. https://doi.org/10.1038/am.2014.48.

[39]   Z. Xu, C. Gao, Graphene fiber: A new trend in carbon fibers, Mater. Today. 18 (2015) 480–492. https://doi.org/10.1016/j.mattod.2015.06.009.

[40] S.H. Aboutalebi, R. Jalili, D. Esrafilzadeh, M. Salari, Z. Gholamvand, S. Aminorroaya Yamini, K. Konstantinov, R.L. Shepherd, J. Chen, S.E. Moulton, P.C. Innis, A.I. Minett, J.M. Razal, G.G. Wallace, High-performance multifunctional Graphene yarns: Toward wearable all-carbon energy storage textiles, ACS Nano 8 (2014) 2456–2466. https://doi.org/10.1021/nn406026z.

[41] W. Cai, T. Lai, J. Lai, H. Xie, L. Ouyang, J. Ye, C. Yu, Transition metal sulfides grown on graphene fibers for wearable asymmetric supercapacitors with high volumetric capacitance and high energy density, Sci. Rep. 6 (2016) 1–9. https://doi.org/10.1038/srep26890.

[42] Q. Chen, Y. Meng, C. Hu, Y. Zhao, H. Shao, N. Chen, L. Qu, MnO_2-modified hierarchical graphene fiber electrochemical supercapacitor, J. Power Sources 247 (2014) 32–39. https://doi.org/10.1016/j.jpowsour.2013.08.045.

[43] K. Gopalsamy, Z. Xu, B. Zheng, T. Huang, L. Kou, X. Zhao, C. Gao, Bismuth oxide nanotubes-graphene fiber-based flexible supercapacitors, Nanoscale 6 (2014) 8595–8600. https://doi.org/10.1039/c4nr02615b.

[44] L. Liu, Y. Yu, C. Yan, K. Li, Z. Zheng, Wearable energy-dense and power-dense supercapacitor yarns enabled by scalable graphene-metallic textile composite electrodes, Nat. Commun. 6 (2015) 1–9. https://doi.org/10.1088/0950-7671/42/8/448.

[45] Q. Yang, Z. Xu, C. Gao, Graphene fiber based supercapacitors: Strategies and perspective toward high performances, J. Energy Chem. 27 (2018) 6–11. https://doi.org/10.1016/j.jechem.2017.10.023.

[46] M.D. Stoller, S. Park, Y. Zhu, J. An, R.S. Ruoff, Graphene-based ultracapacitors, Nano Lett. 8 (2008) 6–10. https://doi.org/10.1021/nl802558y.

[47] L.L. Zhang, X. Zhao, M.D. Stoller, Y. Zhu, H. Ji, S. Murali, Y. Wu, S. Perales, B. Clevenger, R.S. Ruoff, Highly conductive and porous activated reduced graphene oxide films for high-power supercapacitors, Nano Lett. 12 (2012) 1806-1812. https://doi.org/10.1021/nl203903z.

[48] G. Sun, J. Liu, X. Zhang, X. Wang, H. Li, Y. Yu, W. Huang, H. Zhang, P. Chen, Fabrication of ultralong hybrid microfibers from nanosheets of reduced graphene oxide and transition-metal dichalcogenides and their application as supercapacitors, Angew. Chemie Int. Ed. 53 (2014) 12576–12580. https://doi.org/10.1002/anie.201405325.

[49]   Q. Li, X. Guo, Y. Zhang, W. Zhang, C. Ge, L. Zhao, X. Wang, H. Zhang, J. Chen, Z. Wang, L. Sun, Porous graphene paper for supercapacitor applications, J. Mater. Sci. Technol. 33 (2017) 793–799. https://doi.org/10.1016/j.jmst.2017.03.018.

[50]   S. Gan, L. Zhong, T. Wu, D. Han, J. Zhang, J. Ulstrup, Q. Chi, L. Niu, Spontaneous and fast growth of large-area graphene nanofilms facilitated by oil/water interfaces, Adv. Mater. 24 (2012) 3958–3964. https://doi.org/10.1002/adma.201201098.

[51]   G. Wallace, R.B. Kaner, M. Muller, S. Gilje, D. Li, Processable aqueous dispersions of graphene nanosheets, Nat. Nanotechnol. 3 (2008) 101–105. https://doi.org/10.1038/nnano.2007.451.

[52]   F. Gu, H. Shin, C. Biswas, G.H. Han, E.S. Kim, S.J. Chae, Layer-by-layer doping of few-layer graphene film, ACS Nano 4 (2010) 4595–4600. https://doi.org/10.1021/nn1008808

[53]   A. Davies, P. Audette, B. Farrow, F. Hassan, Z. Chen, J.Y. Choi, A. Yu, Graphene-based flexible supercapacitors: Pulse-electropolymerization of polypyrrole on free-standing graphene films, J. Phys. Chem. C. 115 (2011) 17612–17620. https://doi.org/10.1021/jp205568v.

[54]   S. Bose, T. Kuila, A.K. Mishra, R. Rajasekar, N.H. Kim, J.H. Lee, Carbon-based nanostructured materials and their composites as supercapacitor electrodes, J. Mater. Chem. 22 (2012) 767–784. https://doi.org/10.1039/c1jm14468e.

[55]   J.N. Tiwari, R.N. Tiwari, K.S. Kim, Zero-dimensional, one-dimensional, two-dimensional and three-dimensional nanostructured materials for advanced electrochemical energy devices, Prog. Mater. Sci. 57 (2012) 724–803. https://doi.org/10.1016/j.pmatsci.2011.08.003.

[56]   H. Jiang, P.S. Lee, C. Li, 3D Carbon based nanostructures for advanced supercapacitors, Energy Environ. Sci. 6 (2013) 41–53. https://doi.org/10.1039/c2ee23284g.

[57]   L. Liu, Z. Niu, J. Chen, Flexible supercapacitors based on carbon nanomaterials, Chinese Chem. Lett. 29 (2018) 571–581. https://doi.org/10.1016/j.cclet.2018.01.013.

[58]   Y. Ma, Y. Chen, Three-dimensional graphene networks : Synthesis, properties and applications, Natl. Sci. Rev. 2 (2015) 40-53. https://doi.org/10.1093/nsr/nwu072

[59]   X. Cao, Z. Yin, H. Zhang, Three-dimensional graphene materials: Preparation, structures and application in supercapacitors, Energy Environ. Sci. 7 (2014) 1850–1865. https://doi.org/10.1039/c4ee00050a.

[60]   Z. Chen, W. Ren, L. Gao, B. Liu, S. Pei, H.M. Cheng, Three-dimensional flexible and conductive interconnected graphene networks grown by chemical vapour deposition, Nat. Mater. 10 (2011) 424–428. https://doi.org/10.1038/nmat3001.

[61]   J. Sha, C. Gao, S.K. Lee, Y. Li, N. Zhao, J.M. Tour, Preparation of three-dimensional graphene foams using powder metallurgy templates, ACS Nano 10 (2016) 1411–1416. https://doi.org/10.1021/acsnano.5b06857.

[62]   X.H. Xia, D.L. Chao, Y.Q. Zhang, Z.X. Shen, H.J. Fan, Three-dimensional graphene and their integrated electrodes, Nano Today 9 (2014) 785–807. https://doi.org/10.1016/j.nantod.2014.12.001.

[63]   Q. Fang, Y. Shen, B. Chen, Synthesis, decoration and properties of three-dimensional graphene-based macrostructures: A review, Chem. Eng. J. 264 (2015) 753–771. https://doi.org/10.1016/j.cej.2014.12.001.

[64]   Z. Yang, S. Chabi, Y. Xia, Y. Zhu, Preparation of 3D graphene-based architectures and their applications in supercapacitors, Prog. Nat. Sci. Mater. Int. 25 (2015) 554–562. https://doi.org/10.1016/j.pnsc.2015.11.010.

[65]   Y. Ping, Y. Gong, Q. Fu, C. Pan, Preparation of three-dimensional graphene foam for high performance supercapacitors, Prog. Nat. Sci. Mater. Int. 27 (2017) 177–181. https://doi.org/10.1016/j.pnsc.2017.03.005.

[66]   H. Huang, Y. Tang, L. Xu, S. Tang, Y. Du, Direct formation of reduced graphene oxide and 3D lightweight nickel network composite foam by hydrohalic acids and its application for high-performance supercapacitors, ACS Appl. Mater. Interfaces 6 (2014) 10248–10257. https://doi.org/10.1021/am501635h.

[67]   H. Huang, L. Xu, Y. Tang, S. Tang, Y. Du, Facile synthesis of nickel network supported three-dimensional graphene gel as a lightweight and binder-free electrode for high rate performance supercapacitor application, Nanoscale 6 (2014) 2426–2433. https://doi.org/10.1039/c3nr05952a.

[68]   J. Yang, E. Zhang, X. Li, Y. Yu, J. Qu, Z. Yu, Direct reduction of graphene oxide by ni foam as a high-capacitance supercapacitor electrode, ACS Appl. Mater. Interfaces 8 (2016) 2297-2305. https://doi.org/10.1021/acsami.5b11337.

[69]   S. Sivaprakash, P. Sivaprakash, A facile synthesis of graphene foam as electrode material for supercapacitor, Mater. Res. Express 3 (2016) 1–7. https://doi.org/10.1088/2053-1591/3/7/075020.

[70]   P. Miao, J. He, Z. Sang, F. Zhang, J. Guo, D. Su, X. Yan, X. Li, H. Ji, Hydrothermal growth of 3D graphene on nickel foam as a substrate of nickel-cobalt-sulfur for high-performance supercapacitors, J. Alloys Compd. 732 (2018) 613–623. https://doi.org/10.1016/j.jallcom.2017.10.243.

[71]   Z. Niu, J. Chen, H.H. Hng, J. Ma, X. Chen, A leavening strategy to prepare reduced graphene oxide foams, Adv. Mater. 24 (2012) 4144–4150. https://doi.org/10.1002/adma.201200197.

[72]   W.Y. Tsai, R. Lin, S. Murali, L. Li Zhang, J.K. McDonough, R.S. Ruoff, P.L. Taberna, Y. Gogotsi, P. Simon, Outstanding performance of activated graphene based supercapacitors in ionic liquid electrolyte from -50 to 80°C, Nano Energy 2 (2013) 403–411. https://doi.org/10.1016/j.nanoen.2012.11.006.

[73]   Y. Zhu, S. Murali, M.D. Stoller, K.J. Ganesh, W. Cai, P.J. Ferreira, A. Pirkle, R.M. Wallace, K.A. Cychosz, M. Thommes, D. Su, E.A. Stach, R.S. Ruoff, Carbon-based supercapacitors, Science 332 (2011) 1537–1542. https://doi.org/10.1126./science.1200770.

[74]   B.G. Choi, M. Yang, W.H. Hong, J.W. Choi, Y.S. Huh, 3D Macroporous graphene frameworks for supercapacitors with high energy and power densities, ACS Nano 6 (2012) 4020–4028. https://doi.org/10.1021/nn3003345.

[75]   Y. Cai, A. Zhang, Y. Ping Feng, C. Zhang, Switching and rectification of a single light-sensitive diarylethene molecule sandwiched between graphene nanoribbons, J. Chem. Phys. 135 (2011) 1-6. https://doi.org/10.1063/1.3657435.

[76]   M.P. Down, C.E. Banks, Freestanding Three-dimensional graphene macroporous supercapacitor, ACS Appl. Energy Mater. 1 (2018) 891–899. https://doi.org/10.1021/acsaem.7b00338.

[77]   S. Yang, Y. Liu, Y. Hao, X. Yang, W.A. Goddard, X.L. Zhang, B. Cao, Oxygen-vacancy abundant ultrafine Co_3O_4/graphene composites for high-rate supercapacitor electrodes, Adv. Sci. 5 (2018) 1-10. https://doi.org/10.1002/advs.201700659.

[78]   X. Cao, Y. Shi, W. Shi, G. Lu, X. Huang, Q. Yan, Preparation of novel 3d graphene networks for supercapacitor applications, Small 7 (2011) 3163–3168. https://doi.org/10.1002/smll.201100990.

[79]   C. Jiang, B. Zhao, J. Cheng, J. Li, H. Zhang, Z. Tang, J. Yang, Hydrothermal synthesis of $Ni(OH)_2$ nanoflakes on 3D graphene foam for high-performance supercapacitors, Electrochim. Acta 173 (2015) 399–407. https://doi.org/10.1016/j.electacta.2015.05.081.

[80]   J. Gao, H. Xuan, Y. Xu, T. Liang, X. Han, J. Yang, P. Han, D. Wang, Y. Du, Interconnected network of zinc-cobalt layered double hydroxide stick onto rGO/nickel foam for high performance asymmetric supercapacitors, Electrochim. Acta 286 (2018) 92–102. https://doi.org/10.1016/j.electacta.2018.08.043.

[81]   X. Dong, J. Wang, J. Wang, M.B. Chan-Park, X. Li, L. Wang, W. Huang, P. Chen, Supercapacitor electrode based on three-dimensional graphene-polyaniline hybrid, Mater. Chem. Phys. 134 (2012) 576–580. https://doi.org/10.1016/j.matchemphys.2012.03.066.

[82]   J. Ren, R.P. Ren, Y.K. Lv, Stretchable all-solid-state supercapacitors based on highly conductive polypyrrole-coated graphene foam, Chem. Eng. J. 349 (2018) 111–118. https://doi.org/10.1016/j.cej.2018.05.075.

[83]   C. Xiang, M. Li, M. Zhi, A. Manivannan, N. Wu, A reduced graphene oxide/Co_3O_4 composite for supercapacitor electrode, J. Power Sources 226 (2013) 65–70. https://doi.org/10.1016/j.jpowsour.2012.10.064.

[84]   X.C. Dong, H. Xu, X.W. Wang, Y.X. Huang, M.B. Chan-Park, H. Zhang, L.H. Wang, W. Huang, P. Chen, 3D graphene-cobalt oxide electrode for high-performance supercapacitor and enzymeless glucose detection, ACS Nano 6 (2012) 3206–3213. https://doi.org/10.1021/nn300097q.

[85]   Y. He, W. Chen, X. Li, Z. Zhang, J. Fu, C. Zhao, E. Xie, Freestanding three-dimensional graphene/MnO_2 composite networks as ultralight and flexible supercapacitor electrodes, ACS Nano 7 (2013) 174–182. https://doi.org/10.1021/nn304833s.

[86]   J. Ji, L.L. Zhang, H. Ji, Y. Li, X. Zhao, X. Bai, X. Fan, F. Zhang, R.S. Ruoff, Nanoporous $Ni(OH)_2$ thin film on 3D ultrathin-graphite foam for asymmetric supercapacitor, ACS Nano 7 (2013) 6237–6243. https://doi.org/10.1021/nn4021955.

[87]   W. Zhou, X. Cao, Z. Zeng, W. Shi, Y. Zhu, Q. Yan, H. Liu, J. Wang, H. Zhang, One-step synthesis of Ni_3S_2 nanorod@$Ni(OH)_2$ nanosheet core-shell nanostructures on a three-dimensional graphene network for high-performance supercapacitors, Energy Environ. Sci. 6 (2013) 2216–2221. https://doi.org/10.1039/c3ee40155c.

[88]   U.M. Patil, S.C. Lee, J.S. Sohn, S.B. Kulkarni, K. V. Gurav, J.H. Kim, J.H. Kim, S. Lee, S.C. Jun, Enhanced symmetric supercapacitive performance of $Co(OH)_2$ nanorods decorated conducting porous graphene foam electrodes, Electrochim. Acta 129 (2014) 334–342. https://doi.org/10.1016/j.electacta.2014.02.063.

[89] S. Sun, P. Wang, S. Wang, Q. Wu, S. Fang, Fabrication of MnO_2/nanoporous 3D graphene for supercapacitor electrodes, Mater. Lett. 145 (2015) 141–144. https://doi.org/10.1016/j.matlet.2015.01.061.

[90] U.M. Patil, J.S. Sohn, S.B. Kulkarni, H.G. Park, Y. Jung, K.V. Gurav, J.H. Kim, S.C. Jun, A facile synthesis of hierarchical α-MnO_2 nanofibers on 3D-graphene foam for supercapacitor application, Mater. Lett. 119 (2014) 135–139. https://doi.org/10.1016/j.matlet.2013.12.105.

[91] W. Deng, Y. Sun, Q. Su, E. Xie, W. Lan, Porous CoO nanobundles composited with 3D graphene foams for supercapacitors electrodes, Mater. Lett. 137 (2014) 124–127. https://doi.org/10.1016/j.matlet.2014.08.154.

[92] S. Khamlich, T. Khamliche, M.S. Dhlamini, M. Khenfouch, B.M. Mothudi, M. Maaza, Rapid microwave-assisted growth of silver nanoparticles on 3D graphene networks for supercapacitor application, J. Colloid Interface Sci. 493 (2017) 130–137. https://doi.org/10.1016/j.jcis.2017.01.020.

[93] W. Wang, S. Guo, I. Lee, K. Ahmed, J. Zhong, Z. Favors, F. Zaera, M. Ozkan, C.S. Ozkan, Hydrous ruthenium oxide nanoparticles anchored to graphene and carbon nanotube hybrid foam for supercapacitors, Sci. Rep. 4 (2014) 1–9. https://doi.org/10.1038/srep04452.

[94] X. Yu, B. Lu, Z. Xu, Super long-life supercapacitors based on the construction of nanohoneycomb-like strongly coupled $CoMoO_4$-3D graphene hybrid electrodes, Adv. Mater. 26 (2014) 1044–1051. https://doi.org/10.1002/adma.201304148.

[95] X. Hong, B. Zhang, E. Murphy, J. Zou, F. Kim, Three-dimensional reduced graphene oxide/polyaniline nanocomposite film prepared by diffusion driven layer-by-layer assembly for high-performance supercapacitors, J. Power Sources 343 (2017) 60–66. https://doi.org/10.1016/j.jpowsour.2017.01.034.

[96] A.A. Mirghni, D. Momodu, K.O. Oyedotun, J.K. Dangbegnon, N. Manyala, Electrochemical analysis of $Co_3(PO_4)_2 \cdot 4H_2O$/graphene foam composite for enhanced capacity and long cycle life hybrid asymmetric capacitors, Electrochim. Acta 283 (2018) 374–384. https://doi.org/10.1016/j.electacta.2018.06.181.

[97] J. Hao, T. Meng, D. Shu, X. Song, H. Cheng, B. Li, X. Zhou, F. Zhang, Z. Li, C. He, Synthesis of three dimensional N,S co-doped rGO foam with high capacity and long cycling stability for supercapacitors, J. Colloid Interface Sci. 537 (2019) 57–65. https://doi.org/10.1016/j.jcis.2018.11.007.

[98] N.M. Ndiaye, B.D. Ngom, N.F. Sylla, T.M. Masikhwa, M.J. Madito, D. Momodu, T. Ntsoane, N. Manyala, Three dimensional vanadium pentoxide/graphene foam composite as positive electrode for high performance asymmetric electrochemical supercapacitor, J. Colloid Interface Sci. 532 (2018) 395–406. https://doi.org/10.1016/j.jcis.2018.08.010.

[99] K. Halab Shaeli Iessa, Y. Zhang, G. Zhang, F. Xiao, S. Wang, Conductive porous sponge-like ionic liquid-graphene assembly decorated with nanosized polyaniline as active electrode material for supercapacitor, J. Power Sources 302 (2016) 92–97. https://doi.org/10.1016/j.jpowsour.2015.10.036.

[100] P. Yu, X. Zhao, Z. Huang, Y. Li, Q. Zhang, Free-standing three-dimensional graphene and polyaniline nanowire arrays hybrid foams for high-performance flexible and lightweight supercapacitors, J. Mater. Chem. A 2 (2014) 14413–14420. https://doi.org/10.1039/c4ta02721c.

[101] S. Chabi, C. Peng, Z. Yang, Y. Xia, Y. Zhu, Three dimensional (3D) flexible graphene foam/polypyrrole composite: Towards highly efficient supercapacitors, RSC Adv. 5 (2015) 3999–4008. https://doi.org/10.1039/c4ra13743d.

[102] C. Xiong, T. Li, Y. Zhu, T. Zhao, A. Dang, H. Li, X. Ji, Y. Shang, M. Khan, Two-step approach of fabrication of interconnected nanoporous 3D reduced graphene oxide-carbon nanotube-polyaniline hybrid as a binder-free supercapacitor electrode, J. Alloys Compd. 695 (2017) 1248–1259. https://doi.org/10.1016/j.jallcom.2016.10.253.

[103] Z.S. Wu, A. Winter, L. Chen, Y. Sun, A. Turchanin, X. Feng, K. Müllen, Three-dimensional nitrogen and boron co-doped graphene for high-performance all-solid-state supercapacitors, Adv. Mater. 24 (2012) 5130–5135. https://doi.org/10.1002/adma.201201948.

[104] T. Qin, Z. Wan, Z. Wang, Y. Wen, M. Liu, S. Peng, D. He, J. Hou, F. Huang, G. Cao, 3D flexible O/N co-doped graphene foams for supercapacitor electrodes with high volumetric and areal capacitances, J. Power Sources 336 (2016) 455–464. https://doi.org/10.1016/j.jpowsour.2016.11.003.

[105] H. Huang, C. Lei, G. Luo, Z. Cheng, G. Li, S. Tang, Y. Du, Facile synthesis of nitrogen-doped graphene on Ni foam for high-performance supercapacitors, J. Mater. Sci. 51 (2016) 6348–6356. https://doi.org/10.1007/s10853-016-9931-6.

[106] L. Zhang, D. DeArmond, N.T. Alvarez, D. Zhao, T. Wang, G. Hou, R. Malik, W.R. Heineman, V. Shanov, Beyond graphene foam, a new form of three-dimensional graphene for supercapacitor electrodes, J. Mater. Chem. A. 4 (2016) 1876-1886. https://doi.org/10.1039/c5ta10031c.

[107] E. Senthilkumar, V. Sivasankar, B.R. Kohakade, K. Thileepkumar, M. Ramya, G. Sivagaami Sundari, S. Raghu, R.A. Kalaivani, Synthesis of nanoporous graphene and their electrochemical performance in a symmetric supercapacitor, Appl. Surf. Sci. 460 (2018) 17–24. https://doi.org/10.1016/j.apsusc.2017.10.221.

[108] Z. Xing, B. Wang, W. Gao, C. Pan, J.K. Halsted, E.S. Chong, J. Lu, X. Wang, W. Luo, C.H. Chang, Y. Wen, S. Ma, K. Amine, X. Ji, Reducing CO_2 to dense nanoporous graphene by Mg/Zn for high power electrochemical capacitors, Nano Energy 11 (2014) 600–610. https://doi.org/10.1016/j.nanoen.2014.11.

[109] H. Yang, S. Kannappan, A.S. Pandian, J.H. Jang, Y.S. Lee, W. Lu, Graphene supercapacitor with both high power and energy density, Nanotechnology 28 (2017) 1-10. https://doi.org/10.1088/1361-6528/aa8948.

[110] C. Chen, C. Chen, P. Huang, F. Duan, S. Zhao, P. Li, J. Fan, W. Song, Y. Qin, NiO/nanoporous graphene composites with excellent supercapacitive performance produced by atomic layer deposition, Nanotechnology 25 (2014) 1-9. https://doi.org/10.1088/0957-4484/25/50/504001.

[111] Y. Xu, Z. Lin, X. Huang, Y. Liu, Y. Huang, X. Duan, Flexible solid-state supercapacitors based on three-dimensional graphene hydrogel films, ACS Nano 7 (2013) 4042–4049. https://doi.org/10.1021/nn4000836.

[112] G. Gorgolis, D. Karamanis, Solar Energy Materials & Solar Cells, Sol. Energy Mater. Solar Cells 144 (2016) 559–578. https://doi.org/10.1016/j.solmat.2015.09.040.

[113] P. Ngoc Hong, Carbon nanotube and graphene aerogels–The world's 3d lightest materials for environment applications: A review, Int. J. Mater. Sci. Appl. 6 (2017) 277-283. https://doi.org/10.11648/j.ijmsa.20170606.12.

[114] G. Gorgolis, C. Galiotis, Graphene aerogels: A review, 2D Mater. 4 (2017) 1-22. https://doi.org/10.1088/2053-1583/aa7883.

[115] X. Zhang, Z. Sui, B. Xu, S. Yue, Y. Luo, B. Liu, Mechanically strong and highly conductive graphene aerogel and its use as electrodes for electrochemical power sources, J. Mater. Chem. (2011) 6494–6497. https://doi.org/10.1039/c1jm10239g.

[116] S. Ye, J. Feng, Self-assembled three-dimensional hierarchical graphene/polypyrrole nanotube hybrid aerogel and its application for supercapacitors, ACS Appl. Mater. Interfaces 6 (2014) 9671–9679. https://doi.org/10.1021/am502077p.

[117] H. Sun, Z. Xu, C. Gao, Multifunctional, ultra-flyweight, synergistically assembled carbon aerogels, Adv. Mater. 25 (2013) 2554–2560. https://doi.org/10.1002/adma.201204576.

[118] M.A. Worsley, P.J. Pauzauskie, T.Y. Olson, J. Biener, J.H. Satcher, T.F. Baumann, Synthesis of graphene aerogel with high electrical conductivity, J. Am. Chem. Soc. (2010) 14067–14069.

[119] C. Zhu, T. Liu, F. Qian, T.Y.J. Han, E.B. Duoss, J.D. Kuntz, C.M. Spadaccini, M.A. Worsley, Y. Li, Supercapacitors based on three-dimensional hierarchical graphene aerogels with periodic macropores, Nano Lett. 16 (2016) 3448–3456. https://doi.org/10.1021/acs.nanolett.5b04965.

[120] H. Bai, C. Li, G. Shi, A pH-sensitive graphene oxide composite hydrogel, Chem. Commun. 46 (2010) 2376–2378. https://doi.org/10.1039/c000051e.

[121] H. Bai, C. Li, X. Wang, G. Shi, On the gelation of graphene oxide, J. Phys. Chem. C 115 (2011) 5545–5551. https://doi.org/10.1021/jp1120299.

[122] O.C. Compton, Z. An, K.W. Putz, B. Jin, B.G. Hauser, L.C. Brinson, S.T. Nguyen, Additive-free hydrogelation of graphene oxide by ultrasonication, Carbon 50 (2012) 3399–3406. https://doi.org/10.1016/j.carbon.2012.01.061.

[123] W. Chen, S. Li, C. Chen, L. Yan, Self-assembly and embedding of nanoparticles by in situ reduced graphene for preparation of a 3d graphene/nanoparticle aerogel, Adv. Mater. 23 (2011) 5679–5683. https://doi.org/10.1002/adma.201102838.

[124] Y. Tao, X. Xie, W. Lv, D. Tang, D. Kong, Z. Huang, H. Nishihara, T. Ishii, B. Li, D. Golberg, F. Kang, T. Kyotani, Q. Yang, Towards ultrahigh volumetric capacitance : Graphene derived highly dense but porous carbons for supercapacitors, Sci. Rep. 3 (2013) 1–8. https://doi.org/10.1038/srep02975.

[125] J.M. Chem, High-rate capacitive performance of graphene aerogel with a superhigh C/O molar ratio, J. Mater. Chem. 22 (2012) 23186–23193. https://doi.org/10.1039/c2jm35278h.

[126] X. Meng, L. Lu, C. Sun, Green Synthesis of three-dimensional MnO_2/graphene hydrogel composites as a high-performance electrode material for supercapacitors, ACS Appl. Mater. Interfaces 10 (2018) 16474–16481. https://doi.org/10.1021/acsami.8b02354.

[127] C. Wang, H. Chen, S. Lu, Manganese Oxide/Graphene Aerogel Composites as an Outstanding Supercapacitor Electrode Material, Chem. Eur. J. 20 (2014) 517-523. https://doi.org/10.1002/chem.201303483.

[128] J. Chen, J. Song, X. Feng, Facile synthesis of graphene/polyaniline composite hydrogel for high-performance supercapacitor, Polym. Bull. 74 (2017) 27–37. https://doi.org/10.1007/s00289-016-1695-2.

[129] H. Zhou, T. Ni, X. Qing, X. Yue, G. Li, Y. Lu, One-step construction of graphene-polypyrrole hydrogels and their superior electrochemical performance, RSC Adv. 4 (2014) 4134–4139. https://doi.org/10.1039/c3ra44647f.

[130] S. Biswas, L.T. Drzal, Multilayered nanoarchitecture of graphene nanosheets and polypyrrole nanowires for high performance supercapacitor electrodes, Chem. Mater. 22 (2010) 5667–5671. https://doi.org/10.1021/cm101132g.

[131] C. Zhu, T.Y.J. Han, E.B. Duoss, A.M. Golobic, J.D. Kuntz, C.M. Spadaccini, M.A. Worsley, Highly compressible 3D periodic graphene aerogel microlattices, Nat. Commun. 6 (2015) 1–8. https://doi.org/10.1038/ncomms7962.

[132] Q. Zhang, F. Zhang, S.P. Medarametla, H. Li, C. Zhou, D. Lin, 3D Printing of Graphene Aerogels, Small 12 (2016) 1702–1708. https://doi.org/10.1002/smll.201503524.

[133] B. Yao, S. Chandrasekaran, J. Zhang, W. Xiao, F. Qian, C. Zhu, E.B. Duoss, C.M. Spadaccini, M.A. Worsley, Y. Li, Efficient 3D printed pseudocapacitive electrodes with ultrahigh MnO_2 loading, Joule 3 (2019) 458–470. https://doi.org/10.1016/j.joule.2018.09.020.

[134] P. Lv, X. Tang, W. Wei, Graphene/MnO_2 aerogel with both high compression-tolerance ability and high capacitance, for compressible all-solid-state supercapacitors, RSC Adv. 7 (2017) 47116–47124. https://doi.org/10.1039/c7ra08428e.

[135] Y. Zhao, M.P. Li, S. Liu, M.F. Islam, Superelastic pseudocapacitors from freestanding MnO_2-decorated graphene-coated carbon nanotube aerogels, ACS Appl. Mater. Interfaces 9 (2017) 23810-23819. https://doi.org/10.1021/acsami.7b06210.

[136] Z. Yu, M. Mcinnis, J. Calderon, S. Seal, L. Zhai, J. Thomas, Functionalized graphene aerogel composites for high-performance asymmetric supercapacitors, Nano Energy 11 (2015) 611–620. https://doi.org/10.1016/j.nanoen.2014.11.030.

[137] W. Chen, D. Gui, S. Yu, C. Liu, L. Zhao, J. Liu, E.D. Gui, Composites and its electrochemical performance for supercapacitor, Chem. Mater. 18 (2016) 249–252. https://doi.org/10.1109/ICEPT.2016.7583129.

[138] G. Wu, L. Wu, J. Jin, S. Yang, G. Li, Structure and electrochemical performance of melamine/graphene aerogel composite for supercapacitors, Mater. Sci. Forum 898 (2017) 1844–1849. https://doi.org/10.4028/www.scientific.net/MSF.898.1844.

[139] N. Phattharasupakun, J. Wutthiprom, N. Ma, P. Suktha, M. Sawangphruk, High-performance supercapacitors of N-doped graphene aerogel and its nanocomposites with manganese oxide and polyaniline, J. Electrochem. Soc. 165 (2018) A1430–A1439. https://doi.org/10.1149/2.0981807jes.

[140] R. Liu, L. Wan, S. Liu, L. Pan, D. Wu, D. Zhao, An interface-induced co-assembly approach towards ordered mesoporous carbon/graphene aerogel for high-performance supercapacitors, Adv. Funct. Mater. 25 (2015) 526–533. https://doi.org/10.1002/adfm.201403280.

[141] L. Pei, Y. Yang, H. Chu, J. Shen, M. Ye, Self-assembled flower-like FeS$_2$/graphene aerogel composite with enhanced electrochemical properties, Ceram. Int. 42 (2016) 5053–5061. https://doi.org/10.1016/j.ceramint.2015.11.178.

[142] N. Van Hoa, T.T.H. Quyen, N. Van Hieu, T.Q. Ngoc, P.V. Thinh, P.A. Dat, H.T.T. Nguyen, Three-dimensional reduced graphene oxide-grafted polyaniline aerogel as an active material for high performance supercapacitors, Synth. Met. 223 (2017) 192–198. https://doi.org/10.1016/j.synthmet.2016.11.021.

[143] Z. Gao, J. Yang, J. Huang, C. Xiong, Q. Yang, A three-dimensional graphene aerogel containing solvent-free polyaniline fluid for high performance supercapacitors, Nanoscale 9 (2017) 17710–17716. https://doi.org/10.1039/c7nr06847f.

[144] Z. Song, W. Liu, N. Sun, W. Wei, Z. Zhang, H. Liu, G. Liu, Z. Zhao, One-step self-assembly fabrication of three-dimensional copper oxide/graphene oxide aerogel composite material for supercapacitors, Solid State Commun. 287 (2019) 27–30. https://doi.org/10.1016/j.ssc.2018.10.007.

[145] Y. Zhou, L. Le Wen, K. Zhan, Y. Yan, B. Zhao, Three-dimensional porous graphene/nickel cobalt mixed oxide composites for high-performance hybrid supercapacitor, Ceram. Int. 44 (2018) 21848–21854. https://doi.org/10.1016/j.ceramint.2018.08.292.

[146] S. Mao, Z. Wen, H. Kim, G. Lu, P. Hurley, J. Chen, A general approach to one-pot fabrication of crumpled graphene-based nanohybrids for energy applications, ACS Nano 6 (2012) 7505–7513. https://doi.org/10.1021/nn302818j.

[147] J. Luo, H.D. Jang, J. Huang, Effect of sheet morphology on the scalability of graphene-based ultracapacitors, ACS Nano 7 (2013) 1464–1471. https://doi.org/10.1021/nn3052378.

[148] T. Derived, Design of 3D graphene-oxide spheres and their derived hierarchical porous structures for high performance supercapacitors, Small 13 (2017) 1702474. https://doi.org/10.1002/smll.201702474.

[149] E. Hee, J. Choi, S. Park, C. Min, H. Chang, H. Dong, Size and structural effect of crumpled graphene balls on the electrochemical properties for supercapacitor application, Electrochim. Acta 222 (2016) 58–63. https://doi.org/10.1016/j.electacta.2016.11.016.

[150] X. He, H. Zhang, L. Xiaojing, N. Xiao, J. Qiu, Direct synthesis of 3D hollow porous graphene balls from coal tar pitch for high performance supercapacitors, J.Mater. Chem. A 2 (2014) 19633-19640. https://doi.org/10.1039/c4ta03323j.

[151] L. Chen, Y. Liu, Y. Zhao, N. Chen, L. Qu, Graphene-based fibers for supercapacitor applications, Nanotechnology 27 (2016) 1-19.

[152] X. Li, T. Zhao, Q. Chen, P. Li, K. Wang, M. Zhong, J. Wei, D. Wu, B. Wei, H. Zhu, Flexible all solid-state supercapacitors based on chemical vapor deposition derived graphene fibers, Phys. Chem. Chem. Phys. 15 (2013) 17752–17757. https://doi.org/10.1039/c3cp52908h.

[153] S. Chen, L. Wang, M. Huang, L. Kang, Z. Lei, Reduced graphene oxide/Mn_3O_4 nanocrystals hybrid fiber for flexible all-solid-state supercapacitor with excellent volumetric energy density, Electrochim. Acta 242 (2017) 10–18. https://doi.org/10.1016/j.electacta.2017.05.013.

[154] V. Ruiz, C. Blanco, R. Santamaría, J.M. Ramos-Fernández, M. Martínez-Escandell, A. Sepúlveda-Escribano, F. Rodríguez-Reinoso, An activated carbon monolith as an electrode material for supercapacitors, Carbon 47 (2009) 195–200. https://doi.org/10.1016/j.carbon.2008.09.048.

[155] X. Ding, Y. Zhao, C. Hu, Y. Hu, Z. Dong, N. Chen, Z. Zhang, L. Qu, Spinning fabrication of graphene/polypyrrole composite fibers for all-solid-state, flexible fibriform supercapacitors, J. Mater. Chem. A 2 (2014) 12355–12360. https://doi.org/10.1039/c4ta01230e.

[156] V.A. Online, T. Huang, B. Zheng, L. Kou, K. Gopalsamy, Z. Xu, C. Gao, Y. Meng, Z. Wei, Flexible high performance wet-spun graphene fiber supercapacitors, RSC adv. (2013) 23957–23962. https://doi.org/10.1039/c3ra44935a.

[157] G. Qu, J. Cheng, X. Li, D. Yuan, P. Chen, X. Chen, B. Wang, H. Peng, A fiber supercapacitor with high energy density based on hollow graphene/conducting polymer fiber electrode, Adv. Mater. 28 (2016) 3646–3652. https://doi.org/10.1002/adma.201600689.

[158] S. Chen, W. Ma, H. Xiang, Y. Cheng, S. Yang, W. Weng, M. Zhu, Conductive, tough, hydrophilic poly(vinyl alcohol)/graphene hybrid fibers for wearable supercapacitors, J. Power Sources 319 (2016) 271–280. https://doi.org/10.1016/j.jpowsour.2016.04.030.

[159] Scalable non-liquid-crystal spinning of locally aligned graphene fibers for high performance wearble supercapapitors, Nano Energy 15 (2015) 642-653.

[160] X. Zhao, B. Zheng, T. Huang, C. Gao, Graphene-based single fiber supercapacitor with a coaxial structure, Nanoscale 7 (2015) 9399–9404. https://doi.org/10.1039/c5nr01737h.

[161] W. Ma, S. Chen, S. Yang, W. Chen, W. Weng, M. Zhu, Bottom-up fabrication of activated carbon fiber for all-solid-state supercapacitor with excellent electrochemical performance, ACS Appl. Mater. Interfaces 8 (2016) 14622–14627. https://doi.org/10.1021/acsami.6b04026.

[162] Y. Ma, P. Li, J.W. Sedloff, X. Zhang, H. Zhang, J. Liu, Conductive graphene fibers for wire-shaped supercapacitors strengthened by unfunctionalized few-walled carbon nanotubes, ACS Nano 9 (2015) 1352–1359. https://doi.org/10.1021/nn505412v.

[163] X. Jiang, Y. Cao, P. Li, J.Wei, K. Wang, D. Wu, H. Zhu, Polyaniline graphene carbon fiber ternary composites as supercapacitor electrodes, Mater. Lett. 140 (2015) 43-47.

https://doi.org/ 10.1016/j.matlet.2014.10.162

[164] W. Cai, T. Lai, J. Ye, A spinneret as the key component for surface-porous graphene fibers in high energy density micro-supercapacitors, J. Mater. Chem. A 3 (2015) 5060–5066. https://doi.org/10.1039/c5ta00365b.

[165] B.S. Mao, Z. Wen, Z. Bo, J. Chang, X. Huang, J. Chen, Hierarchical nanohybrids with porous cnt-networks decorated crumpled graphene balls for supercapacitors, ACS Appl. Mater. Interfaces 6 (2014) 9881-9889. https://doi.org/10.1021/am502604u.

[166] J.Y. Lee, K. Lee, Y.J. Kim, J.S. Ha, S. Lee, Sea-urchin-inspired 3d crumpled graphene balls using simultaneous etching and reduction process for high- density capacitive energy storage, Adv. Funct. Mater. 25 (2015) 3606–3614. https://doi.org/10.1002/adfm.201404507.

[167] T. Kim, G. Jung, S. Yoo, K.S. Suh, R.S. Ruoff, Activated graphene-based carbons as supercapacitor electrodes with macro- and mesopores, ACS Nano 7 (2013) 6899–6905. https://doi.org/10.1021/nn402077v.

[168] Z. Tang, X. Li, T. Sun, S. Shen, X. Huixin, J. Yang, Microporous and mesoporous materials porous crumpled graphene with hierarchical pore structure and high surface utilization efficiency for supercapacitor, Microporous Mesoporous Mater. 272 (2018) 40–43. https://doi.org/10.1016/j.micromeso.2018.06.020.

[169] T. Ha, S.K. Kim, J. Choi, H. Chang, H.D. Jang, pH controlled synthesis of porous graphene sphere and application to supercapacitors, Adv. Powder Technol. 30 (2019) 18–22. https://doi.org/10.1016/j.apt.2018.10.002.

[170] K. Xia, G. Wang, H. Zhang, Synthesis and characterization of nitrogen-doped graphene hollow spheres as electrode material for supercapacitors, J. Nanopart. Res. 19 (2017) 2-11. https://doi.org/10.1007/s11051-017-3954-z.

[171] C. Zhang, L. Wang, Y. Zhao, Y. Tian, J. Liang, Self-assembly synthesis of graphene oxide double-shell hollow spheres decorated with Mn_3O_4 for electrochemical supercapacitors, Carbon 107 (2016) 100–108. https://doi.org/10.1016/j.carbon.2016.05.057.

Chapter 5

Graphene-Based Materials for Micro-Supercapacitors

Akanksha R. Urade[1], Gurjinder Kaur[2], Indranil Lahiri[1,2] *

[1]Centre of Excellence: Nanotechnology, Indian Institute of Technology Roorkee, Roorkee 247667, India [2]Nanomaterials and Applications Lab., Department of Metallurgical and Materials Engineering, Indian Institute of Technology Roorkee, Roorkee 247667, India

*indrafmt@iitr.ac.in

Abstract

With the increase in the demand for flexible miniaturized electronic devices, there is a rapid development in micro-power sources. Recently developed on-chip micro-supercapacitors are raising possibilities for energy storage because of fast charge-discharge rates, higher power density and long lifetime than their counterparts. Among all the materials, graphene-based materials exhibit great potential in the development of thin, flexible and long-life micro-supercapacitors. This chapter presents fundamental working mechanism and design of planar interdigital micro-supercapacitors with particular emphasis on graphene-based micro-supercapacitors. New trends in the electrode preparations and fabrication techniques are identified. Finally, the future prospects in the development of graphene-based micro-supercapacitors are briefly discussed.

Keywords

Graphene, Carbon, Supercapacitors, Micro-Supercapacitors, Energy Storage

Contents

1.   Introduction

With the changing global landscape and increasing demand for cleaner energy, energy has become the primary focus for the world scientist community. As the availability of fossil fuels decreases, conversion of energy from renewable energy sources like sun and wind is becoming a major demand of the world. But the sun doesn't shine in the night and wind doesn't plow on demand. Hence, the current challenge that we are facing is to construct environmental friendly and highly efficient energy storage devices. Currently batteries are most commonly used power source for portable electronics devices, such as smartphone, laptops, tablets, etc., because of their high energy storage capacity in relatively small volume. As shown in "Ragone Plot" (Fig.1), in spite of exhibiting high energy density (180 Wh/Kg) [1], batteries show moderate power density. On the other hand, the conventional capacitors exhibit a good power density, but suffer from the limitation of low energy density [2]. Unfortunately both the conventional capacitors and batteries cannot provide high energy and power densities at the same time [1].

In search of the solution for a clean and better energy storage device, electrochemical supercapacitor (ES)[3–7] has emerged as an alternate energy storage device with a potential to facilitate major advances in energy storage to fulfill the global energy demand. Supercapacitors, also known as ultracapacitors [2,3], seem to be the best possible candidate to replace the conventional batteries and capacitors. It has advantage such as high energy density (5 Wh/kg) and high power density (10 kW/kg) with fast charging and discharging rates (in seconds) and a long cycle life (>100 000 cycles) [8–11].

As the size of portable electronics devices are becoming smaller, a high demand is observed for on-chip integrated power sources. Micro-supercapacitors are one of the most promising power sources for portable and wearable electronic devices, because of their extended lifetime and high power density [12–17]. Nowadays, micro-power sources have been greatly stimulated by micro-batteries which are dependent on redox reaction between electrode and electrolyte. But they still suffer from low durability and low power density. On the other hand, micro-supercapacitors can provide good durability, fast charge discharge rate and high power density. Furthermore, the specific energy of activated carbon (AC) based micro-supercapacitors is almost two orders of magnitude more than the commercially available traditional supercapacitors and Li-ion batteries, as shown in Fig. 1.

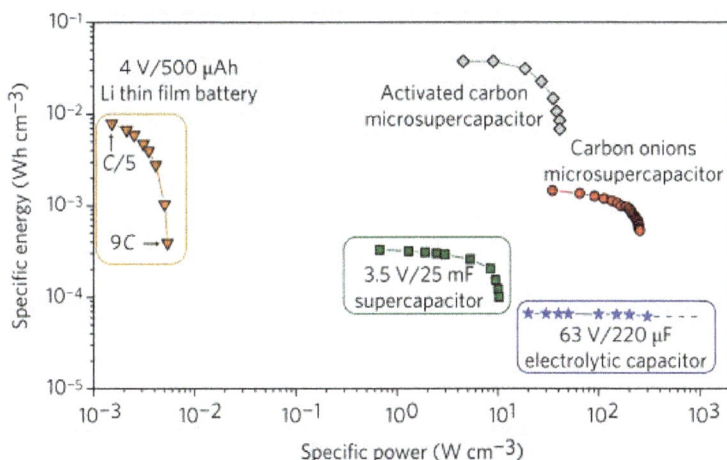

Figure 1. Ragone plot of energy density versus power density for different micro-devices in comparison with the Li thin film battery, electrolytic capacitor, and a micro-supercapacitor. (Reproduced with permission from Ref. [17])

For on-chip applications, areal density is very important because of limited space for integrating components into such system. Therefore, it is very important to consider the areal capacitance of these micro-sized supercapacitors. Till date the reported areal capacitance of electric double layer capacitors is in the range 0.4 to 16 F/cm^2 [17,18].

Hence it is important to consider the pseudocapacitive electrode material such as metal oxide, conducting polymer along with EDL material such as graphene based [16,19] or carbon derived ones [20], in order to achieve high energy density at high charge-discharge rates.

In this chapter, the latest advances in graphene-based micro-supercapacitors will be discussed and the methodology of their synthesis along with the observed capacitance performance will be summarized.

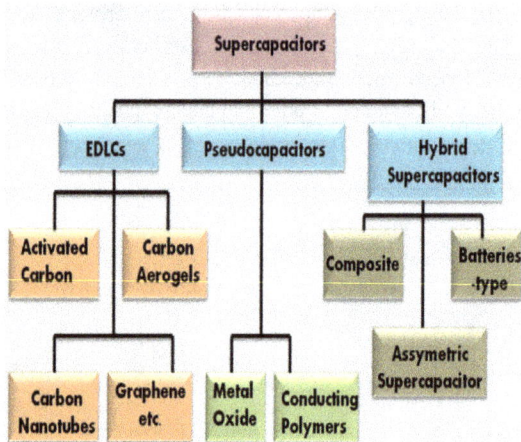

Figure 2. Classification of supercapacitor on the basis of charge storage mechanism

2. Fundamental of supercapacitor

Supercapacitors consist of two electrodes with an electrolyte in between separated by a separator which allow the migration of ions while charging and discharging [21]. While charging, positive ions in the electrolytes travel towards the negative electrode and negative ions move towards the positive electrode and there is formation of two separate layers of electric double layer. Here, each electrode/electrolyte interface indicates a capacitor. So, the entire cell represents two capacitors in series. The cell capacitance can be calculated from

$$\frac{1}{C_{cell}} = \frac{1}{C_1} + \frac{1}{C_2}$$ (1)

where, C_{cell} is the devices capacitance and C_1 and C_2 represent the capacitances of two electrodes, respectively.

Supercapacitors can be divided into several categories depending upon the charge storage mechanism, as shown in Fig. 2. One of the varieties is electrical double layer capacitor (EDLC) which stores the energy by electrostatically separating the charges at the interface of the electrode. Electrode materials in this type are made from highly porous carbon material to increase the charge storage capacity. The second category is pseudo-capacitor which uses metal oxide or conducting polymer as an electrode material where charge transfer occurs through surface redox reaction. The third category is hybrid supercapacitor which is a combination of electric double layer and pseudocapacitor, where one electrode is composed of carbon-based material and the other one is Faradic electroactive materials.

2.1 Electric double layer capacitors (EDLC)

Electric double layer capacitors consist of two electrodes coated with porous carbon materials immersed in an electrolyte, with a separator in between for the migration of ions. The schematic diagram of EDLC is shown in Fig.3. In EDLCs, the charge transfer is non-Faradic, i.e. there is no charge transport beyond the electrode/electrolyte interface. In EDLCs, the specific capacitance is highly dependent on the adsorption of positive and negative ions at the interface of electrode / electrolyte which is also called as the double layer. Hence, it is highly dependent on the surface area of electrode material. The concept of electric double layer was first introduced by Helmhotz [22]. This theory suggests that when a charged surface (either positive or negative) is immersed in an electrolyte, the opposite charges are attracted at the electrode/ electrolyte interface, as shown in Fig. 4a. This layer of electronic charge at the electrode surface and the layer of counter charges at the electrolyte surface, which is separated by an atomic distance, is called electrical double layer. This model was similar to the conventional capacitor. This representation was remodeled by Gouy (1910) and Chapman (1913) [23], as shown in Fig. 4b. The new theory suggests that there is a formation of a diffuse layer which is formed by the electrolyte ions due to thermal motion. However, this model leads to an overestimation of EDL capacitance. Later in 1924, Stern [24] suggested to combine the Helmholtz model with Gouy-Chapman's model. The Stern model proposed the existence of two regions of ion distribution – the inner region, called the compact layer or Stern layer and the diffuse layer. As shown in Fig. 4c, Stern layer distinguished the two types of absorbed ions. These are (i) specifically absorbed ions, which are present in inner Helmholtz plane (IHP) and (ii) solvated ions, which occur in outer Helmholtz plane (OHP). The charging-discharging process of electric double layer capacitor is depicted in Fig.5. During the process of charging, electrons travel from the negative electrode to the positive electrode through an external load. Within electrolyte, cations move towards the negative electrode while anions move towards the positive electrode. Complete reverse process take place during discharge. In the whole during, charge–discharge of EDLCs, transfer of charge takes place across the electrode/electrolyte interface, and no net ion exchange arises between electrolyte and electrode.

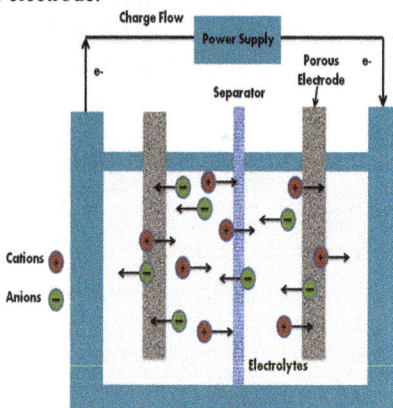

Figure 3. Schematic representation of electric double layer capacitor with porous carbon as an electrode material

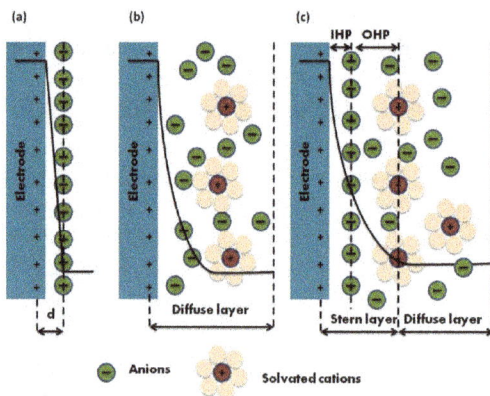

Figure 4. Representation of electrical double layer structures according to (a) the Helmholtz model, (b) the Gouy-Chapman model, and (c) the Gouy Chapman-Stern model.

2.2   Pseudocapacitors

The main difference between pseudo-capacitance and electric double layer capacitance is that the former is Faradic in nature and arises from the fast, reversible redox reaction on the electrode surface. Normally known electrode materials are ruthenium oxide [25–27], vanadium oxide [28], magnesium oxide [29] and some conducting polymers such as polyaniline, polypyrrole, polythiophene [30–33]. Though the pseudo-capacitance is higher than electric double layer capacitance but it suffers from the limitations of cyclic stability and low power density. Ruthenium oxide with theoretical capacitance of 1358 F/g is one of the good examples of material showing pseudo-capacitance. But, from a economical point of view, commercial level production of RuO_2 is not feasible.

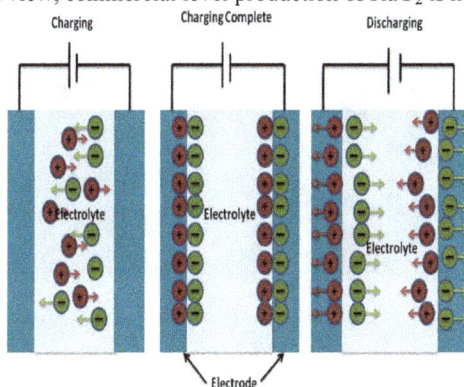

Figure 5. Charging- discharging process in electric double layer capacitors

The low-cost metal oxides such as nickel oxide and hydroxide [34,35], magnesium oxide [29], cobalt oxide and hydroxide [36,37] have attracted much attention during the last few years. The theoretical capacitances of magnesium and nickel oxides has been calculated as high as 1100 F/g and 2580 F/g, respectively [38–40]. The high theoretical capacitance and low cost make them prime candidates to replace ruthenium oxide. Manganese oxide nanowires were prepared by using a porous alumina template immobilized on Ti/Si substrate [41]. Capacitance of about 493 F/g has been achieved at a current density of 4 A/g in 0.5 M Na_2SO_4 aqueous solution. Similarly, nanostructures of nickel oxide and nickel hydroxide have also received tremendous attraction in the last few years. Yang et al., 2008 [42] have reported electrodeposited nickel hydroxide on nickel foam showing capacitance of 3152 F/g at 4 A/g current density in 3 % potassium hydroxide electrolyte.

2.3 Hybrid supercapacitors

Electric double layer capacitors have high power density but it suffers from the drawbacks of low energy density. Hybrid supercapacitors are novel to tackle this problem by increasing the operating voltage window. As the name suggests, hybrid supercapacitor is the combination of EDLC and pseudocapacitor. It uses EDLC carbon electrode with pseudo-capacitive metal oxide electrode as shown in Fig.6.

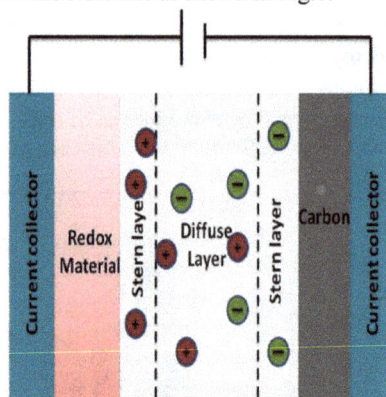

Figure 6. Schematic of hybrid supercapacitor – one electrode is pseudocapacitive and other is porous carbon

Hybrid supercapacitors show improved energy density because of combined effect of Faradic redox reaction and electric double layer. Yat Li et al. [43] have prepared hybrid supercapacitor by hydrothermal method. They have used nickel oxide nanoflakes array on carbon cloth as cathode electrode and reduced graphene oxide on nickel foam as an anode electrode. They achieved a remarkable area capacitance of 248 mF/cm^2 with capacitive retention of 95 % after 3000 cycles. Other oxides reported include the use of ruthenium oxide/titanium oxide as the positive electrode and activated carbon as the negative electrode cell that demonstrated a supercapacitance of 46 F/g and an energy density of 5.7 Wh/kg at a power density of 1207 W/ kg with reported over 90% after 1000 cycles [27].

Moreover, the high specific surface area of positive electrode materials significantly enhances the electric double layer performance, which contributes to increment in power density. Chen and coworkers [44] used a three-dimensional graphene-based porous carbon material with an ultrahigh surface area (3355 m^2/g) as the positive electrode material and Fe_3O_4/graphene nanocomposites as negative electrode. This hybrid supercapacitor demonstrated an ultrahigh energy density of 147 Wh/kg and power density of 150 W/kg, which remained at 86 Wh/kg even for a high power density of 2587 W/kg, so far the highest value was reported for hybrid supercapacitor.

3.   Structural model and working mechanisms of micro supercapacitors

3.1   State-of-art of fabrication

The charge storage mechanism mentioned above is applicable to both conventional supercapacitor and micro-supercapacitor, but the structural model and ionic diffusion performance may vary. Generally, supercapacitors consist of sandwiched design with the electrode, electrolyte and the separator in between. But due to finite space available in the integrated circuits, the micro-supercapacitor uses the planar interdigital design [45–47]. For supercapacitors, major factors that affect the power density are the thickness of electrode and power density. The ion diffusion length increases on increasing the thickness of electrodes which decreases charge/discharge rate and power density. On the contrary, in planar interdigital micro-supercapacitor, the electrodes of width W_e are separated by a gap W_g on an insulating plane. The capacitance is proportional to the width ratio of electrode and the gap W_g/W_e [48]. As the electrode width increases, area available for charge storage increases which helps in improving the areal capacitance. As the gap increases, it results into longer ion diffusion path which leads to a higher electrochemical series resistance (ESR) which results into lower power density [7]. By selecting a suitable fabrication method, both W_e and W_g can be adjusted so that the average migration distance of ions can be controlled. So for micro-supercapacitors, the wider electrode width and minimum gap will be helpful to improve capacitance, power density and energy density of the device [49,50]. The schematic design for interdigital micro-supercapacitor is shown in Fig.7.

The first planar interdigital micro-supercapacitor was reported by Sung et al. in 2003 [51]. Gold and platinum electrodes arrays were fabricated on silicon wafer by photolithography and a wet etching method. After that, conducting polymers such as poly(3-phenylthiophene) (PPT) and polypyrrole (PPy) were synthesized potentiostatically on these microelectrodes. Both the width of electrodes and the gap between them were both 50 μm. But due to the leakage problem of the electrolyte, direct applications of this microcapacitor have been limited. In order to resolve this problem, Sung et al. further fabricated solid state electrochemical microcapacitors [52]. They reported use of a conducting polymer such as polypyrrole as an electrode material, and polymeric ionic conductors such as $PVA–H_3PO_4–H_2O$ and $PANI/LiCF_3SO_3$ as solid electrolytes to form the microcapacitors.

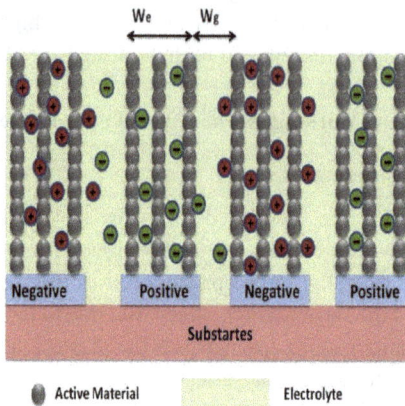

Figure 7. Design of interdigital micro-supercapacitor

After these reports of micro-capacitors, researchers around the world were attracted towards development of novel materials and their fabrication techniques, such as electrochemical and electrophoretic deposition, vapor-based deposition techniques, laser scribing etc. Since then, micro-supercapacitors based on a number of electrode materials, including MnO_2 [53], RuO_2 [26] and PANI [16] have been widely studied. Besides pseudocapacitive micro-supercapacitors, carbon-based [20] and graphene [54][19] based electrode materials have been utilized for EDL micro-supercapacitors. Graphene-based materials are the potential candidates for their application in supercapacitors owing to their amazing electrical, chemical and physical properties, such as high specific surface area, reasonable electrical conductivity and improved theoretical capacitance. In addition, graphene-based electrodes demonstrate additional attractive features, such as transparency, ultrathin structure and flexibility (29). These features can be adapted for the fabrication of multifunctional MSCs. Moreover, graphene-based electrodes allowed faster interaction of electrolyte ions with graphene sheets in horizontal direction compared to conventional supercapacitors.

3.2   Electrode materials and electrolytes

Most commonly used active electrode materials for supercapacitor application are: (i) carbon-based materials, e.g. activated carbon [55–57], carbon nanotubes [58,59] and graphene [60–62]; (ii) electroactive transition metal oxides, e.g. RuO_2, MnO_2 and NiO; (iii) conducting polymers, e.g. polypyrrole and polyaniline. Activated carbons with reported surface area about 2000 m^2/g are the most common materials in electric double layer supercapacitors [63,64]. Graphene is one of the most promising candidates for EDL supercapacitor. Basically a supercapacitor consists of two types of electrode configuration - i) symmetric electrode configuration, in which anode and cathode plates are of same the electrode materials; ii) asymmetric electrode configuration, in which one electrode plate consists of Faradaic electroactive material and the other consists of electrostatic carbon-based materials [9,10,65]. In asymmetric supercapacitors, both Faradaic capacitance mechanisms and electric double layer capacitance occur simultaneously, which lead to a wide voltage window and enhanced energy as well as power density as compared to symmetric supercapacitors.

The first fabrication of micro-supercapacitors based on conducting polymer electrode was reported by Joo-Hwan Sung and his team fabricated by means of photolithography and electrochemical polymerization techniques [51]. Since then, micro-supercapacitors based on a number of electrode materials, including MnO_2 [53], RuO_2 [26] and PANI[16] have been widely studied. Besides this pseudo-capacitive micro-supercapacitors, carbon-based [20] and graphene [19, 54] based electrode materials have been utilized for EDL micro-supercapacitors. Supercapacitor electrolytes can be divided into two major categories: solid-state electrolytes and conventional liquid electrolytes. Requirement for a good electrolyte include low volatility, minimum toxicity, less cost, a high voltage window, high ionic concentration and good electrochemical stability. Conventional liquid electrolytes such as KOH, NaOH and H_2SO_4 have smaller ionic size – hence, the ions can easily access more surface area of electrode materials and also they tend to possess higher ionic conductivity (up to 1 S/cm) [66]. Usually, water molecules decompose at a voltage of 1.23 V and hence regardless of all these advantages discussed above, the cell voltage for aqueous electrolytes is restricted to 1V. Generally, this operating voltage is lower than that for organic electrolytes (2.5 V)[12]. The problem with liquid electrolyte is leakage and hence it requires robust encapsulation to prevent it. On the other hand, solid electrolytes are well dispersed and bound into a polymer matrix and hence they do not show leakage problem [33,51]. Furthermore, solid-state gel electrolyte combines electrolyte and separator into a single layer to offer dual functionality which is totally different from conventional liquid electrolyte, in which a separator is used to avoid any electrical contact between electrodes.

Because of these advantages, solid state electrolytes such as $PVDF/BMIM+BF_4$, PVA/H_3PO_4 and PVA/H_2SO_4, ionic liquid polymer electrolytes and fumed silica nanopowder gel electrolytes are becoming very attractive in the design of flexible micro-supercapacitors [53,67–69].

3.3 Calculation of energy density and power density

The essential characteristics of a micro-supercapacitor are the amount of stored energy and power per unit weight (or, volume). Theoretically, the specific capacitance C (F/g) of electric double layer capacitor is proportional to the surface area of electrode and inverse of charge separation distance [70,71].

$$C = \frac{\varepsilon A}{d} \tag{2}$$

where, ε is dielectric constant, A is the surface area of electrode and d is the charge separation distance.

Experimentally, specific capacitance of electrode can be obtained from following equivalence. Capacitances derived from cyclic voltammetry (CV) curve can be calculated on the basis of following equation as [72]:

$$C = \frac{1}{v \times w \times \Delta V} \int I(V) dV \tag{3}$$

where, v is scan rate (V/s), w is weight (g), ΔV is voltage window (V), integration is for the area under the CV curve.

Capacitances calculated from charge-discharge curve are expressed as [73]:

$$C = \frac{I \times \Delta t}{m \times \Delta V} \times 4 \text{ (for two electrodes)} \tag{4}$$

$$C = \frac{I \times \Delta t}{m \times \Delta V} \times 2 \text{ (for three electrodes)} \tag{5}$$

where, C is the specific/area/volumetric capacitance, I is the constant current applied (mA), Δt is discharge time (s), ΔV is voltage window (V), m is the net mass of the active material on two electrodes (mg).

The energy density (E in Wh/kg) and power density (P in W/kg) delivered for a single cell supercapacitor are given by:

$$E = \frac{1}{2}CV^2 \tag{6}$$

$$P = \frac{E}{\Delta t} \tag{7}$$

where, E is the energy density, Δt is the discharge time C is the capacitance (F/g)

The maximum power density (P_{max}) at the matched impedance condition is calculated as

$$Pmax = \frac{V^2}{8mR_{ESR}} \tag{8}$$

where, $R_{ESR} = V_{drop}/4I$ is the equivalent series resistance, V_{drop} is voltage drop at the start of discharge cycle.

4.   Development of interdigital planar micro-supercapacitor

The electrochemical performance of the supercapacitors (SCs) is predominantly decided by both electrode material and electrolyte solution. However, much attention has been paid, in the recent years, to find suitable electrode materials which have the ability to enhance the specific capacitance, operating cell voltage as well as energy and power densities of supercapacitors. For an electrode to show good supercapacitive performance, it must be electrically conductive and should have a high surface, high porosity with a defined pore size distribution. Transition metal oxides and carbon materials have been most frequently used electrode materials for SC applications. The carbon-based materials are of particular interest for SC applications due to their brilliant properties such as low cost, light weight, high surface area, non-toxicity, high inertness, ease of processability to name *a* few [74]. Because of these fantastic properties, carbon materials possess higher power density and long cyclic life. Carbon nanomaterials may be typically classified on the basis of their dimensions: zero-dimensional *(*0D; carbon quantum dots), one-dimensional (1D, carbon nanotube), two-dimensional *(*2D; graphene*)*.

4.1   Carbon quantum dots (0 D)

Carbon quantum dots (CQDs) are zero-dimensional nanomaterials with exceptionally small size (less than 10 nm results in high surface area) [75]. CQDs' unique characteristics including high surface area, aqueous dispersibility, enhanced electrical conductivity, fast charge transfer, high chemical stability, easy functionalization make them the favorable electrode material for SC applications [76,77]. Another carbon material with 0D configuration is fullerene. However, it has been scarcely used for supercapacitor applications due to its small specific surface area, low electrical conductivity and drastic interaction ability [78]. The uninterrupted evolution of nanostructured material fabrication methods leads to a bouncy stride in the development of supercapacitor technologies. Several synthesis methods have been used to synthesize CQDs including acid dehydration method [79], hydrothermal route [80], pyrolysis process [81], electrochemical oxidation [82,83], ultrasonic method [84] ,laser ablation [85,86], microwave-assisted pyrolysis method [87] and arc-discharge method [88,89]. No report is available on the use of CQDs as single electrode material. These are composited with some other materials, such as metal oxides (MnO_2, RuO_2, Fe_2O_3), graphene, graphene oxide and conducting polymers (polyaniline, polypyrrole). In addition, hetero-atom doped (nitrogen and oxygen) carbon dots have been employed as an electrode material in supercapacitor devices.

Chen et al. [90] have prepared a nanocomposite of MnO_2 and CQDs using sonochemistry technique (Fig.8) and have compared the electrochemical performance of pristine MnO_2 and MnO_2/CQDs composites. It was observed that capacitance retention and specific capacitance were improved by employing a composite of MnO_2 with CQDs, as an electrode material. The specific capacitance values of MnO_2/CQDs and pristine MnO_2 were 330 F/g and 230 F/g current density of 1 A/g, respectively. The capacitance retention reported for the MnO_2/CQDs was 90.3%, while 61.1% for pristine MnO_2, after 10,000 charging/discharging cycles at 1 A/g.

Another hybrid composite of metal oxide RuO_2 (reduced form) and CQDs was reported by Zhu et al. [91]. This hybrid composite possessed a high specific capacitance of 460 F/g at 50 A/g and exceptionally high cycling stability (capacitance retention 96.9% after 5000 cycles at 5 A/g). Feng et al. [92] prepared a 3D composite rGO hydrogel/CQDs by hydrothermal method. The CDQs facilitated the charge transport in rGO hydrogel and also decreased the internal resistance and charge transfer resistance. The device exhibited a specific capacitance of 264 F/g at a current density of 1A/g and excellent cycling stability (capacitance retention of 90.9% after 5000 cycles at 5 A/g), high energy density of 35.3 Wh/kg and large power density of 516 W/kg. Lv et al. [75] also fabricated a 3D network, named 3D CQD aerogel, and demonstrated a specific capacitance of 294.7 F/g at 0.5 A/g and good cyclic stability (capacitance retention of 94% after 1000 cycles at 1 A/g). Conductive polymers are also very attractive materials for *pseudosuperca*pacitors. Mondal et al. [93] reported a CQDs-doped polyaniline composite which displayed a very high specific capacitance value of 1044 F/g at 1 A/g but the cyclic stability (80.1% after 3000 cycles) of this composite was lesser compared to other reports. Another report utilizing the same polymer composite with carbon quantum dot showed a 738.3 F/g at 1.0A/g. Again this hybrid composite also showed a moderate cyclic stability 78.0% after

1000 cycles at 5 A/g[94]. Zhang et al.[95] demonstrated a specific capacitance of 576 F/g at 0.5 A/g for a ternary composite of GO/CDQs/polypyrrole. However, this composite was able to keep a good cyclic stability (capacitance retention of 92.2 % after 5000 cycles at 10A/g).

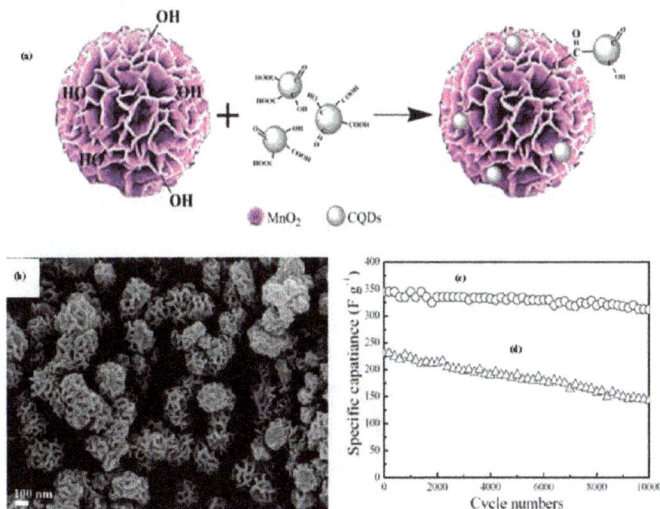

Figure 8. (a) Schematic illustration of the synthetic route of MnO₂/CQDs nanocomposite. (b) FESEM image of MnO₂/CQDs nanocomposite. Cyclic performance of (c) MnO₂/CQDs nanocomposite and (d) pristine MnO₂ at a current density of 1 A/g (17).(Reproduced with permission from Ref.[90])

4.2   Carbon nanotube based materials (1 D)

Carbon nanotubes (CNTs) also attracted the attention of scientists to be used as a possible electrode material for SCs owing to their hollow structure, which leads to large effective surface area, structural stability, excellent electrical properties and mechanical and chemical integrity [96–98]. Vertically aligned CNT (VACNT) was proposed to be superior compared to the randomly aligned CNT (RACNT) because RACNTs might grow into the space regions between two nearby electrodes and may lead to the short circuit in the SC device. Hsia et al. [99] used vertically aligned CNT (VACNT) as an electrode material for flexible planar micro-supercapacitors (MSCs). For electrode fabrication of the device, VACNTs were grown on Si substrate, followed by the deposition of Ni film onto it. The Ni film was deposited to enhance the in-plane electrical conduction of the VACNTs and to improve the power performance of the MSC device.

The Ni-VACNT electrode was patterned and transferred with a laser-assisted dry transfer method onto flexible polycarbonate substrate having 125 μm thickness. Thereafter, an ionic liquid gel (ionogel) used as an electrolyte, was applied onto VACNTs based electrodes to fabricate the MSC devices. The electrochemical tests showed that the synthesised MSC exhibited a specific capacitance of 430 μF/cm² for a scan rate of 0.1 V/s, energy density of 0.1–0.5 μWh/cm² and power density of about 10 mW/cm². The fabricated device demonstrated specific capacitance retention of over 90% after 1000 cycles. Jiang et al. [100] also used VACNT as an electrode for the fabrication of MSC. In this report VACNTs were grown on a conductive Mo/Al/Fe stack substrate. The CNT-covered MSC device demonstrated a specific capacitance of 428 μF/cm². A very cheap, simple and environment friendly extrusion-based 3D printing methodology was also used to prepare CNTs based MSCs[101]. In this technique, printing ink containing the CNTs was moderately solid which was extruded through a micro-nozzle onto a heated base with a pre-programmed printing trajectory. The whole fabrication process of 3D printed MSC is shown in Fig.9.

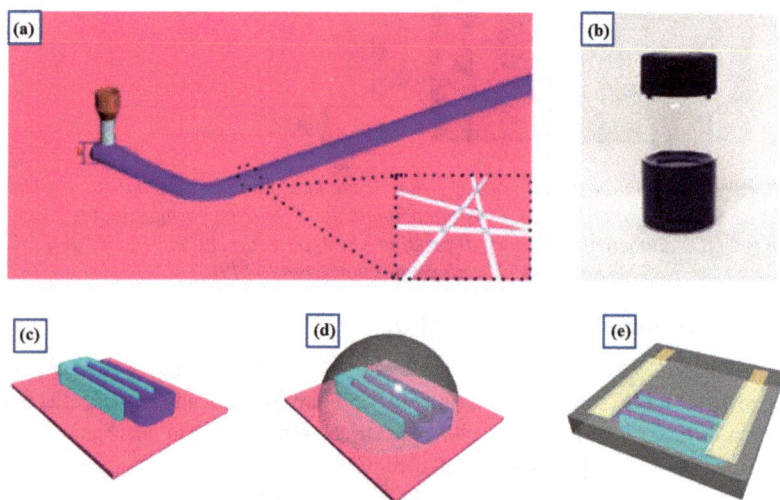

Figure 9. Schematic of the 3D printing process to fabricate CNTs based MSCs. (a) Schematic illustration of the 3D printing method with controllable working distance h, inset shows the the as-obtained electrodes consisting of CNTs inks. (b) Digital photograph of 3D printing inks. (C) The final structure of the interdigital patterned CNT based electrodes. (d) The PVA-H₃PO₄ gel electrolyte was applied and dried. (e) The 3D printed packaged MSCs.(Reproduced with permission from Ref. [101].

Materials Research Forum LLC
https://doi.org/10.21741/9781644900550-5

The important features of 3D printing technique include: (a) in this methodology, various inks can be used with different CNTs contents (b) it is possible to tune the heights and widths of the printed electrode patterns by controlling the diameter of the micronozzle, printing pressure, concentration of CNTs and the working distance. The 3D printed MSCs exhibited a specific capacitance of 2.44 F/cm^2, power density of 3.72 W/cm^3 and energy density of 0.12 mWh/cm^3.

5. Graphene based materials for micro-supercapacitor

Graphene is considered as an unique electrode material due to its 2 dimensional structure with very high electrical conductivity and chemical stability [102] along with the theoretical surface area of 2630 m^2/g [62] and intrinsic capacitance of ~21 $\mu F/cm$. Theoretically, graphene supercapacitors can achieve electric double layer capacitance of ~550 F/g [103]. However, the specific capacitance values for most of chemically exfoliated graphene has been lower than the expected value because of restacking of the graphene sheets [104]. The first publication on the synthesis of graphene based electrochemical supercapacitors was reported by Vivekachand et al. [105]. They achieved the specific capacitance of 117 F/g in aqueous H_2SO_4. The challenge in the synthesis and application of graphene sheets was due to the cohesive Van der Waals force between the π-stacked graphene layers. This was addressed by adding spacers in between graphene sheets such as CNTs [106–108], conducting polymers [109,110] and metal oxide [111–113]. From the time of its discovery, various graphene-based materials such as graphene oxide (GO) and reduced graphene oxide (r-GO) were being manufactured at low cost. Till now, the reported specific surface area values of carbon materials derived from GO have been well below 2630 m^2/g. Here, recent developments in graphene-based electrode materials and their electrochemical performance for micro-supercapacitors are summarized.

Graphene derivative, reduced graphene oxide (rGO) is also used as an electrode material to synthesize MSCs. Gao et al. [114] fabricated micro-scale SCs by laser reduction and patterning of hydrated graphene oxide (GO) films (Fig.10). They discovered that hydrated GO can behave both as an electrical insulator and good ionic conductor simultaneously. Therefore, this group used hydrated GO as both electrolyte and separator to synthesize MSC. In this work, the in-plane geometry was compared with the stacked structure. The capacitance density and volumetric capacitance for in-plane structure with a circular geometry were found to be 0.51 mF/cm^2 and 3.1 F/cm^3, respectively and these values are approximately double of the ones obtained for the conventional sandwich structure. The energy density was also found to be more for in-plane geometry (4.3 × 10^{-4} Wh/cm^3) as compared to sandwich structure (1.9 × 10^{-4} Wh/cm^3). Equivalent series resistance (ESR) of in-plane structure was observed to be more (6.5 kΩ) than the sandwich (126 Ω) structure which has resulted in fast charge-discharge rate and higher power density (9.4 W/cm^3) for sandwich device geometry compared to in-plane device geometry (1.7 W/cm^3).

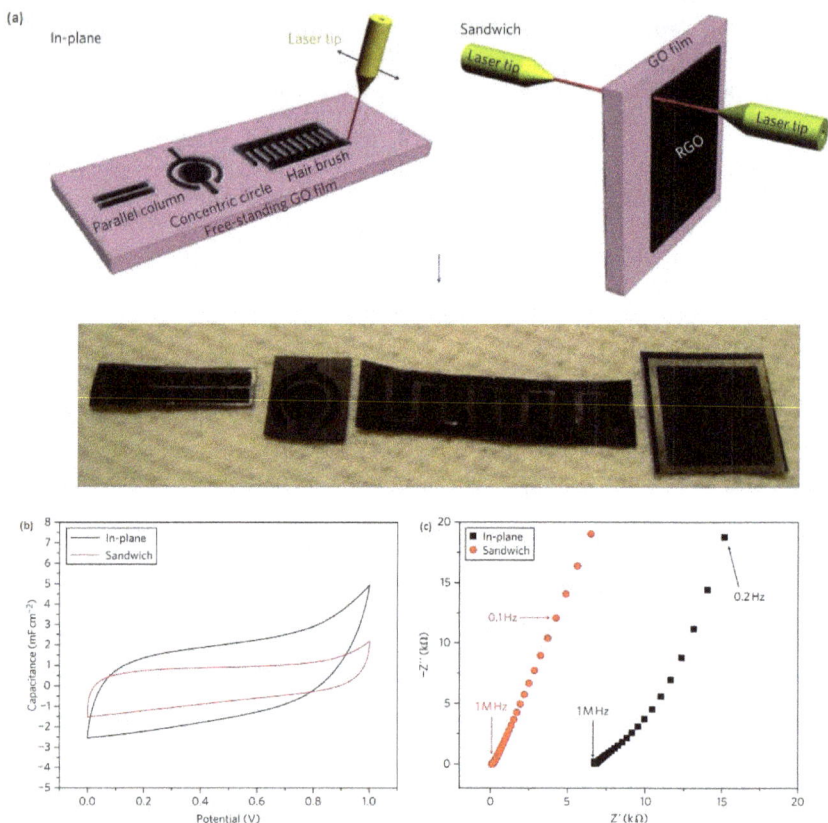

Figure 10. (a) Schematics of CO$_2$ laser-patterning of free-standing hydrated GO films to fabricate RGO–GO–RGO devices with in-plane and sandwich geometries. For in-plane devices, three different geometries were used, and the concentric circular pattern gives the highest capacitance density. The bottom row shows photographs of patterned films. (b) Cyclic Voltametry taken at 40 mV/s. (c) Nyquist plot from 1MHz to 10mHz (Reproduced with permission from Ref.[114])

Focused ion beam (FIB) technology was also used to directly write the rGO micro-electrodes through FIB irradiation reduction of GO films[115]. The rGO electrodes were 40 μm long and the inter-electrode distance was 1 μm. In this method, PVA-H_2SO_4 gel electrolyte was drop-casted on rGO electrodes. The ultra-small electrode dimensions resulted into a good performance of MSC with large capacitance (102 mF/cm^2 at1 mV/s). This capacitance value was much higher than previous reported values for direct laser writing rGO based MSCs (0.51 mF/cm^2) [114], laser scribed graphene based MSCs (2.318 mF/cm^2)[116] and CH_4 plasma reduced graphene oxide MSCs (322 μF/cm^2)[14]. This MSC provided ultra-small time response (0.03 ms) and an ultra-low ESR (0.35 mΩ/cm^2). In addition, this MSC structure was able to retain 95% of the capacitance after 1000 cycles at 45 mA/cm^2. Very recently, Shen et al. [117] have fabricated miniature rGO electrodes based MSC through the reduction of GO films by femtosecond (fs) laser irradiation. The synthesis process of rGO based MSC is shown in Fig. 11. The synthesized rGO pads had 100 μm length, 8 μm width and were separated by 2 μm spacing.

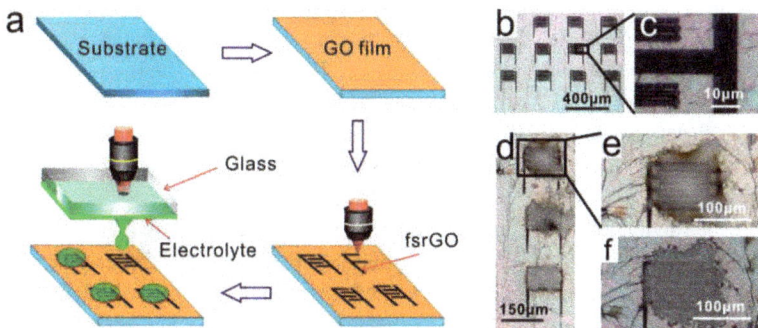

Figure 11. Schematic of the synthesis process for integrated micro-supercapacitor. (a) GO coated Si/SiO$_2$ substrate was subjected to fs laser radiation to producer ultra-thin rGO electrodes. These electrodes were covered by the micro-electrolyte droplets. (b, c) Interlaced rGO electrode arrays with ~2 μm spacing between the fingers. (d) Micro-electrolyte droplets exactly covered the electrode array. (e) MSC after transfer of the electrolyte and (f) after being solidified overnight at RT. (Reproduced with permission from Ref.[117])

After the synthesis of fs-rGO electrodes on GO films, a micro-droplet electrolyte gel was precisely transferred on fs-rGO electrodes using femtosecond laser-induced forward transfer (fsLIFT) methodology. As the electrolyte drops were transferred in a specific way to cover each individual MSC electrodes, this transfer technique can circumvent any interference of the electrolyte droplets with other electronic components. The synthesized MSC exhibited a good electrochemical performance with high capacitance (6.3 mF/cm^2) and ~100% retention after 1000 charge discharge cycles.

Graphene based MSCs were not only through the laser irradiation of graphene oxides, but also by the laser processing of commercial polymers [118–120]. Lin et al. [118] have synthesized a graphene based in-plane MSC – laser induced graphene (LIG) by CO$_2$ infrared laser irradiation of an insulating polyimide (PI) film under ambient conditions. The graphene was formed through the photo-thermal conversion of sp^3 carbon atoms of polymer to sp^2 carbon atoms of graphene by pulsed laser irradiation. The MSC exhibited specific capacitance of 44 mF/cm^2 at a scan rate of 20mV/s, power density of ~9 mW/cm^2 and nearly 100% capacitance after 9000 charge/discharge cycles in aqueous electrolyte.

Zhong-Shuai Wu et al. [14] developed graphene-based in-plane interdigital micro-supercapacitors on a polyethylene terephthalate (PET) substrate. Fig.12. presents the fabrication process of the micro-supercapacitors on a copper substrate. Briefly, a thin film of graphene oxide (GO) was deposited on copper substrate by spin-coating (Fig. 12 a, b). After that, the GO film was rapidly reduced, as shown by the change in colour of film from yellow to grey. (Fig. 12c). Subsequently, the reduced graphene oxide film (MPG film) supported by PMMA film was transferred onto the PET substrate (PMMA/MPG/PET). Next, gold current electrodes were deposited by lithography followed by oxidative etching in oxygen plasma (Figs. 12d, e, f). H$_2$SO$_4$/PVA was then drop-casted onto the interdigital electrodes and kept overnight to solidify, as shown in (Fig.12g). The advantage of gel electrolyte was not only to reduce the thickness of entire microdevice, but also it helped to avoid liquid electrolyte's leakage, and thus simplified the entire fabrication process. The resulting micro-supercapacitors deliver an area capacitance of 78.9 μF/cm^2 with excellent cycling stability of more than 99.9% after 100000 cycles.

6. Fabrication technique for graphene based micro-supercapacitor

Micro-supercapacitors use the same types of material as that of supercapacitors, but additionally, they are generally patterned and often are found to be integrated on chips. Steps of micro-supercapacitor fabrication, thus, mainly include electrode material synthesis, patterning of electrode and packing of the device. Numerous techniques have been developed to construct micro-supercapacitors and the processes can be divided into two categories depending on the patterning technique used for electrode materials. In one of these two types, a colloidal suspension is created and locally deposited onto patterned current collectors via photolithography, inkjet printing or electrophoretic deposition. For the second type, the electrode material is formed by laser irradiation, chemical conversion or electrolytic deposition during the fabrication. This section is aimed to elaborate different techniques used in the fabrication of micro-supercapacitors.

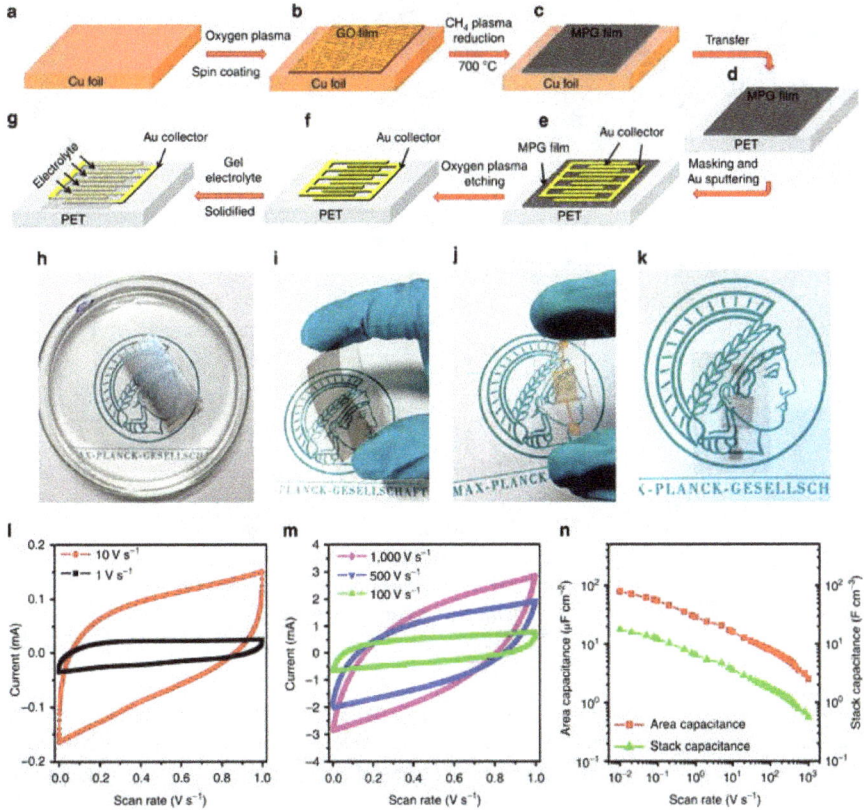

Figure 12. (a–g) Schematic of the fabrication of flexible MPG-MSCs-PET. (h-k) Optical images of MPG film. (l,m) cyclic voltammetry obtained at different scan rates. (n) Area capacitance and stack capacitance of the MPG-MSCs-PET. (Reproduced with permission from Ref.[14])

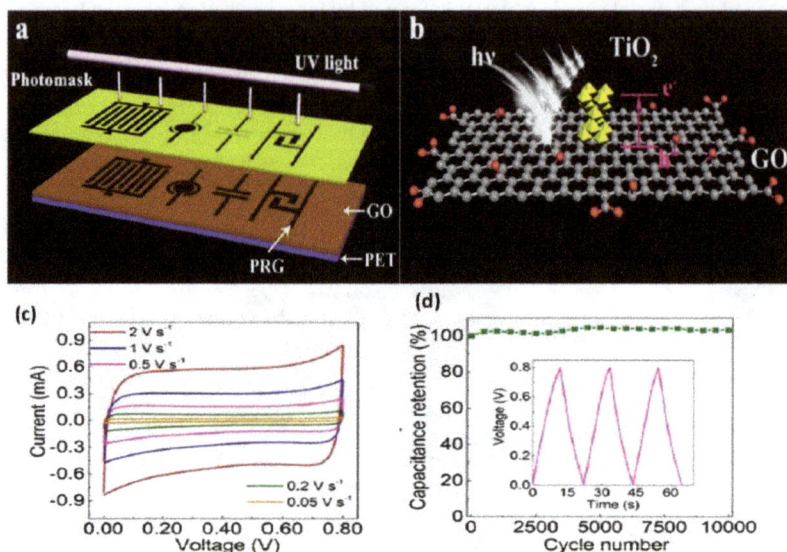

Figure 13. a) Illustration showing the fabrication process for photochemically reduced graphite oxide hybrid film (PRG-MSCs) with different geometries (b) Photoreduction mechanism of titanium dioxide (TiO$_2$) nanoparticles to GO sheets. c) Cyclic Voltametry curves obtained at different scan rates. d) Plot between capacitance retention verses cyclic number. Inset: Galvanostatic charge discharge curve at 0.02 mA/cm^2. (Reproduced with permission from Ref.[121])

6.1 Photolithography

Photolithography is the process which is frequently selected to pattern two very thin and accurately separated adjacent graphene based electrodes. In optical lithography, particular engraving of active materials is carried out by metal masks onto thick active materials as electrodes. By controlling the optical lithography parameters, it is possible to easily adjust the gap between two adjoining electrodes.

Zhiqiang Niu et al. fabricated a flexible graphene interdigital micro-supercapacitor by photolithography using H$_3$PO$_4$ /PVA gel electrolyte [67]. As shown in Fig. 14, a gold film was coated onto PET substrate by thermal evaporation (Fig. 14). The top surface of the gold film was then exposed with resist in the expected interdigitated structures. Afterwards, the r-GO film was deposited onto substrate by electrophoretic deposition. KOH was added drop-wise onto the substrate to remove the resist on the surface of the

Au film. The fabricated device with 400 µm thick electrode and 400 µm separation leads to volumetric energy density of 31.9 mWh/cm^3 and the maximum volumetric power density of 324 W/cm^3. The normalized capacitance by the geometrical area and volume of the rGO micro-supercapacitor were 462µF/cm^2 and 359 F/cm^3, respectively.

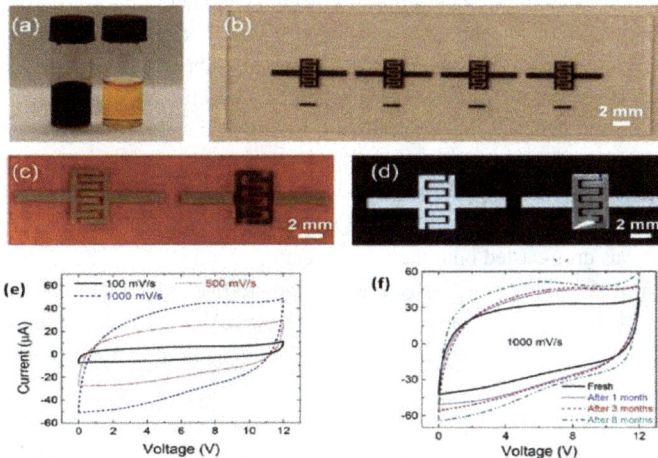

Figure 14. a) Images of PSSH ink (right) and EEG ink (left). (b-d) images of inkjet printed graphene micro-supercapacitor on glass slide, kapton and silicon wafer respectively. e) Cyclic Voltametry curves at 100 mV/s , 500 mV/s and 1000 mV/s f) Cyclic Voltametry curve at the 1000 mV/s after the fabrication. (Reproduced with permission from Ref.[123])

It was possible to follow the process without an actual photoresist. Wang et al. [121] have synthesized a flexible high performance graphite oxide hybrid film (PRG-MSCs) by using a custom made photomask, as shown in Fig.13 a. First, a thin film of GO/TiO$_2$ nanoparticles (graphene oxide/titanium dioxide) was transferred onto PET substrate. Then, the graphene oxide (GO) film was patterned with the help of a custom-made photomask under UV radiation. During this process, photogenerated electrons travel from titanium dioxide nanoparticles to graphene oxide, which result into an efficient reduction of the GO film and restore of the sp2-C network, as shown in figure Fig. 13 b. Lastly, a thin layer of gold was evaporated on PRG film and H$_2$SO$_4$/poly(vinyl alcohol) was carefully drop casted and kept overnight for drying. The as-prepared microdevice exhibited a high volumetric capacitance of 233.0 F/cm^3, as shown in Figs. 13(c & d). It also exhibited a volumetric power density up to 312 W/cm^3 and a high energy density of 7.7 mW h/cm^3.

6.2   Inkjet printing technology

In inkjet printing technology, fine droplets of electrode material ink were sprayed onto a substrate. Inkjet printing technology can avoid a complex photolithography process – thus, simplifying the fabrication process. Le et al. fabricated flexible MSCs using graphene oxide ink and their devices [68] exhibited a specific energy and specific power of 6.74 Wh/kg and 2.19 kW/kg, respectively. Liu et al. developed directly printable in-plane exfoliated graphene (EG) based MSCs on paper and PET substrates [122]. In this work, EG ink was spray-coated onto paper substrates covered by the shadow mask (both electrode and gap widths were 1000 μm, and the length was 2 cm). Then, a gel electrolyte PVA/H_2SO_4 was drop-casted onto the finger electrodes and allowed to solidify overnight. The as-fabricated solid-state EG-based MSC with in-plane geometry showed an areal capacitance of 1080 $μF/cm^2$ at a scan rate of 10 mV/s. The device also exhibited excellent cycling stability with ≈ 90% capacitance retention after 5000 charge/discharge cycles.

Following the same process of inkjet printing, Li et al. [123] fabricated graphene-based MSCs on varieties of substrates, such as glass slide, Kapton and silicon wafer, as shown in Fig. 14. They prepared highly concentrated, electrochemically exfoliated graphene ink (EEG) to print interdigital electrodes with tunable thickness and geometry by using a poly(4-styrenesulfonic acid) (PSSH) electrolyte. The device can sustain stability for about 8 months even after packaging and exhibited an areal capacitance of around 0.7 mF/cm^2. They manufactured a large-scale MSC arrays comprising more than 100 devices on both silicon wafers and flexible substrates (Kapton) and they were charged over 10 V at a scan rate as high as 1 V/s.

6.3   Laser scribing

Laser scribing is the newly developed fabrication technique of MSCs in which the high energy from laser beam is used to increase temperature locally. The resulted localized high-temperature results into breaking of the chemical bonds between carbon, hydrogen, nitrogen and oxygen. Eventually, the carbon atoms rearranged to form graphene and the other light elements escaped as a byproduct gas. By controlling the motion and size of focal point of the laser beam, it is possible to pattern and fabricate the laser scribed graphene (LSG). Laser scribed graphene is a candidate for high energy and power density because of its high porosity and conductivity. Kaner's group [116] illustrated the way to mass production of flexible micro-supercapacitors by direct laser based reduction of GO using a normal DVD writer, as shown in Fig. 15. Briefly, aqueous graphene oxide (GO) dispersion was drop-cast on PET substrate which was attached to a media disc and allowed to dry overnight under suitable conditions to form a uniform graphene oxide (GO) layer. Afterwards the graphene coated disc was put into a DVD drive for laser patterning and a solid-state gel electrolyte PVA-H_2SO_4 was deposited to obtain flexible solid-state micro-supercapacitor. Here, the LSG patterns served as both current collector and an active material. The remaining GO provide spacing between two electrodes. Within a short time period of 30 minutes, more than 100 MSCs could be fabricated on a single disc, in which each of the MSCs delivered an excellent volumetric power density of 200 W/cm^3, or areal density of 152 mW/cm^2.

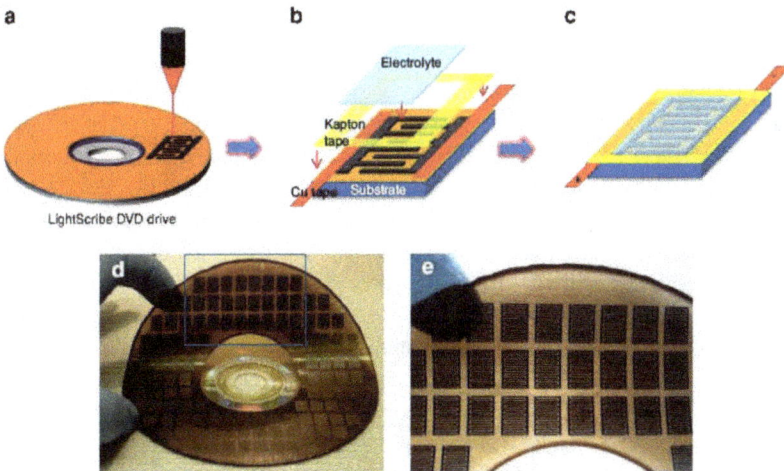

Figure 15. (a–c) Schematic of laser scribed micro-supercapacitor fabrication. (d, e) Images of flexible micro-supercapacitors (Reproduced with permission from Ref.[116])

6.4 Chemical vapor deposition

Chemical vapor deposition is a method which can produce uniform and continuous film of controlled thickness, granting MSCs great capacitive performances. Using 2D graphene, new designs for thin-film planar MSCs with desirable performance have become possible. For instant, Xiong et al. [124] designed high power micro-supercapacitor from microwave plasma chemical vapor deposition (MPCVD) grown graphitic petals (GP) electrodes. Briefly, quartz substrate was subjected to MPCVD conditions. The, quartz substrate with GP was coated with Ti/Au (50/200 nm) by using electron-beam evaporation and Ti/Au patterns were then formed by optical lithography.

Lastly, sixteen interdigitated fingers of the graphitic petal electrodes with Ti/Au as current collectors were formed by using plasma etching. As-prepared microdevices exhibited a volumetric capacitance of 270 F/cm^3 and an areal capacitance of 108 mF/cm^2 at a scan rate of 20 mV/s with a theoretical maximum power density of 292 W/cm^3 at a current of 100 mA. Fabrication of MSC is not only limited to two-dimensional structure. The three-dimensional nanostructured based micro-supercapacitor is also possible through CVD. Recently, Lin et al. [61] fabricated graphene/carbon nanotube carpets based MSC (G/ CNTCs-MCs) on nickel foil by CVD. They reported that the as-fabricated G/ CNTCs-MCs delivered a high volumetric energy density of 2.42 mWh/cm^3 and power density of 115 W/cm^3.

Conclusion

Graphene-based micro-supercapacitors have captivated lots of attention in the last few years. A breakthrough research progress has been made in development of electrode /electrolyte and the design of micro-supercapacitors. Furthermore, enhancement in graphene micro-supercapacitors is possible by addition of capacitive spacers, such as CNTs, nano-size metal oxide and electrically conductive polymers between the graphene sheets, which act as electroactive sites and help to improve porous morphology. The addition of such capacitive spacers prevents the aggregation and restacking of the graphene sheets. The porous matrix between graphene layers provides higher usage of accessible area for charge storage. Another way to obtain high-performance micro-supercapacitors is to develop industrially compatible, novel and cost-effective fabrication techniques. Newly developed methods like photolithography, inkjet printing technology, laser scribing, chemical vapor deposition are quite effective in improving energy density and power density of microdevices. To realize the most desirable features of MSCs, future development of graphene-based micro-supercapacitors should be focused on synthesis of electrode materials, patterning of electrode and packaging of the complete device in a more efficient way.

References

[1]   M. Winter, R.J. Brodd, What are Batteries, Fuel Cells, and Supercapacitors?, Chem Rev. 104 (2004) 4245–4270. https://doi.org/10.1021/cr020730k

[2]   M. Jayalakshmi, K. Balasubramanian, Simple capacitors to supercapacitors-An overview, Int. J. Electrochem. Sci. 3 (2008) 1196–1217.

[3]   A. Burk, Ultracapacitors: why, how, and where is the technology, J. Power Sources 91 (2000) 37–50. https://doi.org/10.1016/S0378-7753(00)00485-7

Materials Research Forum LLC
https://doi.org/10.21741/9781644900550-5

[4]   Z. Yu, L. Tetard, L. Zhai, J. Thomas, Supercapacitor electrode materials: Nanostructures from 0 to 3 dimensions, Energy Environ. Sci. 8 (2015) 702–730. https://doi.org/10.1039/C4EE03229B

[5]   Y. Wang, Y. Xia, Recent progress in supercapacitors: From materials design to system construction, Adv. Mater. 25 (2013) 5336–5342. https://doi.org/10.1002/adma.201301932

[6]   Y. Zhang, H. Feng, X. Wu, L. Wang, A. Zhang, T. Xia, H. Dong, X. Li, L. Zhang, Progress of electrochemical capacitor electrode materials: A review, Int. J. Hydrog. Energy 34 (2009) 4889–4899. https://doi.org/10.1016/j.ijhydene.2009.04.005

[7]   R. Ko, M. Carlen, R. Kotz and M. Carlen, Principles and Applications of electrochemical capacitors, Electrochim. Acta 45 (2000) 1–16. https://doi.org/10.1016/S0013-4686(00)00354-6

[8]   C. Liu, Z. Yu, D. Neff, A. Zhamu, B.Z. Jang, Graphene-based supercapacitor with an ultrahigh energy density, Nano Lett. 10 (2010) 4863–4868. https://doi.org/10.1021/nl102661q

[9]   Z. Fan, J. Yan, T. Wei, L. Zhi, G. Ning, T. Li, F. Wei, Asymmetric supercapacitors based on graphene/MnO$_2$ and activated carbon nanofiber electrodes with high power and energy density, Adv. Funct. Mater. 21 (2011) 2366–2375. https://doi.org/10.1002/adfm.201100058

[10]   J. Yan, Z. Fan, W. Sun, G. Ning, T. Wei, Q. Zhang, R. Zhang, L. Zhi, F. Wei, Advanced asymmetric supercapacitors based on Ni(OH)$_2$/graphene and porous graphene electrodes with high energy density, Adv. Funct. Mater. 22 (2012) 2632–2641. https://doi.org/10.1002/adfm.201102839

[11]   X. Yu, B. Lu, Z. Xu, Super long-life supercapacitors based on the construction of nanohoneycomb-like strongly coupled CoMoO$_4$-3D graphene hybrid electrodes, Adv. Mater. 26 (2014) 1044–1051. https://doi.org/10.1002/adma.201304148

[12]   N. Kurra, Q. Jiang, H.N. Alshareef, A general strategy for the fabrication of high-performance microsupercapacitors, Nano Energy 16 (2015) 1–9. https://doi.org/10.1016/j.nanoen.2015.05.031

[13]   G. Zhang, Y. Han, C. Shao, N. Chen, G. Sun, X. Jin, J. Gao, B. Ji, H. Yang, L. Qu, Processing and manufacturing of graphene-based microsupercapacitors, Mater Chem Front. (2018) 1750–1764. https://doi.org/10.1039/C8QM00270C

[14]   Z.S. Wu, K. Parvez, X. Feng, K. Müllen, Graphene-based in-plane micro-supercapacitors with high power and energy densities, Nat. Commun. 4 (2013) 1-7. https://doi.org/10.1038/ncomms3487

[15]   X. Shi, Z.S. Wu, J. Qin, S. Zheng, S. Wang, F. Zhou, C. Sun, X. Bao, Graphene-based linear tandem micro-supercapacitors with metal-free current collectors and high-voltage output, Adv. Mater. 29 (2017) 1–9. https://doi.org/10.1002/adma.201703034

[16]   W. Liu, X. Yan, J. Chen, Y. Feng, Q. Xue, Novel and high-performance asymmetric micro-supercapacitors based on graphene quantum dots and polyaniline nanofibers, Nanoscale 5 (2013) 6053–6062. https://doi.org/10.1039/c3nr01139a

[17]   D. Pech, M. Brunet, H. Durou, P. Huang, V. Mochalin, Y. Gogotsi, P. Taberna, P. Simon, Ultrahigh-power micrometer-sized supercapacitors based on onion-like carbon. Nat. Nanotechnol. 5 (2010) 651–654. https://doi.org/10.1038/nnano.2010.162

[18]   T. M Dinh, D. Pech, M. Brunet, A. Achour, High-resolution electrochemical micro-capacitors based on oxidized multi-walled carbon nanotubes, J. Phys.: Conf. Ser. 476 2013. https://doi.org/10.1088/1742-6596/476/1/012106

[19]   K. Parvez, X. Feng, K. Mu, Graphene-based in-plane micro-supercapacitors with high power and energy densities, Nat. Commun. 4 (2013) 1-7. https://doi.org/10.1038/ncomms3487

[20]   D. Pech, M. Brunet, P. Taberna, P. Simon, N. Fabre, F. Mesnilgrente, V. Conédéra, H. Durou, Elaboration of a microstructured inkjet-printed carbon electrochemical capacitor, J. Power Sources 195 (2010) 1266–1269. https://doi.org/10.1016/j.jpowsour.2009.08.085

[21]   M. Stoller, S. Park, Y. Zhu, J. An, R. Ruoff, Graphene-based ultracapacitors, Nano Lett. 8 (2008) 3498–3502. https://doi.org/10.1021/nl802558y

[22]   E. Gongadze, S. Petersen, U. Beck, U. Van Rienen, Classical Models of the Interface between an Electrode and an Electrolyte, COMSOL Conference (2009).

[23]   J.P. Valleau, G.M. Torrie, The electrical double layer. III. Modified Gouy-Chapman theory with unequal ion sizes, J. Chem. Phys. 76 (1982) 4623–4630. https://doi.org/10.1016/0079-6816(83)90004-7

[24]   D. Henderson, Recent progress in the theory of the electric double layer, *Prog. Surf. Sci.* 13 (1983) 197–224. https://doi.org/10.1016/0079-6816(83)90004-7

[25]   J.P. Zheng, Hydrous Ruthenium Oxide as an Electrode Material for electrochemical capacitors, J. Electrochem. Soc. 142 (1995) 2699. https://doi.org/10.1149/1.2050077

[26]　S. Makino, Y. Yamauchi, W. Sugimoto, Synthesis of electro-deposited ordered mesoporous RuO x using lyotropic liquid crystal and application toward micro-supercapacitors, J. Power Sources 227 (2013) 153–160. https://doi.org/10.1016/j.jpowsour.2012.11.032

[27]　Y.G. Wang, Z.D. Wang, Y.Y. Xia, An asymmetric supercapacitor using RuO_2/TiO_2 nanotube composite and activated carbon electrodes, Electrochim. Acta 50 (2005) 5641–5646. https://doi.org/10.1016/j.electacta.2005.03.042

[28]　H.Y. Lee, J.B. Goodenough, Amorphous V_2O_5/carbon composites as electrochemical supercapacitor electrodes, Solid State Ionics 153 (2002) 833–841. https://doi.org/10.1016/S0167-2738(02)00383-1

[29]　H. Wang, X. Sun, Z. Liu, Z. Lei, Creation of nanopores on graphene planes with MgO template for preparing high-performance supercapacitor electrodes, Nanoscale 6 (2014) 6577–6584. https://doi.org/10.1039/C4NR00538D

[30]　Q. Xiao, X. Zhou, The study of multiwalled carbon nanotube deposited with conducting polymer for supercapacitor, Electrochim.Acta 48 (2003) 575–580. https://doi.org/10.1016/S0013-4686(02)00727-2

[31]　C. Peng, S. Zhang, D. Jewell, G.Z. Chen, Carbon nanotube and conducting polymer composites for supercapacitors, Prog. Nat. Sci. 18 (2008) 777–788. https://doi.org/10.1016/j.pnsc.2008.03.002

[32]　Q. Meng, K. Cai, Y. Chen, L. Chen, Research progress on conducting polymer based supercapacitor electrode materials, Nano Energy 36 (2017) 268–285. https://doi.org/10.1016/j.nanoen.2017.04.040

[33]　G.A. Snook, P. Kao, A.S. Best, Conducting-polymer-based supercapacitor devices and electrodes, J. Power Sources 196 (2011) 1–12. https://doi.org/10.1016/j.jpowsour.2010.06.084

[34]　D.W. Wang, F. Li, H.M. Cheng, Hierarchical porous nickel oxide and carbon as electrode materials for asymmetric supercapacitor, J. Power Sources 185 (2008) 1563–1568. https://doi.org/10.1016/j.jpowsour.2008.08.032

[35]　Y. Zhu, C. Cao, S. Tao, W. Chu, Z. Wu, Y. Li, Ultrathin nickel hydroxide and oxide nanosheets: Synthesis, characterizations and excellent supercapacitor performances, Sci. Rep. 4 (2014) 1–7. https://doi.org/10.1038/srep05787

[36]　L.Q. Mai, F. Yang, Y.L. Zhao, X. Xu, L. Xu, Y.Z. Luo, Hierarchical $MnMoO_4/CoMoO_4$ heterostructured nanowires with enhanced supercapacitor performance, Nat. Commun. 2 (2011) (1–7). https://doi.org/10.1038/ncomms1387

[37] H. Chen, L. Hu, M. Chen, Y. Yan, L. Wu, Nickel-cobalt layered double hydroxide nanosheets for high-performance supercapacitor electrode materials, Adv. Funct. Mater. 24 (2014) 934–942. https://doi.org/10.1002/adfm.201301747

[38] T. Ohzuku, A. Ueda, Why transition metal (di) oxides are the most attractive materials for batteries, Solid State Ions. 69 (1994) 201–211. https://doi.org/10.1016/0167-2738(94)90410-3

[39] K.-C. Liu, Porous nickel oxide/nickel films for electrochemical capacitors, J. Electrochem. Soc. 143 (1996) 124. https://doi.org/10.1149/1.1836396

[40] B.K. Kim, V. Chabot, A. Yu, Carbon nanomaterials supported $Ni(OH)_2$/NiO hybrid flower structure for supercapacitor, Electrochim. Acta 109 (2013) 370–380. https://doi.org/10.1016/j.electacta.2013.07.119

[41] C. Xu, Y. Zhao, G. Yang, F. Li, H. Li, Mesoporous nanowire array architecture of manganese dioxide for electrochemical capacitor applications, Chem. Comm. (2009) 7575–7577. https://doi.org/10.1039/b915016a

[42] G.W. Yang, C.L. Xu, H.L. Li, Electrodeposited nickel hydroxide on nickel foam with ultrahigh capacitance, Chem. Comm. (2008) 6537–6539. https://doi.org/10.1039/b815647f

[43] F. Luan, G. Wang, Y. Ling, X. Lu, H. Wang, Y. Tong, X.X. Liu, Y. Li, High energy density asymmetric supercapacitors with a nickel oxide nanoflake cathode and a 3D reduced graphene oxide anode, Nanoscale 5 (2013) 7984–7990. https://doi.org/10.1039/c3nr02710d

[44] F. Zhang, T. Zhang, X. Yang, L. Zhang, K. Leng, Y. Huang, Y. Chen, A high-performance supercapacitor-battery hybrid energy storage device based on graphene-enhanced electrode materials with ultrahigh energy density, Energy Environ. Sci. 6 (2013) 1623–1632. https://doi.org/10.1039/c3ee40509e

[45] Y.K.A. Lau, A. V Kvit, A.L. Schmitt, S. Jin, D.J. Chernak, M.J. Bierman, S. Jin, A.R. Harutyunyan, B.I. Yakobson, C. Growth, S. Publishing, N. Cabrera, C. Frank, L.E. Greene, J.C. Johnson, R. Saykally, P.D. Yang, J.H. Song, K. Keis, S.E. Lindquist, A. Hagfeldt, D.S. Boyle, P.B. Kenway, M. Dudley, D. Bliss, M. Callahan, M. Harris, G.M. Fuge, N.A. Fox, D.J. Riley, J.F. Banfield, H.P. Strunk, V.D. Heydemann, G. Pensl, J.D. Eshelby, J.D. Eshelby, C.M. Drum, T.B. Bateman, H.H. Teng, P.M. Dove, J.J. De Yoreo, C. Cottrell, S. Award, D. Young, Monolithic Carbide-Derived Carbon Films for Micro-Supercapacitors, Science 1060 (2010) 480–484.

[46]   W. Sun, X. Chen, Fabrication and tests of a novel three dimensional micro supercapacitor, Microelectron *Eng.* 86 (2009) 1307-1310. https://doi.org/10.1016/j.mee.2008.12.010

[47]   W. Sun, R. Zheng, X. Chen, Symmetric redox supercapacitor based on micro-fabrication with three-dimensional polypyrrole electrodes, J. Power Sources 195 (2010) 7120–7125. https://doi.org/10.1016/j.jpowsour.2010.05.012

[48]   C. Shen, X. Wang, W. Zhang, F. Kang, A high-performance three-dimensional micro supercapacitor based on self-supporting composite materials, J.Power Sources 196 (2011) 10465–10471. https://doi.org/10.1016/j.jpowsour.2011.08.007

[49]   M. Beidaghi, Y. Gogotsi, Capacitive energy storage in micro-scale devices: Recent advances in design and fabrication of micro-supercapacitors, Energy Environ. Sci. 7 (2014) 867–884. https://doi.org/10.1039/c3ee43526a

[50]   M. Beidaghi, C. Wang, Micro-supercapacitors based on interdigital electrodes of reduced graphene oxide and carbon nanotube composites with ultrahigh power handling performance, Adv. Funct. Mater. 22 (2012) 4501–4510. https://doi.org/10.1002/adfm.201201292

[51]   J. Sung, S. Kim, K. Lee, Fabrication of microcapacitors using conducting polymer microelectrodes, J. Power Sources 124 (2003) 343–350. https://doi.org/10.1016/S0378-7753(03)00669-4

[52]   J.H. Sung, S.J. Kim, K.H. Lee, Fabrication of all-solid-state electrochemical microcapacitors, J. Power Sources 133 (2004) 312–319. https://doi.org/10.1016/j.jpowsour.2004.02.003

[53]   M. Xue, Z. Xie, L. Zhang, X. Ma, X. Wu, Y. Guo, W. Song, Microfluidic etching for fabrication of flexible and all-solid-state micro supercapacitor based on MnO_2 nanoparticles, Nanoscale 3 (2011) 2703–2708. https://doi.org/10.1039/c0nr00990c

[54]   M.F. El-kady, R.B. Kaner, Scalable fabrication of high-power graphene micro-supercapacitors for flexible and on-chip energy storage, Nat. Commun. 4 (2013) 1475–1479. https://doi.org/10.1038/ncomms2446

[55]   X. Du, C.Y. Wang, M.M. Chen, Y. Jiao, J. Wang, Electrochemical performances of nanoparticle fe3o4/activated carbon supercapacitor using KOH Electrolyte solution, J Phys Chem. C 113 (2009) 2643–2646. https://doi.org/10.1021/jp8088269

[56]   A. Yuan, Q. Zhang, A novel hybrid manganese dioxide/activated carbon supercapacitor using lithium hydroxide electrolyte, Electrochem. Commun. 8 (2006) 1173–1178. https://doi.org/10.1016/j.elecom.2006.05.018

[57] J. Gamby, P.L. Taberna, P. Simon, J.F. Fauvarque, M. Chesneau, Studies and characterisations of various activated carbons used for carbon/carbon supercapacitors, J. Power Sources 101 (2001) 109–116. https://doi.org/10.1016/S0378-7753(01)00707-8

[58] R.H. Baughman, A.A. Zakhidov, W.A. de Heer, Carbon Nanotubes-the route toward applications, Science 787 (2012) 787–792. https://doi.org/10.1126/science.1060928

[59] M. Kaempgen, C.K. Chan, J. Ma, A. Kiebele, Y. Cui, G. Gruner, Printable Thin film supercapacitors using single-walled carbon nanotubes, Nano Lett. 9 (2009) 1872-1876. https://doi.org/10.1021/nl8038579

[60] B.R. Pan, S.W. Lee, C.J. Tseng, C.L. Chang, W.C. Hung, J.K. Chang, Supercapacitive performance of porous graphene nanosheets in bis(trifluoromethylsulfony)imide and bis(fluorosulfonyl)imide ionic liquid electrolytes, J. *Solid State* Electr. 22 (2018) 2197–2203. https://doi.org/10.1007/s10008-018-3926-y

[61] J. Lin, C. Zhang, Z. Yan, Y. Zhu, Z. Peng, R.H. Hauge, D. Natelson, J.M. Tour, 3-Dimensional graphene-carbon nanotube carpet-based microsupercapacitors with high electrochemical performance, Nano Lett.13 (2013) 72-78. https://doi.org/10.1021/nl3034976

[62] B.Y. Zhu, S. Murali, W. Cai, X. Li, J.W. Suk, J.R. Potts, R.S. Ruoff, Graphene and graphene oxide : Synthesis, properties, and applications, Adv. Mater. 22 (2010) 3906–3924. https://doi.org/10.1002/adma.201001068

[63] L. Li, A. Quinlivan, Patricia, R.U. Knappe, Detlef, Effects of activated carbon surface chemistry and pore structure on the adsorption of organic contaminants from aqueous solutions, Carbon 40 (2AD) 2085–2100. https://doi.org/10.1016/S0008-6223(02)00069-6

[64] E. Frackowiak, Carbon materials for the electrochemical storage of energy in capacitors, Carbon 39 (2001) 937–950. https://doi.org/10.1016/S0008-6223(00)00183-4

[65] Z. Tang, C.H. Tang, H. Gong, A high energy density asymmetric supercapacitor from nano-architectured $Ni(OH)_2$/carbon nanotube electrodes, Adv. Funct. Mater. 22 (2012) 1272–1278. https://doi.org/10.1002/adfm.201102796

[66]   C. Zhong, Y. Deng, W. Hu, J. Qiao, L. Zhang, J. Zhang, A review of electrolyte materials and compositions for electrochemical supercapacitors, Chem. Soc. Rev. 44 (2015) 7484–7539. https://doi.org/10.1039/C5CS00303B

[67]   Z. Niu, L. Zhang, L. Liu, B. Zhu, H. Dong, X. Chen, All-solid-state flexible ultrathin micro-supercapacitors based on graphene, Adv. Mater. 25 (2013) 4035–4042. https://doi.org/10.1002/adma.201301332

[68]   L. Le, H. Qiu et al. Inkjet-printed graphene for flexible micro-supercapacitors, Conference: IEEE Nanotechnology (2011). https://doi.org/10.1109/NANO.2011.6144432

[69]   V. Strong, S. Dubin, M.F. El-Kady, A. Lech, Y. Wang, B.H. Weiller, R.B. Kaner, Patterning and electronic tuning of laser scribed graphene for flexible all-carbon devices, ACS Nano 6 (2012) 1395–1403. https://doi.org/10.1021/nn204200w

[70]   L.L. Zhang, R. Zhou, X.S. Zhao, Graphene-based materials as supercapacitor electrodes, J. Mater. Chem. 20 (2010) 5983–5992. https://doi.org/10.1039/c000417k

[71]   E. Frackowiak, Carbon materials for supercapacitor application, Phys Chem Chem Phys 9 (2007) 1774–1785. https://doi.org/10.1039/b618139m

[72]   L.M. Da Silva, L.A. De Faria, J.F.C. Boodts, Determination of the morphology factor of oxide layers, Electrochim. Acta 47 (2001) 395–403. https://doi.org/10.1016/S0013-4686(01)00738-1

[73]   G. Yu, L. Hu, M. Vosgueritchian, H. Wang, X. Xie, J.R. McDonough, X. Cui, Y. Cui, Z. Bao, Solution-processed graphene/MnO_2 nanostructured textiles for high-performance electrochemical capacitors, Nano Lett. 11 (2011) 2905–2911. https://doi.org/10.1021/nl2013828

[74]   A. Borenstein, O. Hanna, R. Attias, S. Luski, T. Brousse, D. Aurbach, Carbon-based composite materials for supercapacitor electrodes: A review, J. Mater. Chem. A 5 (2017) 12653–12672. https://doi.org/10.1039/C7TA00863E

[75]   L. Lv, Y. Fan, Q. Chen, Y. Zhao, Y. Hu, Z. Zhang, N. Chen, L. Qu, Three-dimensional multichannel aerogel of carbon quantum dots for high-performance supercapacitors, Nanotechnology 25 (2014) 235401. https://doi.org/10.1088/0957-4484/25/23/235401

[76]   Y. Wang, A. Hu, Carbon quantum dots: Synthesis, properties and applications, J. Mater. Chem. C 2 (2014) 6921–6939. https://doi.org/10.1039/C4TC00988F

[77]   Y.-Q. Dang, S.-Z. Ren, G. Liu, J. Cai, Y. Zhang, J. Qiu, Electrochemical and capacitive properties of carbon dots/reduced graphene oxide supercapacitors, Nanomaterials 6 (2016) 212. https://doi.org/10.3390/nano6110212

[78]   X. Chen, R. Paul, L. Dai, Carbon-based supercapacitors for efficient energy storage, Natl. Sci. Rev. 4 (2017) 453–489. https://doi.org/10.1093/nsr/nwx009

[79]   G.E. Lecroy, S.K. Sonkar, F. Yang, L.M. Veca, P. Wang, K.N. Tackett, J.J. Yu, E. Vasile, H. Qian, Y. Liu, P. Luo, Y.P. Sun, Toward structurally defined carbon dots as ultracompact fluorescent probes, ACS Nano 8 (2014) 4522–4529. https://doi.org/10.1021/nn406628s

[80]   Y. Xu, X.H. Jia, X.B. Yin, X.W. He, Y.K. Zhang, Carbon quantum dot stabilized gadolinium nanoprobe prepared via a one-pot hydrothermal approach for magnetic resonance and fluorescence dual-modality bioimaging, Anal. Chem. 86 (2014) 12122–12129. https://doi.org/10.1021/ac503002c

[81]   A.B. Bourlinos, A. Stassinopoulos, D. Anglos, R. Zboril, M. Karakassides, E.P. Giannelis, Surface functionalized carbogenic quantum dots, Small 4 (2008) 455–458. https://doi.org/10.1002/smll.200700578

[82]   Q.L. Zhao, Z.L. Zhang, B.H. Huang, J. Peng, M. Zhang, D.W. Pang, Facile preparation of low cytotoxicity fluorescent carbon nanocrystals by electrooxidation of graphite, Chem. Comm. 41 (2008) 5116–5118. https://doi.org/10.1039/b812420e

[83]   L. Zheng, Y. Chi, Y. Dong, J. Lin, B. Wang, Electrochemiluminescence of water-soluble carbon nanocrystals released electrochemically from graphite, J. Am. Chem. Soc. 131 (2009) 4564–4565. https://doi.org/10.1021/ja809073f

[84]   F.A. Pyatakovich, O. V. Mevsha, T.I. Yakunchenko, K.F. Makkonen, Biotechnical system of automatic classification scattergrams and evaluation of atrial fibrillation outcomes, Int J Pharm Pharm Sci. 8 (2016) 14129–14136.

[85]   H.M.R. Gonçalves, A.J. Duarte, J.C.G. Esteves da Silva, Optical fiber sensor for Hg(II) based on carbon dots, Biosens. Bioelectron. 26 (2010) 1302–1306. https://doi.org/10.1016/j.bios.2010.07.018

[86]   Y.-P. Sun, S.-T. Yang, L. Cao, P.G. Luo, F. Lu, X. Wang, H. Wang, M.J. Meziani, Y. Liu, G. Qi, Carbon dots for optical imaging in vivo, J. Am. Chem. Soc. 131 (2009) 11308–9. https://doi.org/10.1021/ja904843x

[87]   Q. Wang, X. Liu, L. Zhang, Y. Lv, Microwave-assisted synthesis of carbon nanodots through an eggshell membrane and their fluorescent application, Analyst 137 (2012) 5392–5397. https://doi.org/10.1039/c2an36059d

[88]   H. Jiang, F. Chen, M.G. Lagally, F.S. Denes, New strategy for synthesis and functionalization of carbon nanoparticles, Langmuir 26 (2010) 1991–1995. https://doi.org/10.1021/la9022163

[89]   X. Xu, R. Ray, Y. Gu, H.J. Ploehn, L. Gearheart, K. Raker, W.A. Scrivens, Electrophoretic analysis and purification of fluorescent single-walled carbon nanotube fragments, J. Am. Chem. Soc. 126 (2004) 12736–12737. https://doi.org/10.1021/ja040082h

[90]   J. Xu, K. Hou, Z. Ju, C. Ma, W. Wang, C. Wang, J. Cao, Z. Chen, Synthesis and electrochemical properties of carbon dots/manganese dioxide ($CQDs/MnO_2$) nanoflowers for supercapacitor applications, J. Electrochem. Soc. 164 (2017) A430–A437. https://doi.org/10.1149/2.1241702jes

[91]   Y. Zhu, X. Ji, C. Pan, Q. Sun, W. Song, L. Fang, Q. Chen, C.E. Banks, A carbon quantum dot decorated RuO_2 network: Outstanding supercapacitances under ultrafast charge and discharge, Energy Environ. Sci. 6 (2013) 3665–3675. https://doi.org/10.1039/c3ee41776j

[92]   H. Feng, P. Xie, S. Xue, L. Li, X. Hou, Z. Liu, D. Wu, L. Wang, P.K. Chu, Synthesis of three-dimensional porous reduced graphene oxide hydrogel/carbon dots for high-performance supercapacitor, J. Electroanal. Chem. 808 (2018) 321–328. https://doi.org/10.1016/j.jelechem.2017.12.046

[93]   S. Mondal, U. Rana, S. Malik, Graphene quantum dot-doped polyaniline nanofiber as high performance supercapacitor electrode materials, Chem.Comm. 51 (2015) 12365–12368. https://doi.org/10.1039/C5CC03981A

[94]   Z. Zhao, Y. Xie, Enhanced electrochemical performance of carbon quantum dots-polyaniline hybrid, J. Power Sources 337 (2017) 54–64. https://doi.org/10.1016/j.jpowsour.2016.10.110

[95]   X. Zhang, J. Wang, J. Liu, J. Wu, H. Chen, H. Bi, Design and preparation of a ternary composite of graphene oxide/carbon dots/polypyrrole for supercapacitor application: Importance and unique role of carbon dots, Carbon 115 (2017) 134–146. https://doi.org/10.1016/j.carbon.2017.01.005

[96]   L. Ferragut, R. Montenegro, G. Winter, A. Núñez, Accurate extraction of interconnect capacitances by adaptive mixed F.E.M., Microprocessing and Microprogramming 32 (1991) 61–68. https://doi.org/10.1016/0165-6074(91)90324-M

[97]   C. Niu, E.K. Sichel, R. Hoch, D. Moy, H. Tennent, High power electrochemical capacitors based on carbon nanotube electrodes, Appl. Phys. Lett. 70 (1997) 1480–1482. https://doi.org/10.1063/1.118568

[98]   R.Z. Ma J. Liang, D.H. Wu, R.Z. Ma, J. Liang, B.Q. Wei, B. Zhang, C.L. Xu, Study of electrochemical capacitors utilizing carbon nanotube electrodes, J. Power Sources 84 (1999) 126–129. https://doi.org/10.1016/S0378-7753(99)00252-9

[99]   B. Hsia, J. Marschewski, S. Wang, J. Bin In, C. Carraro, D. Poulikakos, C.P. Grigoropoulos, R. Maboudian, Highly flexible, all solid-state micro-supercapacitors from vertically aligned carbon nanotubes, Nanotechnology 25 (2014) 055401. https://doi.org/10.1088/0957-4484/25/5/055401

[100] Y.Q. Jiang, Q. Zhou, L. Lin, Planar mems supercapacitor using carbon nanotube forests, Proceedings of the IEEE International Conference on Micro Electro Mechanical Systems (MEMS), (2009). https://doi.org/10.1109/MEMSYS.2009.4805450

[101] W. Yu, H. Zhou, B.Q. Li, S. Ding, 3D Printing of carbon nanotubes-based microsupercapacitors, ACS Appl. Mater. Interfaces 9 (2017) 4597–4604. https://doi.org/10.1021/acsami.6b13904

[102] A.K. Geim and K.S. Novoselov, The rise of graphene, Nat. Mater. **6** (2007) 183–191. https://doi.org/10.1038/nmat1849

[103] J. Xia, F. Chen, J. Li, N. Tao, Measurement of the quantum capacitance of graphene, Nat. Nanotechnol. 4 (2009) 505–509. https://doi.org/10.1038/nnano.2009.177

[104] X. Cao, Z. Yin, H. Zhang, Three-dimensional graphene materials: Preparation, structures and application in supercapacitors, Energy Environ. Sci. 7 (2014) 1850–1865. https://doi.org/10.1039/C4EE00050A

[105] S.R.C. Vivekchand, C.S. Rout, K.S. Subrahmanyam, A. Govindaraj, C.N.R. Rao, Graphene-based electrochemical supercapacitors, J. Chem. Sci. 120 (2008) 9–13. https://doi.org/10.1007/s12039-008-0002-7

[106] Y. Wu, T. Zhang, F. Zhang, Y. Wang, Y. Ma, Y. Huang, Y. Liu, Y. Chen, In situ synthesis of graphene/single-walled carbon nanotube hybrid material by arc-discharge and its application in supercapacitors, Nano Energy 1 (2012) 820–827. https://doi.org/10.1016/j.nanoen.2012.07.001

[107] J.M. Chem, C.M. Ma, C. Hu, Design and tailoring of a hierarchical graphene-carbon nanotube architecture for supercapacitors J. Mater. Chem. 21 (2011) 2374-2380. https://doi.org/10.1039/C0JM03199B

[108] J. Li, M. Östling, Prevention of graphene restacking for performance boost of supercapacitors—A review, Crystals 3 (2013) 163–190. https://doi.org/10.3390/cryst3010163

[109] Z. Gao, W. Yang, J. Wang, B. Wang, Z. Li, Q. Liu, M. Zhang, L. Liu, A new partially reduced graphene oxide nanosheet/polyaniline nanowafer hybrid as supercapacitor electrode material, Energy Fuels 27 (2013) 568–575. https://doi.org/10.1021/ef301795g

[110] Z. Li, H. Zhang, Q. Liu, L. Sun, L. Stanciu, J. Xie, Fabrication of high-surface-area graphene/polyaniline nanocomposites and their application in supercapacitors, ACS Appl. Mater. Interfaces 5 (2013) 2685–2691. https://doi.org/10.1021/am4001634

[111] J.M. Chem, High performance supercapacitors using metal oxide anchored graphene nanosheet electrodes, J. Mater. Chem. 21 (2011) 16197–16204. https://doi.org/10.1039/c1jm12963e

[112] J. Wang, Z. Gao, Z. Li, B. Wang, Y. Yan, Q. Liu, T. Mann, Journal of Solid State Chemistry Green synthesis of graphene nanosheets / ZnO composites and electrochemical properties, J. Solid State Chem.184 (2011) 1421–1427. https://doi.org/10.1016/j.jssc.2011.03.006

[113] Z. Gao, J. Wang, Z. Li, W. Yang, B. Wang, M. Hou, Y. He, Q. Liu, T. Mann, P. Yang, M. Zhang, L. Liu, Graphene Nanosheet/Ni^{2+}/Al^{3+} Layered Double-Hydroxide Composite as a Novel Electrode for a Supercapacitor, Chem. Mater. 23 (2011) 3509–3516. https://doi.org/10.1021/cm200975x

[114] W. Gao, N. Singh, L. Song, Z. Liu, A.L.M. Reddy, L. Ci, R. Vajtai, Q. Zhang, B. Wei, P.M. Ajayan, Direct laser writing of micro-supercapacitors on hydrated graphite oxide films, Nat. Nanotechnol. 6 (2011) 496–500. https://doi.org/10.1038/nnano.2011.110

[115] D.E. Lobo, P.C. Banerjee, C.D. Easton, M. Majumder, Miniaturized Supercapacitors: Focused Ion Beam Reduced Graphene Oxide Supercapacitors with Enhanced Performance Metrics, Adv. Energy Mater. 5 (2015) (1-10). https://doi.org/10.1002/aenm.201500665

[116] M.F. El-Kady, R.B. Kaner, Scalable fabrication of high-power graphene micro-supercapacitors for flexible and on-chip energy storage, Nat. Commun. 4 (2013) (1-7). https://doi.org/10.1038/ncomms2446

[117] D. Shen, G. Zou, L. Liu, W. Zhao, A. Wu, W.W. Duley, Y.N. Zhou, Scalable High-Performance Ultraminiature Graphene Micro-Supercapacitors by a Hybrid Technique Combining Direct Writing and Controllable Microdroplet Transfer, ACS Appl. Mater. Interfaces 10 (2018) 5404–5412. https://doi.org/10.1021/acsami.7b14410

[118] J. Lin, Z. Peng, Y. Liu, F. Ruiz-Zepeda, R. Ye, E.L.G. Samuel, M.J. Yacaman, B.I. Yakobson, J.M. Tour, Laser-induced porous graphene films from commercial polymers, Nat. Commun. 5 (2015) 5–12. https://doi.org/10.1038/ncomms6714

[119] J. Luo, F.R. Fan, T. Jiang, Z. Wang, W. Tang, C. Zhang, M. Liu, G. Cao, Z.L. Wang, Integration of micro-supercapacitors with triboelectric nanogenerators for a flexible self-charging power unit, Nano Res. 8 (2015) 3934–3943. https://doi.org/10.1007/s12274-015-0894-8

[120] Z. Peng, R. Ye, J.A. Mann, D. Zakhidov, Y. Li, P.R. Smalley, J. Lin, J.M. Tour, Flexible boron-doped laser-induced graphene microsupercapacitors, ACS Nano 9 (2015) 5868–5875. https://doi.org/10.1021/acsnano.5b00436

[121] S. Wang, Z. Wu, S. Zheng, F. Zhou, C. Sun, H. Cheng, X. Bao, Scalable fabrication of photochemically reduced graphene-based monolithic micro-supercapacitors with superior energy and power densities, ACS Nano 11 (2017) 4283–4291. https://doi.org/10.1021/acsnano.7b01390

[122] Z. Liu, Z. Wu, S. Yang, R. Dong, X. Feng, K. Müllen, Ultraflexible in-plane micro-supercapacitors by direct printing of solution-processable electrochemically exfoliated graphene, Adv. Mater. 28 (2016) 2217–2222. https://doi.org/10.1002/adma.201505304

[123] J. Li, S.S. Delekta, P. Zhang, S. Yang, M.R. Lohe, X. Zhuang, X. Feng, Scalable fabrication and integration of graphene microsupercapacitors through full inkjet printing, ACS Nano 11 (2017) 8249–8256. https://doi.org/10.1021/acsnano.7b03354

[124] G. Xiong, C. Meng, R.G. Reifenberger, Graphitic petal micro-supercapacitor electrodes for ultra-high power graphitic petal micro-supercapacitor electrodes for ultra-high power density, Energy Technol. 2 (2014) 897-905. https://doi.org/10.1002/ente.201402055

Materials Research Forum LLC
https://doi.org/10.21741/9781644900550-6

Chapter 6

Graphene-Based Materials for Self-Healable Supercapacitors

Ramyakrishna Pothu[1], Sashivinay Kumar Gaddam[2], S. Vadivel[3], Rajender Boddula[4]*

[1]College of Chemistry and Chemical Engineering, Hunan University, Changsha 410082, China

[2]LAMTUF Plastics Private Ltd, Phase III, IDA Pashamylaram, Sangareddy-502307, Telangana state, India

[3]Department of Chemistry, PSG College of Technology, Coimbatore, 641004, Tamilnadu, India

[4]CAS Key Laboratory of Nanosystem and Hierarchical Fabrication, National Center for Nanoscience and Technology, Beijing 100190, PR China

*research.raaj@gmail.com

Abstract

The self-healing supercapacitor (SHS) based on graphene has interesting physical and chemical properties, which has attracted wide attention and has been proven to have great application potential in the field of energy conversion and storage. In this chapter, we will focus on recent important aspects of the advances in the fabrication and application of graphene-based self-healable electrodes and electrolytes for flexible SHS. And also discusses the challenges and the direction needed for future development of graphene hybrids.

Keywords

Graphene, Self-Healable Capacitor, Electrodes, Electrolytes, Composites

Contents

Graphene as Energy Storage Material for Supercapacitors   Materials Research Forum LLC
Materials Research Foundations **64** (2020) 167-180   https://doi.org/10.21741/9781644900550-6

1.   Introduction

The rising world population has led to an increase in energy, and global demand for electricity is expected to increase over the next few decades. Therefore, it is necessary for scientists and researchers to develop sustainable, clean, and renewable energy technologies, economies, and environmental protection that can meet the challenges of our society in the growing demand for energy. In this energy dependence of the world, energy storage electrochemical devices play a vital role in overcoming the depletion of fossil fuels. Among various explored electrochemical energy storage devices, the supercapacitor (SC) or ultracapacitor are promising power supplies due to its high power density, fast charging/discharging rates and long cycle life performance [1]. They are considered to be one of the most promising electrochemical energy storage devices, with the potential to replenish or eventually replace the battery with the ability to install and portable applications for electronic, electrical and hybrid vehicles.

Based on energy storage mechanism, SCs can be classified into two categories, namely, i) electrochemical double-layer capacitors (EDLC), which store energy anion/cation on surface electrodes by adsorption; ii) The pseudocapacitor (PC) that produces the capacitor responds to the redox reaction from the charge storage in the redox material. However, poor conductivity of the well-known PC electrodes can limit Faraday's reaction and thus lead to unsatisfactory electrochemical properties and life cycles [2].

2.   Graphene

Graphene is a new class of 2D "aromatic" single-atom thick carbon layer atoms that accumulate intensively in honeycomb lattices. Graphene has attracted great attention since its discovery in 2004, as the next generation SCs energy's promising material storage device due to its excellent performance, including large theoretical specific surface area (2,630 m^2 g^{-1}), which is significantly higher than 1D carbon (i.e. carbon nanotubes, 1315 m^2 g^{-1}) and 3D carbon (10 m^2 g^{-1}) counterpart, high carrier migration rate of 200,000 $cm^2V^{-1}S^{-1}$, excellent electrical conductivity (6.6 Ms m^{-1} vs. 0.35 MS m^{-1}), significant thermal conductivity at room temperature for 3,000-5,000 W m^{-1} K^{-1}, excellent optical transparency of 97.3%, which leads to an unexpectedly high opacity of the atomic single layer, excellent Li^+ chemical diffusion coefficient of 10^{-7}-10^{-6} cm^2 s^{-1} [3]. In addition, graphene has high tensile strength and excellent flexibility and is very beneficial to build flexible nanodevices [4]. A number of physical and chemical methods can be used to synthesize graphene including micro-mechanical cutting (original transparent tape method), chemical vapor deposition, liquid phase stripping and so on. As a precursor to graphene tablets, graphene oxide or functionalized graphene have a rich functional surface with highly reactive groups, to help design the preparation of composite structures. So far, various methods have been developed for the integration of graphene tables divided into 1D and 3D macroscopic assemblies, including wet spinning, filtration, and hydrothermal treatment methods.

3. Graphene-based SHS

Graphene sheets can be considered a building block and can be reasonably arranged into a specific architecture, especially for large 2D construction films and three-dimensional (3D) frames, providing a good opportunity to transform the microscopic characteristics of excellent graphene into macroscopic characteristics. Active electrode materials for EDLC are usually porous carbon, such as activated carbon, carbon nanotubes, graphene, and carbide-derived carbon. For example, the monolayer-graphene is theoretical specific capacitance of ~21 μF cm^{-2}, corresponding to the capacitance of ~550 F g^{-1}, when the entire surface area is fully utilized [5]. However, the actual capacitance behavior of pure graphene content below the expected value in the preparation and production process of serious aggregation application procedures. Therefore, the improvement of the overall electrochemical properties of graphene materials is still a huge challenge. Graphene-based materials have been extensively studied as conductive networks to support the redox reaction of transition metal oxides, hydroxides and conductive polymers. In fact, these hybrid electrodes, consisting of graphene and transition metal oxides/hydroxides or conductive polymer nanoparticles, exhibit excellent electrochemical properties [6–10], which are the result of the synergistic effect of graphene layer promoting the dispersion of metal oxides/hydroxides nanomaterials and play a high role. Conductive substrates for enhanced conductivity, metal oxides/hydroxides/conductive polymers provide the required pseudocapacitance. So far, flexible supercapacitors based on graphene components have been widely studied. While, tremendous efforts have been devoted to developing graphene-based flexible supercapacitors, the separation of the separator and electrodes is even inadequacy. Conventional preparation methods of graphene macroscopic assemblies require severe conditions, and, once formed, the assemblies can't be edited, reshaped, or recycled [4].

Traditional flexible supercapacitors can work continuously with bending, folding, or even distortion without compromising performance. However, these devices are difficult to use under large tensile strain. Traditional rigid SC or mainly flexible bendable and twistable SCs is far from practical application requirements of wearable electronic devices [11]. Because they have many potential applications in electronic devices, they are in great need of flexible stretchable and repairable supercapacitors. However, it is challenging to

Materials Research Forum LLC
https://doi.org/10.21741/9781644900550-6

manufacture supercapacitors that can withstand tensile and self-penetrating nature. SHS with strong deformation or healing capabilities has been developed to meet these stringent requirements and has recently become a hot topic. By benefiting from the function of self-healing, organisms can restore mechanical injuries to wounds, thereby improving their viability [12]. Inspired by this, a SHS concept is proposed to repair cracks and ensure that crack SC works well in the event of mechanical fractures.

Fiber materials play a vital role in our lives and extensively used as SC structural materials are woven into lightweight, highly numerous advantages of flexible, soft, and low-cost textiles. Graphene is a typical auspicious fiber material widely used to assemble their high electrical and thermal conductivity, extraordinary toughness and elasticity, with strong stability. Graphene-carbon nanotubes fiber composites have been further proposed and manufactured to prevent the interaction of graphene π-π by sheet stacking. However, when these graphene fibers based SCs in practical applications, the operation of electrode materials and substrates can be caused by mechanical deformation or accidental cutting damage. The reliability and stability of the SCs are limited through these failures, leading to the collapse of electronic devices of the entire scale. Self-healing materials have the ability to partially or completely cure damage and restore mechanical and structural properties inflicted to them. Therefore, self-repairing materials can be used in SCs to prevent structural cracks and restore structural integrity after mechanical damage to the electrical properties of the device. Unfortunately, conventional graphene-based fiber electrodes are thin, so reconnecting tiny batteries is unrealistic by visually checking the exact combination of broken fibers. In order to obtain the recovery of electrochemical properties after damage, a good reconnection of the broken yarn (with a larger diameter of the graphene fiber supercapacitor) is necessary [13].

Development of stretchable and self-repairing materials is prepared through multi-walled carbon nanotubes with reduced graphene oxide (RGO/MWCNT) based spring electrodes for SC. RGO/MWCNT fibers are decorated with polypyrrole (PPy) for different electrodeposition times, where stretchable carboxymethyl polyurethane (PU) is used as a self-healing material. Once the structure is damaged, these fiber springs (diameters of 295 μm thickness) electrodes can be recovered by accurately breaking the electrodes (Fig. 1). These fiber springs exhibit a 54.2% retention capacitor after the third healing, showing 82.4% (100%) after large stretching. Even after 3,000 charge/discharge cycles, the device was 84% of the capacitance was restored at the third healing time without degradation [14]. It is worth noting that the flexible and post-assembly structure of polymer hydrogel films (SWCNT-PANI) SHS interlayer is obtained from physical crosslinked hydrogel with a large number of hydrogen bonds. The hydrogel film is made from in-situ polymerization and nano-composite electrode material deposited on both sides of the self-repairing hydrogel electrolyte separator. When the current density is 0.044 mA cm^{-2}, the specific capacitance of the SWCNT-PANI is 15.8 mF cm^{-2}, much larger than the original SWCNT for 0.16 mF cm^{-2} and PANI 3.85 mF cm^{-2}. The SHS has a cyclic stability of up to 2,400 cycles under 0.13 mA cm^{-2}, reliable self-healing capability, 80% retention after the fifth repair cycle, and high flexibility [15].

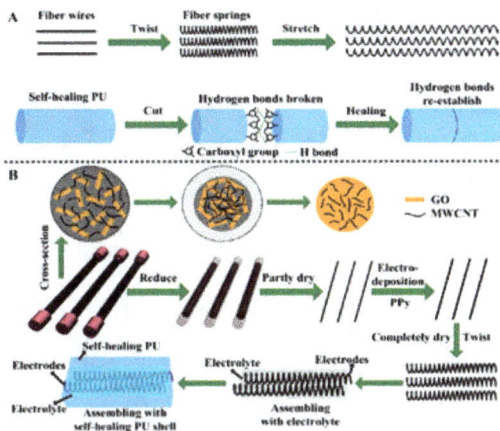

Figure 1. Fabrication procedure of stretchable SHS composed of PPy/RGO/MWCNT electrodes, hydrogel solid electrolyte and PU shell self-healable substrate; self-repairing property generate from interfacial hydrogen bonding. Reproduced with permission [14]. Copyright, American Chemical Society.

For practical applications, flexible SCs can work in a continuous deformation state; However, many flexible SCs are studied in static form, rather than fully reflecting the true performance of stretchable energy storage devices. Therefore, it is necessary to investigate the effect of dynamic deformation on electrochemical properties. In addition, self-healing electrodes have healing capacity damage and can be used in SCs to prevent structural breakage, restore structural integrity and electrical mechanical damage after the characteristics of the devices. PANI nanowires decorative graphene nanosheets grown by in-situ polymerization on the surface of CNT yarn and the resultant film (CNT@Graphene@PANI=CGP) is attached to PDMS substrate buckle structure is formed as a stretchable electrode. The tensile carboxymethyl polyurethane (PU) composite membrane electrode coated as a self-repairing material ensures the healing performance of SC. The elastic PDMS substrate is pre-stretched to 180% strain, and then the connecting electrode is photographed on top of it. The release of the pre-strain substrate causes the perfect wrinkle structure to form with the electrode to bend firmly on the PDMS (Fig. 2a). Then, two identical PDMS support electrodes are PVA/H3PO4 gel clips in the middle to obtain electrolytes and flexible solid-state SC (Fig. 2b). The device's photo strain under 0% to 180% (Fig. 2c) [16].

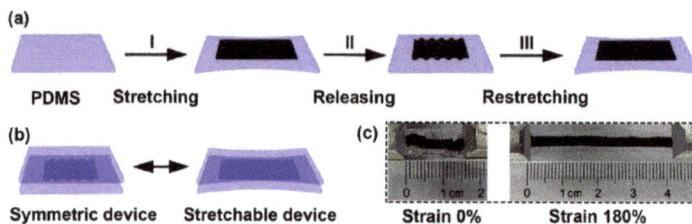

Figure 2. (a) Schematic diagram of the manufacturing process of a stretchable electrode, (b) A schematic diagram based on the CGP stretchable device, and (c) Photographs of flexible equipment based on strain 0% and 180%. Reproduced with permission [16]. Copyright, RSC Publishers.

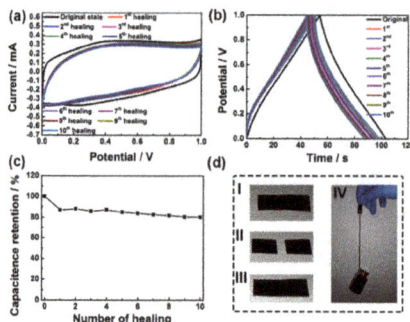

Figure 3. (a) CV and (b) GCD curves before and after different healing times, (c) Capacitance retention as a function of different healing times, (dI–III) Photographs of the supercapacitor before and after self-healing, (dIV) the healable supercapacitor is used to lift a 200 g weight. Reproduced with permission [16]. Copyright, RSC publishers.

Constructed SHS shows that there is no significant reduction in CV profile even after 10th self-repairing cycles (Fig. 3a), indicates after cutting and healing, the device capacitance was still maintained at high level and charge-discharge profile (Fig. 3b) is consistent with CV profile, reveals good recovering SHS. The specific capacitance can be well maintained after different cut/healing cycles (Fig. 3c). After the 10th cutting/healing cycle, the programmable SHS can be used to lift the weight of 200 grams (Fig. 3d), proving the high strength and self-tightness of the device. In addition, the manufactured supercapacitor has a capacitance retention force after 80.2% of the 10th healing cycle. The successful manufacture of dynamic stretchable and self-repairing supercapacitors can inspire the design and preparation of a variety of stretchable, self-repairing and wearable electronic products in the near future [16].

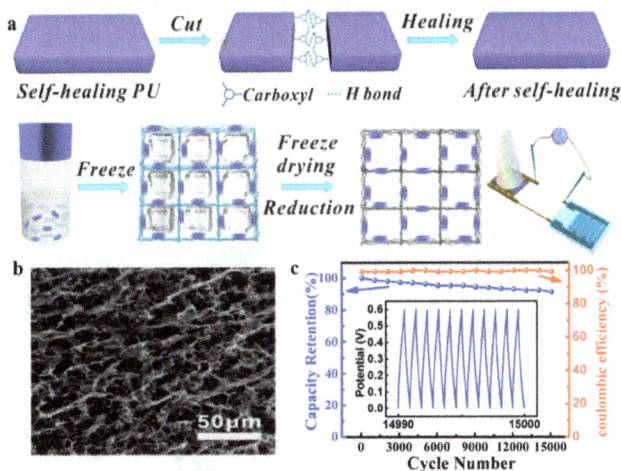

Figure 4. (a) Self-repairing mechanism of MXene-RGO aerogel for MSC, (b) SEM image of MXene-RGO aerogel, and (c) Cycle stability of the MXene-RGO aerogel MSC at the 2 mA cm^{-2} current density (inset, charge-discharge curve from the 14990th to the 15000th cycle). Reproduced with permission [18]. Copyright, American Chemical Society.

Over the past decade, graphene and MXene are the two normal representative of the 2D material "Goldrush". Different technologies in material preparation and device manufacturing for electronic applications have been extensively prepared on graphene and MXenes [17]. A 3D high-performance micro-supercapacitor (MSC) consisting of MXene (Ti$_3$C$_2$T$_x$)-Graphene (3D-MXene-rGO) composite aerogel (Fig. 4) was synthesized by freeze-drying and laser cutting technology. This composite aerogel electrode is wrapped with a self-repaired carboxymethyl polyurethane (PU) as a shell. This aerogel composite with high conductivity of the MXene and the high surface area of the RGO large nanosheets not only can avoid the self-restacking of the lamellar structure

173

but also additionally oppose the MXene from poor oxidization to a degree. The composite has capacitance of 34.6 mF cm^{-2} at the scanning rate of 1mV s^{-1} and an abnormal cyclic execution with a capacity retention rate of 91% after 15,000 cycles [18].

Xinda et al. fabricated waterborne flexible SHS biochar (BC) based electrodes and poly-amphoteric electrolyte gel electrolytes for cryogenic appliances (Fig. 5). Electrodes based on BC are combined into RGO to obtain electrical conductivity and mechanical integrity to form the resulting BC-RGO. When the current density is 0.5 Ag^{-1}, the specific capacitance of the BC-RGO electrode is 216 Fg^{-1}, the power density is 50 W kg^{-1}, and the high energy density 30 Wh kg^{-1} with ~90% retention capacity after 5000 charge-discharge cycles at room temperature. Under very low temperatures (-30 °C), BC-RGO electrode shows specific capacitance, energy dentistry and power density of 193 Fg^{-1}, 10.5 Wh kg^{-1} and 500 W kg^{-1} respectively. This promising method shows the actual power grid-scale energy applications in low-temperature environments [19].

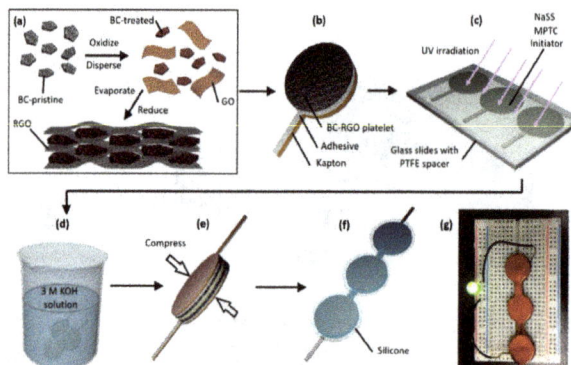

Figure 5. Schematic fabrication for BC-RGO SHS (a) synthesis of BC-RGO, (b) BC-RGO supported on Kapton substrate, (c) polyampholyte electrolyte synthesized on the BC-RGO electrodes, (d) then dialyzed in 3 M KOH, (e) compressing KOH electrolyte with polyampholyte@BC-RGO electrode for SC, (f) two symmetric SC compressed in silicone to green LED light. Reproduced with permission *[19]*. Copyright, Nature Publishing Group.

Conventional polymer electrolytes are often limited by ionic conductivity and lack additional functionality, which hampers their use in the development of flexible energy-related devices. Motivated by this challenge, methacrylated graphene oxide-polyacrylic acid (MGO-PAA) polyelectrolytes spotlighted high stretchability and excellent self-healing capability as MGO adopted the formation of hydrogen-bonded networks between

PAA chains and acted as a cross-linking mediator to expand the mechanical properties. And also MGO-PAA have shown that up to 950% of the ionic strain conductivity is much higher than the commonly used PVA based electrolytes. Coupled with excellent self-healing efficacy, MGO-PAA can be directly applied to the construction of SHS (80% capacity to maintain after 3 self-healing cycles) and Stretchable SC (with approximately 1.8 times the capacitance enhancement 300% strain) without additional addition to other self-healing or stretchable materials [20].

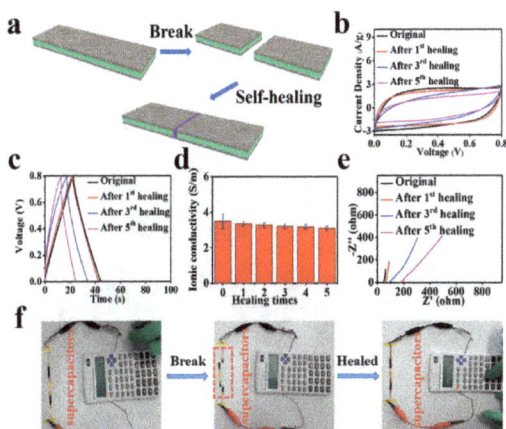

Figure 6. Self-healing performance of the SC based on the MGO-PAA polyelectrolyte. (a) The self-healing diagram of the MWCNT-PPy/MGO-PAA/MWCNT-PPy SC. (b) CV curves (at 100 mV s^{-1}), (c) GCD profiles (at 1 A g^{-1}) and (e) EIS curves of the device in repetitive cutting/self-healing cycles. (d) The ionic conductivity of the self-healable MGO-PAA polyelectrolyte in repetitive cutting/self-healing cycles. (f) Photographs showing that two series-connected SCs that experienced two cutting/self-healing cycles could still power a commercial calculator after recharging. Reproduced with permission [20]. Copyright, RSC Publishers.

For self-repair tests, the prepared MWCNT-PPy/MGO-PAA/MWCNT-PPy SHS is cut in half, and then the broken half is exposed to each other under environmental conditions (Fig. 6a), the MWCNT-PPy films reconnected without additional stimuli and materials, this process exclusively different from other self-repairing SC prepared using additional self-sealing polymers. Prior to the comparison of CV curves, SHS device worked up to 5th self-repairing cycles (Fig. 6b). In Fig. 6c, charge-discharge profiles of self-repairing cycles (1st, 3rd and 5th cycles) represents 51.1, 44.8 and 33.6 F g^{-1} at a current density of 1

Ag^{-1}, respectively. This result shows 80% and 60% capacitance retention after the 3rd and 5th healing process compared to the original capacitance (56 F g^{-1}) indicating that excellent self-healing with good self-healable behavior due to the ionic conductivity of polyelectrolyte (Fig. 6d) increases the series resistance of the device relatively small (Fig. 6e) in the entire repetitive cutting/self-healing cycle. For practical application, commercial calculator connected in series through the two identical MWCNT-PPy/MGO-PAA/MWCNT-PPy SHS could be well powered by the SHS even after the self-healing process (no charge required). At the same time, the calculator is fully functional, keeping its screen brightness even after three self-repairing cycles (Fig. 6f) verifies that the SHS with polyelectrolytes are reliable power sources to drive electronic devices [20].

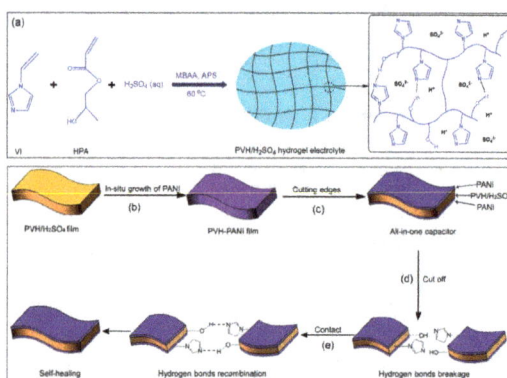

Figure 7. Schematics illustrate the fabrication of the all-in-one capacitor and its self-healing behavior. (a) Synthesis of a covalently cross-linked PVH/H$_2$SO$_4$ hydrogel film, (b) in situ growth of PANI on the hydrogel to obtain the PVH–PANI film, (c) the edges of the PVH–PANI film were cut to fabricate the all-in-one PVH–PANI capacitor, (d) the capacitor was cut into halves and hydrogen bonds were broken at the cut interfaces, and (e) the cut pieces were brought into contact allowing the broken hydrogen bonds to recombine, leading to the self-healing of the capacitor. Reproduced with permission [21]. Copyright, RSC Publishers.

Liu et al. SHS manufactured copolymer hydroxypropyl acrylate and vinyl imidazole (PVH) into integrating PANI and H$_2$SO$_4$ solution into the capacitor in situ comprising a single network. H$_2$SO$_4$ solution is incorporated in situ into the covalent crosslinking copolymer (PVH) in vinyl imidazole (VI) and Hydroxypropyl acrylate (HPA), forms PVH hydrogel film has high ion conductivity. The conductive polyaniline (PANI) is then integrated into the hydrogel network in situ to form all-in-one the functional configuration (Fig. 7).

The resulting prototype of the PVH-PANI supercapacitor shows a high area capacitance, an ultra-long cycle life and excellent self-healability, which is superior to the existing self-healing energy. Even if self-repairing itself 10 times, it delivers areal capacitance of 341.7 mF cm^{-2} at 2mA cm^{-2} and shows high cycle stability 1000 cycles with retention rate of 96%. Superior cycle life indicates that the flexible PVH network can accommodate swelling and shrink the PANI chain thus alleviating the structural pulverization of the PANI layers in the charge/discharge process. All functions in all of the con group configuration combine fast charge carrier transport, good structural durability and excellent self-healing ability. As a result, capacitors are produced to provide a high specific areal capacitance and rate capacity compare to its counterparts, but even after maintaining an extremely long cycle life even after self-repairing 10 cycles. This configuration model engineering offers a promising way to design SHS with excellent performance [21].

4.   Challenges and perspectives

SHS using solid-state polymer electrolytes is an ideal source due to its high power density and long cycle life. SHS is direct assembly of polymer electrolyte between two active electrodes. Since traditional polymer electrolytes and most electrode materials simply do not heal themselves, adding self-repairing materials, capacitive differential characteristics (such as polyurethane or TiO_2 in supramolecular network) is essential for the manufacture of SHS, but unnecessarily reduced its specific capacitance owes an increase in the weight and volume of the device. To date, attempts to synthesize self-repairing polymer-based electrolytes (vinyl alcohol, vinyl imidazole, hydroxypropyl acrylate, crosslinked-acrylate, etc.), have been successful, but the SCs based on these electrolytes have no tensile property. Appropriate scaling and improved ionic conductivity nature self-healable polymer electrolytes remain in the development of the next generation of multifunctional solid-state SC with high specific capacitance is a challenge for energy storage. As a result, SHS may be assisted in repairing its own conductive patches from graphene to improve the conductivity of the polyelectrolyte and stretchable substrates to show high electrochemical properties and a wide voltage range of polyelectrolytes. There are still many challenges to the application of SHS in a wide and practical application that must be investigated and developed through in-depth research and development. For example, many self-healing materials require external stimulation (moisture, pH and heat to trigger or accelerate the process. Clearly, there are huge challenges to the prospect of exploring self-healing materials with functional features, new materials, scalable preparation of self-repairing materials requires the design and manufacture of self-repairing supercapacitor devices. And also the integration of self-healing electrodes and electrolytes into a single device like all-in-one SC indispensable requirement for wearable SCs.

Conclusions

In this chapter, we summarize the advanced progress of graphene electrodes and electrolytes as active electrodes for SC applications. It is envisaged to cover the field of SHS technology based on graphene, and to systematically review the most advanced full range of graphene hybrids in SHS, as well as the views and challenges. Graphene is known for its excellent electrical, mechanical and thermal properties. In particular, the energy generation and storage is one of the hot topics that have attracted more and more attention from scientists. As for energy storage, a range of improvements with special functions based on graphene, such as deformability, abrasion resistance, stimulating response, self-protection and self-adaptability, self-healing, integration and miniaturization are essential for researchers to be more attractive and practical for energy storage devices. SHS feature high tensile elongation, arbitrary deformability and even self-healing ability enable them to withstand repetitive deformations or sudden rupture of long-term operation. SHS based on graphene is important may have the greatest potential and can have a significant impact in addressing energy storage technology challenges. We believe that more breakthroughs in graphene-based intelligent energy generation and storage devices can be achieved in the near future.

References

[1]   M. Winter, R.J. Brodd, What are Batteries, Fuel Cells, and Supercapacitors?, Chem. Rev. 104 (2004) 4245–4270. https://doi.org/10.1021/cr020730k.

[2]   Y. Shao, M.F. El-Kady, J. Sun, Y. Li, Q. Zhang, M. Zhu, H. Wang, B. Dunn, R.B. Kaner, Design and mechanisms of asymmetric supercapacitors, Chem. Rev. 118 (2018) 9233–9280. https://doi.org/10.1021/acs.chemrev.8b00252.

[3]   M. Ye, Z. Zhang, Y. Zhao, L. Qu, Graphene platforms for smart energy generation and storage, Joule. 2 (2018) 245–268. https://doi.org/10.1016/j.joule.2017.11.011.

[4]   W.K. Chee, H.N. Lim, Z. Zainal, N.M. Huang, I. Harrison, Y. Andou, Flexible graphene-based supercapacitors: A review, J. Phys. Chem. C. 120 (2016) 4153–4172. https://doi.org/10.1021/acs.jpcc.5b10187.

[5]   J. Xia, F. Chen, J. Li, N. Tao, Measurement of the quantum capacitance of graphene, Nat. Nanotechnol. 4 (2009) 505–509. https://doi.org/10.1038/nnano.2009.177.

[6]   B. Ravi, B. Rajender, S. Palaniappan, Improving the electrochemical performance by sulfonation of polyaniline-graphene-silica composite for high performance supercapacitor, Int. J. Polym. Mater. Polym. Biomater. 65 (2016) 835–840. https://doi.org/10.1080/00914037.2016.1171221.

[7]   R. Bolagam, R. Boddula, P. Srinivasan, Design and synthesis of ternary composite of polyaniline-sulfonated graphene oxide-TiO_2 nanorods: A highly stable electrode material for supercapacitor, J. Solid State Electrochem. 22 (2018) 129–139. https://doi.org/10.1007/s10008-017-3732-y.

[8]   R. Bolagam, R. Boddula, P. Srinivasan, One-step preparation of sulfonated carbon and subsequent preparation of hybrid material with polyaniline salt: a promising supercapacitor electrode material, J. Solid State Electrochem. 21 (2017) 1313–1322. https://doi.org/10.1007/s10008-016-3487-x.

[9]   R. Boddula, R. Bolagam, P. Srinivasan, Incorporation of graphene-Mn_3O_4 core into polyaniline shell: supercapacitor electrode material, Ionics 24 (2018) 1467–1474. https://doi.org/10.1007/s11581-017-2300-x.

[10]   P. Ramyakrishna, B. Rajender, G. Sadanandam, P. Srinivas, Ultrasonic Assisted Synthesis of 2D-functionalized grapheneoxide@PEDOT composite thin films and its application in electrochemical capacitors, in: Inamuddin M.F.A. Abdullah M. Asiri (Ed.), Electrochem. Capacit. Theory, Mater. Appl., Volume 26, Materials Research Foundations, 2018: pp. 93–106. https://doi.org/10.21741/9781945291579-4.

[11]   Y. Huang, C. Zhi, Functional flexible and wearable supercapacitors, J. Phys. D. Appl. Phys. 50 (2017) 273001. https://doi.org/10.1088/1361-6463/aa73b8.

[12]   K. Guo, N. Yu, Z. Hou, L. Hu, Y. Ma, H. Li, T. Zhai, Smart supercapacitors with deformable and healable functions, J. Mater. Chem. A. 5 (2017) 16–30. https://doi.org/10.1039/C6TA08458C.

[13]   Q. Yang, Z. Xu, C. Gao, Graphene fiber based supercapacitors: Strategies and perspective toward high performances, J. Energy Chem. 27 (2018) 6–11. https://doi.org/10.1016/j.jechem.2017.10.023.

[14]   S. Wang, N. Liu, J. Su, L. Li, F. Long, Z. Zou, X. Jiang, Y. Gao, Highly stretchable and self-healable supercapacitor with reduced graphene oxide based fiber springs, ACS Nano 11 (2017) 2066–2074. https://doi.org/10.1021/acsnano.6b08262.

[15]   Y. Guo, K. Zheng, P. Wan, A Flexible Stretchable Hydrogel Electrolyte for Healable All-in-One Configured Supercapacitors, Small 14 (2018) 1704497. https://doi.org/10.1002/smll.201704497.

[16]   X. Liang, L. Zhao, Q. Wang, Y. Ma, D. Zhang, A dynamic stretchable and self-healable supercapacitor with a CNT/graphene/PANI composite film, Nanoscale (2018) 22329–22334. https://doi.org/10.1039/C8NR07991A.

[17]   C. (John) Zhang, V. Nicolosi, Graphene and MXene-based transparent conductive electrodes and supercapacitors, Energy Storage Mater. 16 (2019) 102–125. https://doi.org/10.1016/j.ensm.2018.05.003.

[18]   Y. Yue, N. Liu, Y. Ma, S. Wang, W. Liu, C. Luo, H. Zhang, F. Cheng, J. Rao, X. Hu, J. Su, Y. Gao, Highly Self-Healable 3D Microsupercapacitor with MXene-graphene composite aerogel, ACS Nano. 12 (2018) 4224–4232. https://doi.org/10.1021/acsnano.7b07528.

[19]   X. Li, L. Liu, X. Wang, Y.S. Ok, J.A.W. Elliott, S.X. Chang, H.-J. Chung, Flexible and self-healing aqueous supercapacitors for low temperature applications: polyampholyte gel electrolytes with biochar electrodes, Sci. Rep. 7 (2017) 1685. https://doi.org/10.1038/s41598-017-01873-3.

[20]   X. Jin, G. Sun, H. Yang, G. Zhang, Y. Xiao, J. Gao, Z. Zhang, L. Qu, A graphene oxide-mediated polyelectrolyte with high ion-conductivity for highly stretchable and self-healing all-solid-state supercapacitors, J. Mater. Chem. A. 6 (2018) 19463–19469. https://doi.org/10.1039/c8ta07373b.

[21]   F. Liu, J. Wang, Q. Pan, An all-in-one self-healable capacitor with superior performance, J. Mater. Chem. A. 6 (2018) 2500–2506. https://doi.org/10.1039/c7ta10323a.

Chapter 7

Graphene-Based ZnO Nanocomposites for Supercapacitor Applications

Udaya Bhat K.[*], Sunil Meti

Department of Metallurgical and Materials Engineering, National Institute of Technology Karnataka, Surathkal, India, 575025

*udayabhatk@gmail.com

Abstract

In the recent past, the increase in energy needs and the development of efficient energy storage devices have opened a new area of research. Batteries and supercapacitors are generally used as energy storage devices. Graphene-based nanocomposite materials have proven to be promising material for supercapacitor applications. Graphene-based ZnO nanocomposites are known to have high specific capacitance. In this chapter, many approaches for graphene and graphene-based ZnO nanocomposite synthesis are explained. Electrochemical characterization and specific capacitance of graphene-based ZnO nanocomposites are also reviewed.

Keywords

Graphene, ZnO Nanoparticles, Synthesis Methods, Specific Capacitance, Crystal Growth

Contents

1.   Introduction

The dependency on fossil fuels for the production of electricity adversely affects the environment due to the increase in greenhouse gas emissions. Electricity production and efficient energy utilization are the challenging task for most of the countries. Increase in energy needs and energy management have been prime concerns for many countries [1-3]. Energy production, distribution and storage devices are important components of energy management. The requirement of efficient energy systems drives continuous research on new materials, design and development of efficient energy storage devices [4,5].

2.   Energy storage devices

Important energy storage devices are grouped into batteries and supercapacitors. The batteries store energy in chemical form and generate higher voltages. Supercapacitors store energy in the form of an electric field and are limited to low voltage applications. The batteries are mostly used due to their high energy density whereas supercapacitors have high power density [6,7]. These two devices are complementary to each other when they are hybridized. The need for designing new electrode materials, electrolytes and electrochemical concepts for supercapacitors are of prime concern to improve the energy and power densities of the supercapacitors.

2.1   Supercapacitors

Supercapacitors are also called ultracapacitors or electrochemical supercapacitors. Supercapacitors have characteristic features, like, long life cycle, higher energy density than capacitors, higher power density than batteries and are environmental friendly [2,8,9]. Since the invention of supercapacitor by General Electric, it has drawn the attention of many researchers [8]. Supercapacitors are mainly used in pulse power and potential energy backup applications. Based on the charge storage mechanism, supercapacitors are classified into two types as electric double layer capacitors (EDLCs) and pseudocapacitors. The EDLCs store the charges at the interface of the electrode/electrolyte. Pseudocapacitors store the charge by Faradaic reactions on the surface of the electrode materials [3,10].

2.1.1 Components of supercapacitors

Recently, there has been a few reports on the uses of graphene and its derivatives as a supercapacitor electrode material [8,9,11–29]. The important limitation of the supercapacitors is their low specific capacitance. This limitation could be overcome by hybridizing the graphene with the pseudocapacitor materials [30]. The hybridization of the graphene with the inorganic structures improves the specific capacitance and electrochemical stability when compared with the individual counterparts. The pseudocapacitive contributions and double-layer contributions improve the effective supercapacitive behaviour of the graphene hybrid materials [5,31–42].

2.2 Inorganic nanostructures as electrode materials

During the last two decades, there are many reports on the inorganic nanostructures. The inorganic nanostructures have various advantages compared to their bulk counterpart materials and are listed in the references. The main demerit associated with the inorganic nanostructure particles is stacking of the nanoparticles due to the Van der Waals forces. This phenomenon causes the limitation in the usage of the inorganic nanoparticles and results in structural instability, thus minimizes their applicability. One approach to overcome this limitation is the use of graphene sheets along with the nanoparticles in the form of nanocomposites [41,43,44]. The graphene matrix prevents the clustering of the inorganic nanoparticles. The chemical interaction between the graphene sheets and the inorganic nanoparticles produces the hybrid materials with certain unique features. These inorganic nano-hybrid composites have special features in most of the applications. Graphene is a well-known electron acceptor material, whereas the inorganic nanoparticles are good electron donors. Graphene having large surface area is qualified material as the best matrix for inorganic nanomaterials [9,31,41,42].

The nanostructures deposited onto the graphene matrix potentially display the unique properties of the hybrid materials, additional novel properties and functionalities. The planar structure of the graphene with the sp^2 hybridized carbon atoms in the graphene sheet provides nucleation sites for the inorganic nanoparticles to grow on the graphene surface. The nanostructures grown on graphene sheets are considered to be having tiny sized particles and narrow size distribution [34,38,42]. These characteristic features of the graphene attracted many researchers to focus on the material to incorporate the inorganic nanoparticles in the graphene matrix to improve the ability of individual materials.

2.2.1 Benefits of the zinc oxide nanoparticles for supercapacitor applications

Zinc oxide (ZnO) is one among the semiconductors with a band gap of 3.37 eV (which is similar to that of the GaN), high electron mobility and large binding energy of 60 meV at room temperature. The optical band gap of the ZnO is tuned by the variables, like, morphology, composition and size. The ZnO is also a good semiconducting material which is utilized for fabrication of the light emitting diodes, transparent electrodes, anti-corrosive and antifungal surfaces, pH sensors, biosensors, gas sensors, acoustic wave devices, UV filters, photocatalytic and laser diodes applications [45–47].

A new nanocomposite synthesized using ZnO and graphene is expected to display unique properties and functionalities due to synergic interaction between the components. There are many innovative approaches to synthesize graphene-based zinc oxide (G/ZnO, GO/ZnO or rGO/ZnO) nanocomposites and their possible beneficial characteristics [11,18,47]. Graphene-based ZnO materials improve the properties of the host ZnO nanomaterial [48,49]. Hybrid graphene-based ZnO nanocomposites display properties, like, ultrafast nonlinear optical switching, optoelectronic behaviour and energy storage properties [50].

There are multiple reports on the synthesis of the graphene-based ZnO nanocomposites. A few of them are hydrothermal, sol-gel, electrochemical deposition and vapour-phase processes. It is also possible to tune morphologies, such as, nanowires, nanorods, nanoneedles, hollow structures, and self-assembled architectures of graphene-ZnO nanocomposites [11,17,18,21,22,24,27,28,47,49,51–60]. Complex reaction procedures are necessary requirements for the growth of ZnO on to the graphene oxide nanosheets. Therefore, there is a need for producing graphene-based ZnO nanocomposites with improved properties for their potential applications.

3.   Importance of graphene/graphene oxide for the assembly of nanostructures

Single atomic layer graphite is called graphene. Fig. 1 shows a schematic representation of the single layer graphene. Graphene has unique characteristic features, like, extraordinary carrier mobility ($20~m^2V^{-1}s^{-1}$), large specific surface area ($\sim2600~m^2g^{-1}$), Young's modulus 1 TPa, the thermal conductivity of $\sim5000~Wm^{-1}K^{-1}$, the optical absorption of $\approx2.3\%$ and complete impermeability to any gas [41]. The bandgap between the highest occupied molecular orbital and the lowest unoccupied orbital for the graphene is zero. The two-dimensional (2-D) structure of the graphene offers unique features, like fast electron transport with distance between consecutive carbon-carbon atoms as 0.142 nm. The lattice units of the graphene sheets are represented as a_1 and a_2, as shown in Fig.1. There are various reports on the graphene used in the electronic device applications [12,41,61–63]. The discovery of 2-D graphene structure has opened a new area for electronic devices due to its superb electrical, thermal, optical and mechanical properties. Graphene provides a significant number of possible centres for rapid electron transfer through the four edges of the sheet [64]. Graphene sheets are widely applicable in the field of electrochemistry. Graphite is brittle compared to that of the graphene. The flexibility of the graphene is an important feature in the case of electro-mechanical and energy storage devices [12,41,65,66]. Graphene boosts the electrochemical catalytic activity of materials by predominantly increasing the surface area. The fascinating features of the graphene, such as optical, electronic, mechanical and thermal properties, make it a necessary material for diverse applications [43,63].

3.1   GO and rGO

The oxidized form of graphene is called graphene oxide (GO). The commonly used GO synthesis approach is called as Hummers method. The process involves exfoliation and intercalation of graphite sheets with potassium permanganate and sulphuric acid. The GO contains reactive oxygen functional groups presented on the basal plane and edges. This phenomenon enables the graphene to interact with the inorganic nanoparticles through covalent, electrostatic and hydrogen bonding [41]. Graphene synthesized by chemical oxidation of the graphite is known as reduced graphene oxide (rGO). Reduced graphene oxide (rGO) is obtained by thermal, chemical and UV treatment of GO. The chemical process is carried out in the presence of reducing agents (ex: hydrazine, etc). The rGO restores the electrical conductivity and optical absorbance of the GO whereas the oxygen content, surface charge and hydrophilicity are reduced [67].

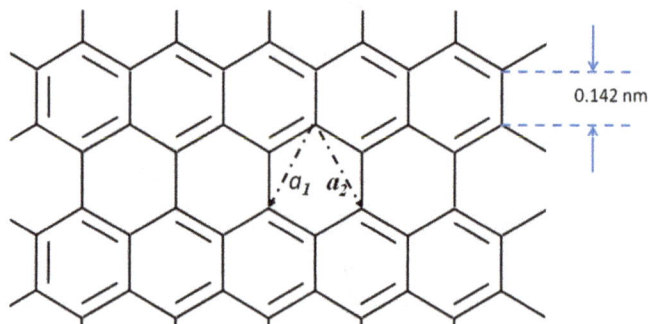

Figure 1. A schematic image of single atomic graphene layer.

3.2   Various synthesis methods for graphene, GO and rGO

There are different methods to synthesize graphene, GO and rGO. The aim of this section is to review different synthesis techniques for graphene, GO and rGO. The methods of graphene, GO and rGO syntheses are discussed as follows.

3.2.1   Tang Lau method

Tang Lau method is also called as hydrothermal synthesis of GO. It suggests that using table sugar and water, graphene oxide nanosheets (GON) can be synthesized. This method is environmental friendly (because reagents used are only sugar and water), facile, low-cost and capable to scale up for mass production [64]. Glucose, sucrose, fructose or sugar is used as a source of carbon, which is having C: H: O in the ratio of 1: 2: 1. This method is used to synthesize GO with thickness ranging from ($\sim$ 1 nm (monolayer) to $\sim$ 1500 nm). In our previous work, we have synthesized GO nanosheets by following Tang Lau method [68]. Fig. 2 shows the SEM micrograph of the GO nanosheets synthesized by using Tang Lau method.

Figure 2. SEM micrograph of the GO synthesized by following Tang Lau method.

3.2.2 Scotch tape method

This method is used for micromechanical exfoliation of the graphene from the graphite. An adhesive tape is used to detach graphene layers from a graphite crystal. The method is cheap and easy to perform as no special equipment and specialization are needed. However, this method produces very uneven graphene films [69,70].

3.2.3 Modified Hummers method

This method is widely used for the production of GO. The method utilizes sulphuric acid and potassium permanganate ($KMnO_4$) as reducing agents. The method produces GO from the graphite by exfoliation and intercalation [71]. The method is cost effective and can produce large quantity of GO with minimal utilization of instrumentation. The graphene or rGO can be obtained by treating the GO with hydrazine or other reducing agents [71].

3.2.4 Graphene film growth by sputtering

The RF magnetron sputtering machine, with three-target RF diode, is used to deposit the graphene film. The deposition is carried out at a base pressure of 2×10^{-7} Tor in presence of the argon gas. The samples are subjected to rapid thermal processing (RTP) either *in-situ* or *ex-situ*. The heating power and thermal mass of the sample holder are the two variables controlling the heating and cooling rates of *in-situ* RTP process.

The quality of the graphene by *ex-situ* RTP is better compared to *in-situ* RTP. The *ex-situ* RTP requires a quartz chamber and commercial RTP apparatus. The heating and cooling rates in this apparatus are much faster than the *in-situ* system. The rapid cooling rate is also possible in the apparatus with the help of cold chamber wall design and quartz sample holder with a small thermal mass. The process is carried out in the presence of an inert atmosphere using (argon gas) to avoid the oxidation of the sample [72].

3.2.5 Catalyst-assisted and catalyst-free growth of the graphene by chemical vapour deposition (CVD)

A commercial system (Black Magic, AIXTRON Nanoinstruments Ltd) with copper foil, as a catalyst and substrate, is used for the catalyst-assisted graphene growth by CVD process. A mixture of 5 % volume of high purity methane and argon is used as the precursor [73]. The catalyst-free growth of the graphene is directly grown onto the silica substrate without the use of any catalyst. The single-layer of graphene is achieved by this method. The layers of the graphene are tuned by controlling the deposition time and the precursor concentration [70,74].

3.2.6 E-Beam deposition of the graphene

The AVAC HVC600 system is used for the deposition of the carbon films at room temperature. A polycrystalline graphite rod is used as the source material and the thickness of the deposited carbon films is kept as 20 nm [70,75].

According to the literature, the Modified Hummer's method has been widely used for the production of GO and rGO. GO could be reduced to graphene using reducing agents. GO contains graphene sheets having oxy-functional groups on the surface. The functional groups are the main source to attract nanoparticles. Many researchers have used GO as a base material for capturing metallic ions and to develop various nanocomposites [62].

4. Various synthesis methods for producing ZnO nanostructures and ZnO nanoparticles on the rGO nanosheets

4.1 Hydrothermal synthesis of the graphene-based ZnO

Hydrothermal method is widely used for the production of a variety of inorganic nanoparticles under controlled pressure and temperature. This method has various advantages including low processing temperatures and less reaction time. The

Materials Research Forum LLC
https://doi.org/10.21741/9781644900550-7

hydrothermal method is used to produce chemically homogeneous, various single and multi-component powders of ultrafine size exhibiting high purity as well as good crystallinity. If the solutions are non-aqueous, the method is called solvothermal process. The closed system of Teflon-lined stainless steel autoclave is used in the hydrothermal synthesis. When the temperature of the vessel is increased, the pressure inside the vessel also increased and if it exceeds the critical pressure of the water, the disintegration of the thermodynamically unstable compounds intensifies. The macromolecules in the aqueous solution are broken down to form nano-sized particles [76]. Absolute ethanol has been utilized as a solvent for the ZnO production. The advantage of using ethanol is to improve the dispersion of gel-like GO [24,49].

One pot synthesis of ZnO nanoparticles and the ZnO/graphene nanocomposites has been performed using Zn salt and GO, where the ZnO nanoparticles chemically bonded to the graphene sheets during the reaction. ZnO nanoparticles with the dimensions of 10 - 15 nm are covered with the graphene sheets [13]. Zhou et al. have utilized hydrothermal method to synthesize the ZnO-rGO nanocomposite using zinc nitrate hexahydrate and ammonia as the raw materials. The resulted nanocomposite exhibited high degree of photocatalytic degradation performance [77]. Li and Cao et al. [78] have synthesized ZnO-graphene composite. They reported that graphene sheets were incorporated onto ZnO by chemically reducing the mixture of graphite oxide and zinc acetate in aqueous solution in the presence of $NaBH_4$. Xu et al. [79] have synthesized graphene–ZnO nanocomposites by reducing GO with the help of hydrazine. The graphene sheets were coated on the surface of the ZnO nanoparticles.

4.2   Microwave-assisted hydrothermal synthesis of the graphene-based ZnO

Generally, microwave heating is done using the microwave frequency of 2.45 GHz. The heating depends on the molecular properties and reaction conditions. The complex binary or ternary metal oxides including zeolites can be synthesized by using this method. The nanoparticles produced using this method exhibited uniform particle size due to minimal synthesis time and highly concentrated localised heating. The particle sizes of ZnO ranged from 10 - 100 nm [80]. Microwave-assisted graphene-based ZnO nanocomposites synthesis has also reported [12,80]. Graphene sheets constrain the growth of the ZnO nanostructures. T. Lu et al. [17] carried out significant work on the synthesis of graphene-ZnO nanocomposites with high specific capacitance, via microwave-assisted reduction. Zinc sulphate, NaOH and GO were used as the target materials. A. Kajbafvala et al. [54] has worked on the synthesis of the ZnO wires (sword shape) by using microwave energy. Methanol, tri-ethanol amine, zinc acetate dehydrate and sodium hydroxides were used as the raw materials. The width of the ZnO nanowires was in the range of 80 – 250 nm and

Materials Research Forum LLC
https://doi.org/10.21741/9781644900550-7

length ~ 1-4 μm. S. Meti et al. [48] synthesized GO by following Tang Lau method and utilized the GO in the preparation of rGO/ZnO nanocomposite by microwave-assisted hydrothermal method.

4.3 Electrochemical growth of the ZnO

Effective control on the ZnO dimensions is possible through electrochemical growth. The dimensions of the ZnO can be tuned by controlling the growth variables, like, solution concentration and applied potential. The deposition of ZnO is carried out at lower temperatures (<100 °C) from an aqueous salt solution saturated with the oxygen. The variation in growth conditions results in a change from the planar film growth to the nanosized islands. The growth of the ZnO microstructure was found to depend on the substrate lattice parameter, the electrolyte used (nitrate, chloride or perchlorate), the amount of the oxygen content in the electrolyte, the applied potential and pre-activation surface treatment [81].

Wong et al. [81] have studied the electrochemical growth of the ZnO nanorods. The ZnO nanorods were grown on the polycrystalline Zn foil following cathodic electrodeposition method. Aqueous zinc chloride/calcium chloride was used in this reaction at 80 °C. Zeng et al. [82] synthesized ZnO nanorods array on the GaAs substrate by electrochemical deposition (ECD) without any buffer layer. Aqueous zinc nitrate was used as the electrolyte solution. Peulon et al. [83] deposited the ZnO using ECD approach. Zinc chloride and potassium chloride were used as electrolytes. Liu et al. [84] deposited ZnO nanopillars on the single crystal of gold following ECD method using $ZnCl_2$/KCl solution. M. Izaki [85] has deposited transparent ZnO films by ECD method using zinc nitrate aqueous solution.

4.4 Sol-gel synthesis of the graphene-based ZnO

This method involves the use of GO, zinc precursors, additives and solvents in the synthesis of nanocomposites. The basic reactions involved in sol-gel methods are hydrolysis and condensation reactions.

Pourshaban et al. [86] have used zinc acetate as the zinc precursor with amino-additives in alcoholic solvents, like methanol and ethanol. The ZnO nanorods were synthesized by following the calcination technique at 500 °C for 60 min. Alvarado et al. [87] have synthesized ZnO nanoparticles by using zinc acetate and KOH as the target materials with an annealing temperature of 600 °C. Singh et al. [88] used zinc acetate and ethanol at calcination temperature of 395 °C. Zhou et al. [89] developed the ZnO porous film with wood template method. Zinc acetate and ethanol were used as the targeting agents at sintering temperatures of 600, 800 and 1000 °C for 3 hours. Chebil et al. [90] synthesized the ZnO thin films using zinc acetate, isopropanol and monoethanolamine as a precursor, solvent and stabilizer, respectively. Haarindraprasad et al. [91] have developed the ZnO nanowires using zinc acetate, alcohol and amine as the target chemicals. Para et al. [92] synthesized the ZnO nanoparticles using zinc acetate, glacial acetic acid and ammonium acetate as target materials. H. Li et al. [93] have developed the ZnO/graphene nanocomposite film as the anode material for the lithium-ion batteries using zinc acetate, ethanol, graphene and lithium hydroxide aqueous solution. Demes et al. [94] have studied the effect of different amine agents, like mono and diethanolamine on the morphology of the ZnO nanowires. Ayana et al. [95] have developed multilayer ZnO thin films using ethanol, zinc acetate monoethanolamine as target materials. They have also studied the effect of different substrates on the development of ZnO films. Zimmermann et al. [96] have synthesized ZnO quantum dots and examined the effect of the ethylene glycol on their growth. They have used zinc acetate, NaOH, 2-propanol and ethylene glycol.

4.5   Biological synthesis of the ZnO nanoparticles

Biological synthesis method of ZnO nanoparticles is economical, environmentally friendly and safe compared to other methods. This method uses microorganisms, enzymes and plant extracts for nanoparticles production. The nanoparticles synthesized by this method have been used in the applications of biosensors, catalysts, antibacterial, antiviral and water purification applications. ZnO nanoparticles have been produced from plant extracts, plant sources and microbes.

Al-Dhabi et al. [46] synthesized the ZnO nanoparticles using aqueous extract of *Scadoxus Multiflorus* (*S. multiflorus*) leaves powder. Khalil et al. [97] have used *Sageretia thea* (*Osbeck*) mediated synthesis of the ZnO nanoparticles. Venkatesan et al. *[98]* produced the ZnO nanoparticles from the extract of *Ipomoea pes-caprae* leaves.

5.   The effect of synthesis methods and parameters on the ZnO nanostructures and rGO/ZnO nanocomposites

5.1   Hydrothermal method
5.1.1   Zinc nitrate and NaOH as the target chemicals

The effects of various precursors and hydrothermal heating temperatures on the synthesis of the ZnO nanostructures have been studied by many authors. Fig. 3 shows the SEM micrograph of the ZnO, where zinc nitrate along with NaOH was used. The prepared solution was hydrothermally heated to 80 °C in Teflon lined stainless steel autoclave.

The effect of hydrothermal temperature on the ZnO morphology was also investigated. The SEM image (Fig. 4) shows the ZnO nanostructures with rod-like morphology. The increase in temperature provided sufficient activation energy and nucleation sites for the production of rod-like morphology of ZnO nanostructures.

Figure 3. SEM micrograph of the hydrothermally synthesized ZnO using zinc nitrate and NaOH aqueous solution as reagents at 80 °C.

Figure 4. SEM micrograph of the hydrothermally synthesized ZnO using zinc nitrate and NaOH aqueous solution as reagents at 180 °C.

5.2   ZnO synthesis by microwave-assisted hydrothermal method

The ZnO nanostructures were synthesized by following microwave-assisted hydrothermal method. Zinc nitrate and NaOH were added into deionized water (DIW). The mixture was then transferred to a Parr acid digester vessel. The Parr microwave autoclave was heated in a microwave oven at 700 W power for few minutes. The resulting constituent was then washed with ethanol and DIW several times. The final powder was dried in an oven at 60±1 °C for 12 hours. Fig. 5 shows the SEM micrograph of the ZnO nanoparticles.

Figure 5. SEM micrograph of the ZnO nanoparticles, synthesized by microwave assisted hydrothermal method using NaOH as reagent at 700 W power.

5.3 rGO/ZnO nanocomposite synthesis by microwave-assisted hydrothermal method

There are many reports on the ZnO nanoparticles decorated on the graphene/GO/rGO nanosheets. The ZnO nanostructures anchored onto the graphene sheets improve the electron transferability of the ZnO. In our previous work, we have synthesized rGO/ZnO nanocomposites using microwave assisted hydrothermal method. We used hydrothermally synthesized GO nanosheets using zinc nitrate and NaOH as the target chemicals [48]. The synthesis of the GO was carried out using the hydrothermal method [64]. The GO was dispersed in DIW. Zinc nitrate and NaOH were added into the same solution. The solution was then heated in Parr microwave acid digester vessel at 700 W power for few minutes. The FESEM micrograph (Fig.6) shows the ZnO nanorods anchored onto rGO nanosheets. The properties of the ZnO were improved by anchoring the ZnO onto the rGO nanosheets.

6. Interaction of the ZnO with the graphene/GO/rGO

The graphene interacts with the decorated inorganic nanostructures through following mechanism. The functional groups, such as carboxyl, hydroxyl and epoxy on the GO sheets attract the positively charged metal ions. The metal ions attached on the edges and surface of the GO sheets involve in reduction and oxidation reactions. The metal ion attachment sites form the nucleation sites for the growth of the nanostructures on the graphene sheets. The simultaneous reduction of metal ions and GO produces metal-graphene nanocomposites. Graphene oxide produces photogenerated electrons which reduce the metal ions. This phenomenon is explained through the chemical reactions from 1–4,

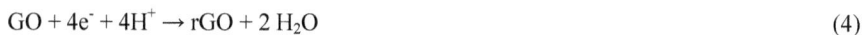

$$GO + hv \rightarrow GO\ (h^+ + e^-) \tag{1}$$

$$4h^+ + 2\ H_2O \rightarrow O_2 + 4H^+ \tag{2}$$

$$M^+ + GO\ (e^-) \rightarrow GO + M \tag{3}$$

$$GO + 4e^- + 4H^+ \rightarrow rGO + 2\ H_2O \tag{4}$$

The nanoparticles are incorporated onto the graphene sheets through the mechanisms such as chemisorption, physisorption, electrostatic interaction, van der Waals or covalent bonding. The aggregation and restacking of the graphene sheets reduced due to the attachments of the nanostructures onto the surface of the graphene sheets [99].

Figure 6. FESEM image of the rGO/ZnO synthesized by microwave assisted hydrothermal method using NaOH as reagent [48].

7. Supercapacitors

The graphene has been extensively used as the electrode material in case of the supercapacitors [12]. As mentioned in section 2.1, supercapacitors store electric charge through EDLCs and pseudocapacitance. When graphene is hybridized with ZnO nanoparticles, pseudocapacitance increases predominantly. The pseudocapacitors store the electrical energy Faradically by electron charge transfer between electrode and electrolyte. Graphene sheets due to higher surface area and electrical conductivity improves the electron charge transfer kinetics. Also, ZnO nanoparticles attached to the graphene sheets individually increase the quantum of charge available for transfer. Added to that, the sp^2 hybridized carbon atoms in graphene sheet contribute for the transfer of additional charges. The specific capacitance of the graphene/ZnO nanostructure hybrids was found to be extremely increased compared to the individual materials.

To find the electrochemical (EC) performance of the graphene-based ZnO nanocomposites, the cyclic voltammetry (CV) and galvanostatic charge-discharge test in an aqueous medium have to be carried out. A sequence of EC experiments, like cyclic voltammetry (CV), chronopotentiometry (CP), electrochemical impedance spectroscopy (EIS) and cycling stability have been used to find the specific capacitance and the stability of the nanocomposites.

The EC test can be performed using two-electrode system or three-electrode system. The three electrode system uses the synthesized graphene-based nanocomposites as the working electrode, platinum foil as the counter electrode and Ag/AgCl as the reference electrode in aqueous electrolyte solution of Na_2SO_4, $NaNO_3$ or KOH. The two-electrode system uses working electrode and counter/reference electrode. The cyclic voltammetry and charge-discharge measurements were recorded for various scan rates and current densities to evaluate the performance of nanocomposites.

Graphene within the nanocomposites enhances the specific capacitance by influencing the electrical-double-layer capacitance, whereas the ZnO enhances the specific capacitance by improving the pseudocapacitance nature [26–28,41,59]. When Na_2SO_4 or KOH is used as the electrolyte, the Na^+ or K^+ ions from the electrolyte intercalate and deintercalate the ZnO, producing redox peaks. The absence of redox peaks indicates the non-contribution of ZnO towards the capacitance. In such cases, it is believed that the capacitance in the ZnO sample originates from the carbon species produced from the combustion of the binders.

The EIS measurement is used for examining the behaviour of the electrode materials of the supercapacitors. The EIS data are analyzed by using the Nyquist plots. The impedance data is collected by sweeping across the frequencies. Each data in the Nyquist plot is at a different frequency. The equivalent series resistance (ESR) is plotted for the measured Nyquist plots. The ESR is the effective resistance of the combination of resistance of the electrolyte solution, the intrinsic resistance of the electroactive material and contact resistance at the interface between the electrode and the current collector.

Table 1. Table showing reported specific capacitance of the ZnO nanoparticles with the graphene/GO/rGO nanocomposites.

Si. No.	Methods	Materials	Specific capacitance [Fg^{-1}]	Power density [kW kg^{-1}]	References
1	Low temperature *in-situ* wet chemistry method	G-ZnO	786		[28]
2	One-pot method	G-ZnO	192	20.8	[13]
3	Hydrothermal	rGO/ZnO	280		[14]
4	Hydrothermal	G-ZnO	122.4		[15]
5	Ultrasonic spray pyrolysis	G-ZnO	61.7		[16]
6	Microwave assisted synthesis	G-ZnO	146		[17]
7	Ultrasonic assisted solution process	rGO-ZnO	312		[18]
8	Two-step method	G-ZnO	135		[19]
9	In-situ crystallization	G-ZnO	62.2	8.1	[21]
10	Sonochemical	G-ZnO	109		[20]
11	Hydrothermal	G-ZnO	156		[22]
12	Ultrasonic spray pyrolysis	G-ZnO	11.3		[23]
13	Hydrothermal	G-ZnO	719.2		[24]
14	Chemical vapour deposition reaction	3D-G-ZnO	275		[25]
15	Hydrothermal	rGO-ZnO	112		[58]
16	Hydrothermal	rGO-ZnO	96.1		[26]
17	Sol-gel deposition	rGO-ZnO	95		[11]
18	CVD and hydrothermal	3D-G-ZnO	554.23		[27]
19	One-step green approach using supercritical CO$_2$	rGO-ZnO	314		[100]

Table 1 lists the reported specific capacitance of the graphene/GO/rGO based ZnO nanocomposites. Fang et al. [28] reported the specific capacitance of the graphene-ZnO nanocomposites. They reported high cycle life with capacitance retention of 92% after 500 cycles.

7.1 Different formulae for calculating the specific capacitance

Many authors have used different formulas (equations) to measure the specific capacitance (C_{sp}) of supercapacitors. Few equations have been listed as follows:
The C_{sp} is determined from galvanostatic charge-discharge measurements using the Eq. (1) to Eq. (5)

$$C_{sp} = \frac{i_{average}}{v \times m} \text{-------- Eq. (1)}$$

Where C_{sp} is specific capacitance, $i_{average}$ is the average current measured in amperes (A), v is scan rate measured in (Vs^{-1}) and m is the mass of the electrode.

$$C_{sp} = \frac{i \times t}{\Delta V \times m} \pi r^2 \text{-------- Eq. (2)}$$

where i is the current measured in (A), t is the time required for charge-discharge in seconds and ΔV is the potential difference.

$$C_{sp} = \frac{1}{mV(V_f - V_i)} \int_{V_i}^{V_f} I(V)\, dV \text{ --------- Eq. (3)}$$

where m is the mass of the active electrode, V is the scan rate measured in (Vs^{-1}), V_f and V_i are the higher and lower limits of the integration potential in volumetric curve, $I(V)$ is the volumetric current measured in (A) [27].

$$C_{sp} = \frac{(I_+ - I_-)}{mv} \text{-------- Eq. (4)}$$

where I_+ is the maximum current in the positive scan (A) and I_- is the maximum current in the negative scan (A).

$$C_{sp} = \frac{\int i\, dV}{S\, \Delta V\, m} \text{ --------- Eq. (5)}$$

where $\int i dV$ indicates the area under one complete cycle of the CV curve, ΔV is the potential measured in (V) and S is the scan rate measured in (mVs^{-1}).

7.2 Factors affecting the specific capacitance

The scan rate affects specific capacitance. Increase in the scan rate decreases the specific capacitance. This is due to the reason that at higher scan rates, the diffusion limits the movement of the electrolyte ions and only the immediate surface participates in the charge storage. At lower scan rates, the entire active surface centers are utilized to store the charge. The specific capacitance also remains high at higher current densities.

The amount of graphene content in the nanocomposites affects the specific capacitance. The graphene/GO in the nanocomposites prevents the restacking of the ZnO in the nanocomposites. Graphene/GO provides the intercalated space between individual graphene sheets for ZnO nanoparticles growth. The superior performance is also due to the aggregation of faster electron movement provided by the graphene sheets.

Summary

This article basically introduces various techniques available for the production of graphene/GO/rGO based ZnO nanocomposites. Each one has its own merits and demerits. The general methods available for supercapacitor performance are discussed. The domain is relatively young and sufficient scope for performance improvement and optimization exists.

List of Abbreviations

ZnO	-Zinc oxide
GO	-Graphene oxide
rGO	-Reduced graphene oxide
EDLCs	-Electric double layer capacitors
eV	-Electron volts
TPa	-Terapascal
nm	-Nanometer
GON	-Graphene oxide nanosheets
KMnO$_4$	-Potassium permanganate
RTP	-Rapid thermal processing
CVD	-Chemical vapour deposition
W	-Watt
GHz	-Gigahertz
μm	-Micrometer
°C	-Degree Celsius
GaAs	-Gallium arsenide
ZnCl$_2$	-Zinc chloride
KCl	-Potassium chloride
GaN	-Gallium nitride
NaBH$_4$	-Sodium borohydride
KOH	-Potassium hydroxide
NaOH	-Sodium hydroxide
DIW	-Deionized water
CV	-Cyclic voltammetry
EC	-Electrochemical
Ag	-Silver
AgCl	-Silver chloride
Na$_2$SO$_4$	-Sodium sulfate
C_{sp}	-Specific capacitance
NaNO$_3$	-Sodium nitrate
EIS	-Electrochemical impedance spectroscopy

ESR	-Equivalent series resistance
SEM	-Scanning electron microscope
FESEM	-Field emission scanning electron microscope

References

[1]   B.E. Conway, Electrochemical Supercapacitors, Springer US, Boston, MA, 1999. https://doi.org/10.1007/978-1-4757-3058-6.

[2]   L.L. Zhang, X.S. Zhao, Carbon-based materials as supercapacitor electrodes, Chem. Soc. Rev. 38 (2009) 2520-2531. https://doi.org/10.1039/b813846j.

[3]   P. Simon, Y. Gogotsi, Materials for electrochemical capacitors, Nat. Mater. 7 (2008) 845–854. https://doi.org/10.1038/nmat2297.

[4]   M. Jayalakshmi, K. Balasubramanian, Simple capacitors to supercapacitors-An overview, Int. J. Electrochem. Sci. 3 (2008) 1196-1217.

[5]   R. Kötz, M. Carlen, Principles and applications of electrochemical capacitors, Electrochim. Acta 45 (2000) 2483–2498. https://doi.org/10.1016/S0013-4686(00)00354-6.

[6]   Y. Bar-Cohen, Electroactive polymer (EAP) actuators as artificial muscles : reality, potential, and challenges, SPIE Press, 2004. https://doi.org/10.1117/3.547465

[7]   H. Wang, J.T. Robinson, G. Diankov, H. Dai, Nanocrystal growth on graphene with various degrees of oxidation, J. Am. Chem. Soc. 132 (2010) 3270–3271. https://doi.org/10.1021/ja100329d.

[8]   J. Yan, Q. Wang, T. Wei, Z. Fan, Recent advances in design and fabrication of electrochemical supercapacitors with high energy densities, Adv. Energy Mater. 4 (2014) 1300816. https://doi.org/10.1002/aenm.201300816.

[9]   F. Béguin, V. Presser, A. Balducci, E. Frackowiak, Carbons and electrolytes for advanced supercapacitors, Adv. Mater. 26 (2014) 2219–2251. https://doi.org/10.1002/adma.201304137.

[10] X. Lang, A. Hirata, T. Fujita, M. Chen, Nanoporous metal/oxide hybrid electrodes for electrochemical supercapacitors, Nat. Nanotechnol. 6 (2011) 232–236. https://doi.org/10.1038/nnano.2011.13.

[11] I.Y.Y. Bu, R. Huang, One-pot synthesis of ZnO/reduced graphene oxide nanocomposite for supercapacitor applications, Mater. Sci. Semicond. Process. 31 (2015) 131–138. https://doi.org/10.1016/J.MSSP.2014.11.037.

[12] Q. Ke, J. Wang, Graphene-based materials for supercapacitor electrodes – A

Materials Research Forum LLC
https://doi.org/10.21741/9781644900550-7

review, J. Mater. (2016). https://doi.org/10.1016/j.jmat.2016.01.001.

[13]   Y.Z. Liu, Y.F. Li, Y.G. Yang, Y.F. Wen, M.Z. Wang, A one-pot method for producing ZnO–graphene nanocomposites from graphene oxide for supercapacitors, Scr. Mater. 68 (2013) 301–304. https://doi.org/10.1016/j.scriptamat.2012.10.048.

[14]   M. Raja, A.B.V.K. Kumar, N. Arora, J. Subha, Studies on electrochemical properties of zno/rgo nanocomposites as electrode materials for supercapacitors, Fullerenes Nanotub. Carbon Nanostructures. 23 (2015) 691–694. https://doi.org/10.1080/1536383X.2014.971117.

[15]   M. Saranya, R. Ramachandran, F. Wang, Graphene-zinc oxide (G-ZnO) nanocomposite for electrochemical supercapacitor applications, J. Sci. Adv. Mater. Devices J. (2016). https://doi.org/10.1016/j.jsamd.2016.10.001.

[16]   T. Lu, Y. Zhang, H. Li, L. Pan, Y. Li, Z. Sun, Electrochemical behaviors of graphene–ZnO and graphene–SnO$_2$ composite films for supercapacitors, Electrochim. Acta. 55 (2010) 4170–4173. https://doi.org/10.1016/J.ELECTACTA.2010.02.095.

[17]   T. Lu, L. Pan, H. Li, G. Zhu, T. Lv, X. Liu, Z. Sun, T. Chen, D.H.C. Chua, Microwave-assisted synthesis of graphene–ZnO nanocomposite for electrochemical supercapacitors, J. Alloys Compd. 509 (2011) 5488–5492. https://doi.org/10.1016/J.JALLCOM.2011.02.136.

[18]   J. Jayachandiran, J. Yesuraj, M. Arivanandhan, A. Raja, S.A. Suthanthiraraj, R. Jayavel, D. Nedumaran, Synthesis and electrochemical studies of rgo/zno nanocomposite for supercapacitor application, J. Inorg. Organomet. Polym. Mater. 28 (2018) 2046–2055. https://doi.org/10.1007/s10904-018-0873-0.

[19]   Y.-L. Chen, Z.-A. Hu, Y.-Q. Chang, H.-W. Wang, Z.-Y. Zhang, Y.-Y. Yang, H.-Y. Wu, Zinc Oxide/Reduced graphene oxide composites and electrochemical capacitance enhanced by homogeneous incorporation of reduced graphene oxide sheets in zinc oxide matrix, J. Phys. Chem. C. 115 (2011) 2563–2571. https://doi.org/10.1021/jp109597n.

[20]   A. Ramadoss, S.J. Kim, Facile preparation and electrochemical characterization of graphene/ZnO nanocomposite for supercapacitor applications, Mater. Chem. Phys. 140 (2013) 405–411. https://doi.org/10.1016/J.MATCHEMPHYS.2013.03.057.

[21]   J. Wang, Z. Gao, Z. Li, B. Wang, Y. Yan, Q. Liu, T. Mann, M. Zhang, Z. Jiang, Green synthesis of graphene nanosheets/ZnO composites and electrochemical

properties, J. Solid State Chem. 184 (2011) 1421–1427.
https://doi.org/10.1016/J.JSSC.2011.03.006.

[22] Z. Li, Z. Zhou, G. Yun, K. Shi, X. Lv, B. Yang, High-performance solid-state
supercapacitors based on graphene-ZnO hybrid nanocomposites, Nanoscale Res.
Lett. 8 (2013) 473. https://doi.org/10.1186/1556-276X-8-473.

[23] Y. Zhang, H. Li, L. Pan, T. Lu, Z. Sun, Capacitive behavior of graphene–ZnO
composite film for supercapacitors, J. Electroanal. Chem. 634 (2009) 68–71.
https://doi.org/10.1016/J.JELECHEM.2009.07.010.

[24] V. Rajeswari, R. Jayavel, A. Clara Dhanemozhi, Synthesis and characterization of
graphene-zinc oxide nanocomposite electrode material for supercapacitor
applications, Mater. Today Proc. 4 (2017) 645–652.
https://doi.org/10.1016/J.MATPR.2017.01.068.

[25] Z. Li, P. Liu, G. Yun, K. Shi, X. Lv, K. Li, J. Xing, B. Yang, 3D (Three-
dimensional) sandwich-structured of ZnO (zinc oxide)/rGO (reduced graphene
oxide)/ZnO for high performance supercapacitors, Energy. 69 (2014) 266–271.
https://doi.org/10.1016/J.ENERGY.2014.03.003.

[26] L. Huang, G. Guo, Y. Liu, Q. Chang, Y. Xie, Reduced graphene oxide-ZnO
nanocomposites for flexible supercapacitors, J. Disp. Technol. 8 (2012) 373–376.
https://doi.org/10.1109/JDT.2011.2173158.

[27] X. Li, Z. Wang, Y. Qiu, Q. Pan, P. Hu, 3D graphene/ZnO nanorods composite
networks as supercapacitor electrodes, J. Alloys Compd. 620 (2015) 31–37.
https://doi.org/10.1016/J.JALLCOM.2014.09.105.

[28] L. Fang, B. Zhang, W. Li, J. Zhang, K. Huang, Q. Zhang, Fabrication of highly
dispersed ZnO nanoparticles embedded in graphene nanosheets for high
performance supercapacitors, Electrochim. Acta. 148 (2014) 164–169.
https://doi.org/10.1016/J.ELECTACTA.2014.10.065.

[29] A.G. Pandolfo, A.F. Hollenkamp, Carbon properties and their role in
supercapacitors, J. Power Sources 157 (2006) 11–27.
https://doi.org/10.1016/J.JPOWSOUR.2006.02.065.

[30] D. Pech, M. Brunet, H. Durou, P. Huang, V. Mochalin, Y. Gogotsi, P.-L. Taberna,
P. Simon, Ultrahigh-power micrometre-sized supercapacitors based on onion-like
carbon, Nat. Nanotechnol. 5 (2010) 651–654.
https://doi.org/10.1038/nnano.2010.162.

[31] H. Wang, Y. Liang, T. Mirfakhrai, Z. Chen, H.S. Casalongue, H. Dai, Advanced

asymmetrical supercapacitors based on graphene hybrid materials, Nano Res. 4 (2011) 729–736. https://doi.org/10.1007/s12274-011-0129-6.

[32] M. Winter, R.J. Brodd, What are Batteries, Fuel Cells, and Supercapacitors?, Chem. Rev. 104 (2004) 4245–4270. https://doi.org/10.1021/cr020730k.

[33] M. Inagaki, H. Konno, O. Tanaike, Carbon materials for electrochemical capacitors, J. Power Sources. 195 (2010) 7880–7903. https://doi.org/10.1016/j.jpowsour.2010.06.036.

[34] A.S. Aricò, P. Bruce, B. Scrosati, J.-M. Tarascon, W. van Schalkwijk, Nanostructured materials for advanced energy conversion and storage devices, Nat. Mater. 4 (2005) 366–377. https://doi.org/10.1038/nmat1368.

[35] J. Li, F. Gao, Analysis of electrodes matching for asymmetric electrochemical capacitor, J. Power Sources 194 (2009) 1184–1193. https://doi.org/10.1016/j.jpowsour.2009.06.017.

[36] E. Yoo, J. Kim, E. Hosono, H. Zhou, T. Kudo, I. Honma, Large reversible Li storage of graphene nanosheet families for use in rechargeable lithium ion batteries, Nano Lett. 8 (2008) 2277–2282. https://doi.org/10.1021/nl800957b.

[37] T. Bhardwaj, A. Antic, B. Pavan, V. Barone, B.D. Fahlman, Enhanced electrochemical lithium storage by graphene nanoribbons, J. Am. Chem. Soc. 132 (2010) 12556–12558. https://doi.org/10.1021/ja106162f.

[38] M.D. Stoller, S. Park, Y. Zhu, J. An, R.S. Ruoff, Graphene-based ultracapacitors, Nano Lett. 8 (2008) 3498–3502. https://doi.org/10.1021/nl802558y.

[39] Y. Wang, Z. Shi, Y. Huang, Y. Ma, C. Wang, M. Chen, Y. Chen, Supercapacitor devices based on graphene materials, J. Phys. Chem. C 113 (2009) 13103–13107. https://doi.org/10.1021/jp902214f.

[40] X. An, T. Simmons, R. Shah, C. Wolfe, K.M. Lewis, M. Washington, S.K. Nayak, S. Talapatra, S. Kar, Stable aqueous dispersions of noncovalently functionalized graphene from graphite and their multifunctional high-performance applications, Nano Lett. 10 (2010) 4295–4301. https://doi.org/10.1021/nl903557p.

[41] Y. Wang, Z. Shi, Y. Huang, Y. Ma, C. Wang, M. Chen, Y. Chen, Supercapacitor devices based on graphene materials, J. Phys. Chem. C. 113 (2009) 13103–13107. https://doi.org/10.1021/jp902214f.

[42] T. Palaniselvam, J.-B. Baek, Graphene based 2D-materials for supercapacitors, 2D Mater. 2 (2015) 32002. https://doi.org/10.1088/2053-1583/2/3/032002.

[43] J. Basu, J.K. Basu, T.K. Bhattacharyya, The evolution of graphene-based electronic devices, Int. J. Smart Nano Mater. 1 (2010) 201–223. https://doi.org/10.1080/19475411.2010.510856.

[44] V. Singh, D. Joung, L. Zhai, S. Das, S.I. Khondaker, S. Seal, Graphene based materials: Past, present and future, Prog. Mater. Sci. 56 (2011) 1178–1271. https://doi.org/10.1016/j.pmatsci.2011.03.003.

[45] A. Kolodziejczak-Radzimska, T. Jesionowski, Zinc oxide-from synthesis to application: A review, Materials 7 (2014) 2833–2881. https://doi.org/10.3390/ma7042833.

[46] N. Al-Dhabi, M. Valan Arasu, Environmentally-friendly green approach for the production of zinc oxide nanoparticles and their anti-fungal, ovicidal, and larvicidal properties, Nanomaterials 8 (2018) 500. https://doi.org/10.3390/nano8070500.

[47] Y. Wang, X. Xiao, H. Xue, H. Pang, Zinc oxide based composite materials for advanced supercapacitors, ChemistrySelect 3 (2018) 550–565. https://doi.org/10.1002/slct.201702780.

[48] S. Meti, M.R. Rahman, M.I. Ahmad, K.U. Bhat, Chemical free synthesis of graphene oxide in the preparation of reduced graphene oxide-zinc oxide nanocomposite with improved photocatalytic properties, Appl. Surf. Sci. 451 (2018) 67–75. https://doi.org/10.1016/j.apsusc.2018.04.138.

[49] A.R. Marlinda, N.M. Huang, M.R. Muhamad, M.N. An'Amt, B.Y.S. Chang, N. Yusoff, I. Harrison, H.N. Lim, C.H. Chia, S.V. Kumar, Highly efficient preparation of ZnO nanorods decorated reduced graphene oxide nanocomposites, Mater. Lett. 80 (2012) 9–12. https://doi.org/10.1016/j.matlet.2012.04.061.

[50] B. Saravanakumar, R. Mohan, S.-J. Kim, Facile synthesis of graphene/ZnO nanocomposites by low temperature hydrothermal method, Mater. Res. Bull. 48 (2013) 878–883. https://doi.org/10.1016/J.MATERRESBULL.2012.11.048.

[51] Y.. b Zhao, G.. Chen, Y.. Wang, Facile synthesis of graphene/ZnO composite as an anode with enhanced performance for lithium ion batteries, J. Nanomater. 2014 (2014). https://doi.org/10.1155/2014/964391.

[52] X. Bai, L. Wang, R. Zong, Y. Lv, Y. Sun, Y. Zhu, Performance enhancement of ZnO photocatalyst via synergic effect of surface oxygen defect and graphene hybridization, Langmuir 29 (2013) 3097–3105. https://doi.org/10.1021/la4001768.

[53] X. Liu, L. Pan, Q. Zhao, T. Lv, G. Zhu, T. Chen, T. Lu, Z. Sun, C. Sun, UV-assisted photocatalytic synthesis of ZnO–reduced graphene oxide composites with

enhanced photocatalytic activity in reduction of Cr(VI), Chem. Eng. J. 183 (2012) 238–243. https://doi.org/10.1016/J.CEJ.2011.12.068.

[54] A. Kajbafvala, M.R. Shayegh, M. Mazloumi, S. Zanganeh, A. Lak, M.S. Mohajerani, S.K. Sadrnezhaad, Nanostructure sword-like ZnO wires: Rapid synthesis and characterization through a microwave-assisted route, J. Alloys Compd. 469 (2009) 293–297. https://doi.org/10.1016/j.jallcom.2008.01.093.

[55] A. Ashkarran, B. Mohammadi, ZnO nanoparticles decorated on graphene sheets through liquid arc discharge approach with enhanced photocatalytic performance under visible-light, Appl. Surf. Sci. 342 (2015) 112–119. https://doi.org/10.1016/j.apsusc.2015.03.030.

[56] J. Wu, X. Shen, L. Jiang, K. Wang, K. Chen, Solvothermal synthesis and characterization of sandwich-like graphene/ZnO nanocomposites, Appl. Surf. Sci. 256 (2010) 2826–2830. https://doi.org/10.1016/j.apsusc.2009.11.034.

[57] W. Zou, J. Zhu, Y. Sun, X. Wang, Depositing ZnO nanoparticles onto graphene in a polyol system, Mater. Chem. Phys. 125 (2011) 617–620. https://doi.org/10.1016/j.matchemphys.2010.10.008.

[58] G. Du, X. Wang, L. Zhang, Y. Feng, Y. Li, Controllable synthesis of different ZnO architectures decorated reduced graphene oxide nanocomposites, Mater. Lett. 96 (2013) 128–130. https://doi.org/10.1016/J.MATLET.2013.01.063.

[59] I.Y.Y. Bu, R. Huang, One-pot synthesis of ZnO/reduced graphene oxide nanocomposite for supercapacitor applications, Mater. Sci. Semicond. Process. 31 (2015) 131–138. https://doi.org/10.1016/J.MSSP.2014.11.037.

[60] M. Jayalakshmi, M. Palaniappa, K. Balasubramanian, Single step solution combustion synthesis of ZnO/carbon composite and its electrochemical characterization for supercapacitor application, Int. J. Electrochem. Sci. 3 (2008) 96-103.

[61] L. Liao, X. Duan, Graphene-dielectric integration for graphene transistors, Mater. Sci. Eng. R Reports 70 (2010) 354–370. https://doi.org/10.1016/j.mser.2010.07.003.

[62] N. Hashim, Z. Muda, M.Z. Hussein, I.M. Isa, A. Mohamed, A. Kamari, S.A. Bakar, M. Mamat, A. Jaafar, A brief review on recent graphene oxide-based material nanocomposites: Synthesis and applications, J. Mater. Environ. Sci. 7 (2016) 3225–3243.

[63] T.-H. Han, H. Kim, S.-J. Kwon, T.-W. Lee, Graphene-based flexible electronic

devices, Mater. Sci. Eng. R Reports. 118 (2017) 1–43.
https://doi.org/10.1016/J.MSER.2017.05.001.

[64] L. Tang, X. Li, R. Ji, K.S. Teng, G. Tai, J. Ye, C. Wei, S.P. Lau, Bottom-up synthesis of large-scale graphene oxide nanosheets, J. Mater. Chem. 22 (2012) 5676-5683. https://doi.org/10.1039/c2jm15944a.

[65] M.M. Benameur, F. Gargiulo, S. Manzeli, G. Autès, M. Tosun, O. V. Yazyev, A. Kis, Electromechanical oscillations in bilayer graphene, Nat. Commun. 6 (2015) 8582. https://doi.org/10.1038/ncomms9582.

[66] Z. Shi, H. Lu, L. Zhang, R. Yang, Y. Wang, D. Liu, H. Guo, D. Shi, H. Gao, E. Wang, G. Zhang, Studies of graphene-based nanoelectromechanical switches, Nano Res. 2012 (n.d.) 82–87. https://doi.org/10.1007/s12274-011-0187-9.

[67] M. Gu, Y. Liu, T. Chen, F. Du, X. Zhao, C. Xiong, Y. Zhou, Is graphene a promising nano-material for promoting surface modification of implants or scaffold materials in bone tissue engineering?, Tissue Eng. Part B. Rev. 20 (2014) 477–91. https://doi.org/10.1089/ten.TEB.2013.0638.

[68] S. Meti, U.B. K, M.R. Rahman, M. Jayalakshmi, Photocatalytic behaviour of nanocomposites of sputtered titanium oxide film on graphene oxide nanosheets, Am. J. Mater. Sci. 5 (2015) 12–18. https://doi.org/10.5923/c.materials.201502.03.

[69] K.S. Novoselov, A.K. Geim, S. V Morozov, D. Jiang, Y. Zhang, S. V Dubonos, I. V Grigorieva, A.A. Firsov, Electric field effect in atomically thin carbon films, Science 306 (2004) 666-669. https://doi.org/10.1126/science.1102896.

[70] R. Sellappan, J. Sun, A. Galeckas, N. Lindvall, A. Yurgens, A.Y. Kuznetsov, D. Chakarov, Influence of graphene synthesizing techniques on the photocatalytic performance of graphene–TiO_2 nanocomposites, Phys. Chem. Chem. Phys. 15 (2013) 15528–15537. https://doi.org/10.1039/C3CP52457D.

[71] N.I. Zaaba, K.L. Foo, U. Hashim, S.J. Tan, W.W. Liu, C.H. Voon, Synthesis of graphene oxide using modified Hummers method: Solvent influence, Procedia Eng. 184 (2017) 469–477. https://doi.org/10.1016/J.PROENG.2017.04.118.

[72] G. Pan, B. Li, M. Heath, D. Horsell, M.L. Wears, L. Al Taan, S. Awan, Transfer-free growth of graphene on SiO_2 insulator substrate from sputtered carbon and nickel films, Carbon 65 (2013) 349–358.
https://doi.org/10.1016/J.CARBON.2013.08.036.

[73] C.-C. Hsu, J.D. Bagley, M.L. Teague, W.S. Tseng, K.L. Yang, Y. Zhang, Y. Li, Y. Li, J.M. Tour, N.C. Yeh, High-yield single-step catalytic growth of graphene

nanostripes by plasma enhanced chemical vapor deposition, Carbon 129 (2018) 527–536. https://doi.org/10.1016/J.CARBON.2017.12.058.

[74] S. Zheng, G. Zhong, X. Wu, L. D'arsì, J. Robertson, Metal-catalyst-free growth of graphene on insulating substrates by ammonia-assisted microwave plasma-enhanced chemical vapor deposition, RSC Adv. 7 (2017) 33185-33193. https://doi.org/10.1039/c7ra04162d.

[75] S. Hari, A.M. Goossens, L.M.K. Vandersypen, C.W. Hagen, Electron Beam Induced Deposition on graphene on silicon oxide and hexagonal boron nitride: A comparison of substrates, Microelectron. Eng. 121 (2014) 122–126. https://doi.org/10.1016/J.MEE.2014.04.037.

[76] M. Shandilya, R. Rai, J. Singh, Review: hydrothermal technology for smart materials, Adv. Appl. Ceram. 115 (2016) 354–376. https://doi.org/10.1080/17436753.2016.1157131.

[77] X. Zhou, T. Shi, H. Zhou, Hydrothermal preparation of ZnO-reduced graphene oxide hybrid with high performance in photocatalytic degradation, Appl. Surf. Sci. 258 (2012) 6204–6211. https://doi.org/10.1016/j.apsusc.2012.02.131.

[78] B. Li, H. Cao, ZnO@graphene composite with enhanced performance for the removal of dye from water, J. Mater. Chem. 21 (2011) 3346–3349. https://doi.org/10.1039/C0JM03253K.

[79] T. Xu, L. Zhang, H. Cheng, Y. Zhu, Significantly enhanced photocatalytic performance of ZnO via graphene hybridization and the mechanism study, Appl. Catal. B Environ. 101 (2011) 382–387. https://doi.org/10.1016/J.APCATB.2010.10.007.

[80] Y.J. Zhu, F. Chen, Microwave-assisted preparation of inorganic nanostructures in liquid phase, Chem. Rev. 114 (2014) 6462–6555. https://doi.org/10.1021/cr400366s.

[81] M.H. Wong, A. Berenov, X. Qi, M.J. Kappers, Z.H. Barber, B. Illy, Z. Lockman, M.P. Ryan, J.L. MacManus-Driscoll, Electrochemical growth of ZnO nano-rods on polycrystalline Zn foil, Nanotechnology. 14 (2003) 968–973. https://doi.org/10.1088/0957-4484/14/9/306.

[82] H.B. Zeng, Y. Bando, Xi, J. Xu, L. Li, Tian, Y. Zhai, X.S. Fang, D. Golberg, Heteroepitaxial growth of ZnO nanorod arrays on GaAs (111) substrates by electrochemical deposition, (2010). https://doi.org/10.1002/ejic.201000527.

[83] S. Peulon, D. Lincot, Cathodic electrodeposition from aqueous solution of dense or

open-structured zinc oxide films, Adv. Mater. 8 (1996) 166–170.
https://doi.org/10.1002/adma.19960080216.

[84] Run Liu, Alexey A. Vertegel, Eric W. Bohannan, and Thomas A. Sorenson, J.A.
Switzer, Epitaxial Electrodeposition of zinc oxide nanopillars on single-crystal
gold, Chem. Mater. 13 (2001) 2508-2512. https://doi.org/10.1021/CM000763L.

[85] M. Izaki, T. Omi, Transparent zinc oxide films prepared by electrochemical
reaction, Appl. Phys. Lett. 68 (1998) 2439. https://doi.org/10.1063/1.116160.

[86] E. Pourshaban, H. Abdizadeh, M.R. Golobostanfard, A close correlation between
nucleation sites, growth and final properties of ZnO nanorod arrays: Sol-gel
assisted chemical bath deposition process, Ceram. Int. 42 (2016) 14721–14729.
https://doi.org/10.1016/J.CERAMINT.2016.06.098.

[87] J.A. Alvarado, A. Maldonado, H. Juarez, M. Pacio, Synthesis of colloidal ZnO
nanoparticles and deposit of thin films by spin coating technique, J. Nanomater.
2013 (2013). https://doi.org/10.1155/2013/903191.

[88] M. Singh, M.Y. Mulla, M.V. Santacroce, M. Magliulo, C. Di Franco, K. Manoli,
D. Altamura, C. Giannini, N. Cioffi, G. Palazzo, G. Scamarcio, L. Torsi, Effect of
the gate metal work function on water-gated ZnO thin-film transistor performance,
J. Phys. D. Appl. Phys. 49 (2016) 275101. https://doi.org/10.1088/0022-
3727/49/27/275101.

[89] M. Zhou, D. Zang, X. Zhai, Z. Gao, W. Zhang, C. Wang, Preparation of
biomorphic porous zinc oxide by wood template method, Ceram. Int. 42 (2016)
10704–10710. https://doi.org/10.1016/J.CERAMINT.2016.03.188.

[90] W. Chebil, M.A. Boukadhaba, A. Fouzri, Epitaxial growth of ZnO on quartz
substrate by sol-gel spin-coating method, Superlattices Microstruct. 95 (2016) 48–
55. https://doi.org/10.1016/J.SPMI.2016.04.033.

[91] R. Haarindraprasad, U. Hashim, S.C.B. Gopinath, V. Perumal, W.-W. Liu, S.R.
Balakrishnan, Fabrication of interdigitated high-performance zinc oxide nanowire
modified electrodes for glucose sensing, Anal. Chim. Acta. 925 (2016) 70–81.
https://doi.org/10.1016/J.ACA.2016.04.030.

[92] T.A. Para, H.A. Reshi, S. Pillai, V. Shelke, Grain size disposed structural, optical
and polarization tuning in ZnO, Appl. Phys. A. 122 (2016) 730.
https://doi.org/10.1007/s00339-016-0256-8.

[93] H. Li, Y. Wei, Y. Zhang, C. Zhang, G. Wang, Y. Zhao, F. Yin, Z. Bakenov, In situ
sol-gel synthesis of ultrafine ZnO nanocrystals anchored on graphene as anode

material for lithium-ion batteries, Ceram. Int. 42 (2016) 12371–12377. https://doi.org/10.1016/J.CERAMINT.2016.05.010.

[94] T. Demes, C. Ternon, D. Riassetto, H. Roussel, L. Rapenne, I. Gélard, C. Jimenez, V. Stambouli, M. Langlet, New insights in the structural and morphological properties of sol-gel deposited ZnO multilayer films, J. Phys. Chem. Solids. 95 (2016) 43–55. https://doi.org/10.1016/J.JPCS.2016.03.017.

[95] D.G. Ayana, R. Ceccato, C. Collini, L. Lorenzelli, V. Prusakova, S. Dirè, Sol-gel derived oriented multilayer ZnO thin films with memristive response, Thin Solid Films 615 (2016) 427–436. https://doi.org/10.1016/J.TSF.2016.07.025.

[96] L.M. Zimmermann, P. V Baldissera, I.H. Bechtold, Stability of ZnO quantum dots tuned by controlled addition of ethylene glycol during their growth, Mater. Res. Express. 3 (2016) 75018. https://doi.org/10.1088/2053-1591/3/7/075018.

[97] A.T. Khalil, M. Ovais, I. Ullah, M. Ali, Z.K. Shinwari, S. Khamlich, M. Maaza, *Sageretia thea* (Osbeck.) mediated synthesis of zinc oxide nanoparticles and its biological applications, Nanomedicine 12 (2017) 1767–1789. https://doi.org/10.2217/nnm-2017-0124.

[98] A. Venkateasan, R. Prabakaran, V. Sujatha, Phytoextract-mediated synthesis of zinc oxide nanoparticles using aqueous leaves extract of Ipomoea pes-caprae (L).R.br revealing its biological properties and photocatalytic activity, Nanotechnol. Environ. Eng. 2 (2017) 8. https://doi.org/10.1007/s41204-017-0018-7.

[99] V. Singh, D. Joung, L. Zhai, S. Das, S.I. Khondaker, S. Seal, Graphene based materials: Past, present and future, Prog. Mater. Sci. 56 (2011) 1178–1271. https://doi.org/10.1016/j.pmatsci.2011.03.003.

[100] Y. Haldorai, W. Voit, J.-J. Shim, Nano ZnO@reduced graphene oxide composite for high performance supercapacitor: Green synthesis in supercritical fluid, Electrochim. Acta. 120 (2014) 65–72. https://doi.org/10.1016/J.ELECTACTA.2013.12.063.

Materials Research Forum LLC
https://doi.org/10.21741/9781644900550-8

Chapter 8

Defect Engineered Graphene Materials for Supercapacitors

Madhabi Devi[1,2] and Ashok Kumar[1*]

[1]Department of Physics, Tezpur University, Napaam, Tezpur-784028, Assam, India

[2]Department of Physics, Majuli College, Kamalabari-785106, Assam, India

*ask@tezu.ernet.in

Abstract

Growing global demand for efficient and sustainable energy storage systems can be met through the development of electrochemical capacitors or supercapacitors owing to their high power performance, long cycle life and environmental benignity. Graphene and its derivatives have been widely used as electrode materials for supercapacitors because of their high specific surface area, low density, high electrical conductivity, thermal and mechanical stability. However, their re-aggregation reduces specific surface area and inhibits their electrochemical performance. Introducing defects in graphene has gained importance to address this issue, which can change its density of states, crystal symmetry and porosity enhancing electrode-electrolyte interfacial interaction. The present chapter deals with defect engineering on graphene-based materials for improved supercapacitor performance. Different defect generation and characterization techniques have been discussed in detail.

Keywords

Supercapacitors, Graphene, Defect Engineering, Doping, Functionalization, Irradiation, Swift Heavy Ions

Contents

1.   Introduction

The exhausting of energy resources due to the growing population worldwide and the accelerating global warming and environmental pollution due to increased combustion of fossil fuels have caused major concerns for the survival of life on earth. At the rate of present energy consumption, the global energy needs are expected to almost double by 2050 [1]. Predictions are that the global energy demand will be dominated by fossil fuels accounting for 78% by 2030 [2]. To meet the high energy demand and to mitigate environmental pollution, there has been a high demand for clean, sustainable and renewable alternative energy sources that can replace fossil fuels. Renewable energy sources that include solar and wind energies are available in abundance and are potentially cheaper technologies. However, these energy sources are intermittent and cannot fulfill the energy demands at peak hours leading to power fluctuations. Utilizing energy storage devices to regulate their supply is a chief requirement to improve their usability for mass production. The other renewable energy sources like biomass, biofuel and hydropower are comparatively stable means of energy production, but energy storage is essential for their effective usage during depletion time. Hence all renewable energy sources suffer from power fluctuations and share a common requirement of the need of reliable energy storage systems. In this regard, supercapacitors are receiving increasing importance as energy storage devices over batteries due to their high power performance, excellent cyclic stability and environmental benignity [3]. However, their low energy density poses significant challenge in their development and applications. Enormous research has been focused on synthesizing new electrode materials and tuning their properties to achieve improved specific capacitance (C_{sp}) and energy density [4-7]. Carbon materials are potential candidates for electrochemical supercapacitors because of their large specific surface area (SSA) and electronic conductivity. The charge storage mechanism in carbon materials is by electrostatic separation and formation of electric double layer between electrolyte ions and high surface area electrodes, which is termed as electric double layer capacitance (EDLC). The cycle life and power delivery obtained with EDL electrodes could be very high as the charge is stored by physical adsorption and no faradaic charge transfer processes are involved between the electrode and electrolyte. Amongst the carbon materials, graphene and reduced graphene oxide (ReGO) have gained immense attention as EDL capacitor electrodes owing to their unique mechanical and thermal stability, high electrical conductivity, large aspect ratio and low density [8]. However, in spite of their large specific surface areas, their specific capacitance is limited and the energy density of commercial carbon-based supercapacitors is not more than 10 Wh/kg, which is sufficiently low to meet the demand of high-performance energy storage devices. To develop advanced electrode materials with high specific energy and power density, the overall performance of supercapacitor electrodes needs to be optimized, which could be achieved by combining graphene with secondary component to form nanocomposite electrodes. Hybridizing graphene and its derivatives with conducting polymers and transition metal oxides has been found to be a promising approach to utilize the merits and overcome the shortcomings of both the individual components in the nanocomposite electrode. In such hybrid nanocomposite electrode, graphene can serve as an underlying conductive backbone with fast charge transport and long cycle life, while the secondary component can provide a high and stable electrochemical capacitance. Moreover, the tendency of graphene sheets to agglomerate due to their interlayer van der Waals interaction could be minimized by the presence of secondary component as spacers between them. However, the experimentally obtained capacitances are not high enough for potential applicability, since not all the SSA of the electrode material is available for the electrolyte ions and therefore the capacitive performance was not up to expectation. To address this limitation and to enhance the electrochemical performance of graphene-based electrodes, defect engineering by a variety of techniques have been developed.

This chapter presents a brief overview of supercapacitors, their charge storage principle and their classifications. Graphene-based electrode materials investigated for supercapacitor applications have been highlighted. Different defect-engineering methods implemented on graphene derived electrode materials for supercapacitor applications have been discussed. Further, the chapter deals with defect engineering of graphene-based polypyrrole nanocomposite electrodes by swift heavy ion (SHI) irradiation investigated by the authors. The effect of SHI irradiation on the electrochemical performance of graphene-based polypyrrole nanocomposite electrodes has been presented in detail.

2. Supercapacitors

Electrochemical supercapacitors have developed as efficient, environmentally benign and low-cost energy storage devices that are capable of storing charge at the interfaces between the electrode and the electrolyte [9]. They are promising candidates for energy storage applications due to their pulse power supply, fast charge-discharge rates, long cycle life, low self-discharge and safe operation [10]. The charge storage principle of a supercapacitor is similar to conventional capacitor except that the two electrodes separated by a porous dielectric membrane are dipped in a common electrolyte. The electrode-electrolyte contact area or interface serves as the dielectric in supercapacitors which is of the order of a few angstroms, much smaller than a conventional capacitor [11]. This gives rise to very high capacitance values in the range of farads and therefore they are referred to as supercapacitors. Filling the gap between capacitors and batteries, supercapacitors exhibit energy densities in the range of 8-10 Wh/kg and power densities greater than 10 kW/kg [12]. The cycle life of supercapacitors is very high upto 10^6 cycles as there are no chemical conversions of the electrode materials involved with non-faradaic processes. They are safer and their internal resistance is lower than that of capacitors and batteries i.e. the energy loss to heat is less during each cycle. The charge-discharge rates of supercapacitors are highly reversible and fast and can be charged within seconds. Supercapacitors store charge by two different mechanisms: (i) electrostatically by accumulation of charges across the electrode-electrolyte interface and (ii) Faradically by reversible redox reactions between the electrode and the electrolyte

[13]. Depending on these two charge storage mechanisms and the electrode materials used, supercapacitors are categorized into two types: Electric double layer capacitors and pseudocapacitors. The combination of these two charge storage mechanisms gives rise to the third classification of supercapacitors known as hybrid capacitors [14]. The block diagram of the classification of supercapacitors based on their types and electrode materials is depicted in Fig. 1.

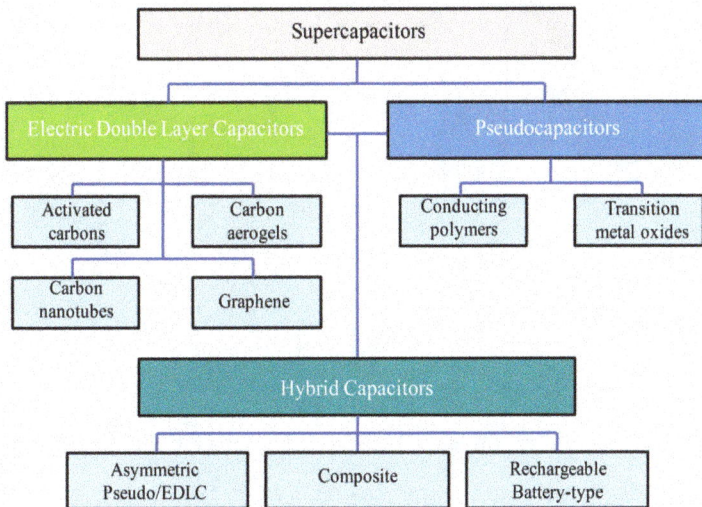

Fig. 1. Block diagram of classification of supercapacitors.

2.1 Electric double layer capacitors

Electric double layer (EDL) capacitors store charge by electrostatic attraction between the electrolyte ions and the charges on the electrode surface forming two electric double layers at the electrode-electrolyte interface. The thickness of the double layer is of the order of a few Angstroms. The charge storage process is purely electrostatic and no faradaic reactions occur across the interface in EDL capacitors [15]. The energy storage process is associated with the movement of charge to the interfaces and no chemical changes are involved, therefore the process is fast and highly reversible. For this reason, EDL capacitors have a high power density and can operate upto 10^6 charge-discharge cycles [16]. The schematic diagram of EDL capacitors is shown in Fig. 2 comprising of two porous carbon electrodes, an electrolyte and a separator that allows ions to pass through as well as provides electric insulation between the electrodes. When an external potential is applied, positive and negative charges of the electrodes accumulate on the surface, which attracts electrolyte ions of opposite polarity. This forms a double layer at both the electrode-electrolyte interface separated by few Angstroms. The double layer formed at each electrode-electrolyte interface represents two capacitors connected in series. Owing to the short distance of charge separation and large surface area of porous electrodes, EDL capacitors achieve high capacitances than that of normal capacitors. Electrode materials for EDL capacitors are generally carbon-based materials with high SSA such as carbon aerogels, activated carbon, carbon nanotubes, graphene etc. [17, 18].

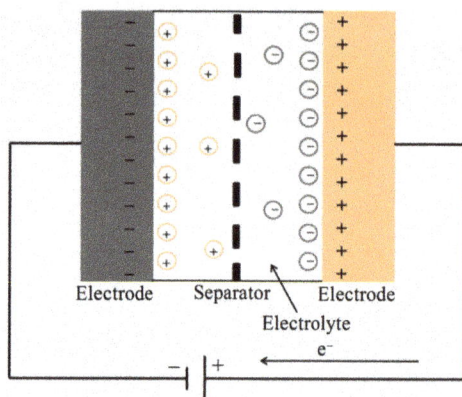

Fig. 2. Schematic of charge storage process in electric double layer capacitors.

2.2   Pseudocapacitors

The charge storage in pseudocapacitors occurs by repeated redox reactions at the electrode-electrolyte interface in addition to the electrostatic charge accumulation. When a potential is applied, charge transfer occurs between the electrode and electrolyte associated with oxidation state changes of the electrode, which is referred to as a faradaic process [14]. Due to the presence of faradaic reactions, they are also known as redox supercapacitors. Unlike EDL capacitors, the charge storage mechanism of pseudocapacitors is indirect, where the amount of charge transferred depends on the potential of the electrode, which is similar to a capacitor and hence is designated as pseudocapacitance ("pseudo" implies "not real"). Pseudocapacitance is exhibited by redox-active materials like conducting polymers [polypyrrole, polyaniline, poly(3,4-ethylenedioxythiophene], transition metal oxides (MnO_2, RuO_2, NiO), metal hydroxides [$Co (OH)_2$,$Ni(OH)_2$] etc. [19]. The schematic of the working principle of pseudocapacitor using polypyrrole (PPr) as electrode and KCl as electrolyte is shown in Fig. 3. The faradaic reactions between the PPr electrode and KCl electrolyte during charging at the negative and positive electrode could be expressed as given in Eq. 1.

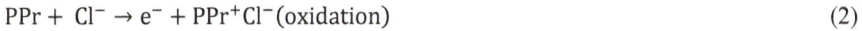

$$PPr + e^- + K^+ \rightarrow PPr^-(K^+)(\text{reduction}) \tag{1}$$

$$PPr + Cl^- \rightarrow e^- + PPr^+Cl^-(\text{oxidation}) \tag{2}$$

It can be seen that the surface as well as the bulk of the material contributes to pseudocapacitive charge storage. Therefore, pseudocapacitance can be higher by 10-100 times than that of EDL capacitors [20]. The faradaic reactions are reversible and faster than that of rechargeable batteries, but are slower than the ion adsorption process in EDL capacitors. During the faradaic charge transfer reactions, volumetric changes are experienced by the electrode, which degrades their performance with time. Hence, pseudocapacitors are characterized by a high specific capacitance and energy density than that of EDL capacitors but have a low power density and cycle life [21].

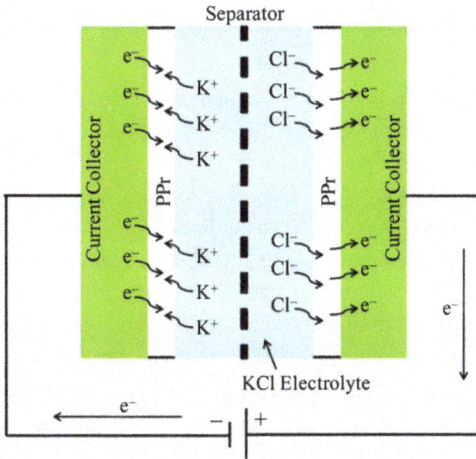

Fig. 3. Schematic of charge storage process in pseudocapacitors.

2.3 Hybrid capacitors

Due to the relative disadvantages of EDL capacitors and pseudocapacitors, hybrid capacitors have been developed to combine the best properties of individual electrode materials and mitigate their drawbacks to produce a synergistic effect for improved electrochemical performance. In hybrid capacitors, both faradaic and non-faradaic charge storage processes are utilized to achieve high specific capacitance and energy density than that of EDL capacitors without sacrificing the power density and cycle life that limit the performance of pseudocapacitors [22]. Depending on their electrode configuration, hybrid capacitors are classified into three categories: (i) composite hybrid, (ii) asymmetric hybrid and (iii) battery-type hybrid supercapacitors, respectively. The supercapacitors that use composite materials as electrodes are called composite hybrid supercapacitors. Here, carbon-based materials are integrated with conducting polymers or metal oxides and both the charge storage mechanisms of ion adsorption and faradaic redox reactions are utilized in a single electrode [23]. Asymmetric hybrid supercapacitors are fabricated by using an EDLC material as one electrode and a pseudocapacitive material as the other electrode. Both the electrodes are of different materials and store charge by two different mechanisms which mitigate the problems of low operating voltage range and cycle life [24]. Battery-type hybrids are similar to asymmetric hybrids except that they couple two different technologies of battery and supercapacitor in a single device. This device configuration is designed to achieve a high specific energy of batteries combined with the fast power delivery and cycle life of supercapacitors [25].

3. Graphene

Graphene, a 2D single layer sheet of sp^2 bonded carbon atoms in a hexagonal lattice has recently gained utmost attention as novel supercapacitor electrodes. The one-atomic thick sheet of carbon arranged in a honeycomb structure was a ground-breaking discovery of Geim and Novoselov in 2004 by mechanical exfoliation of a monolayer of graphite by 'scotch-tape' method, which bagged the Nobel Prize in Physics in 2010 [26]. Graphene is a single sheet of graphite as shown in Fig. 4 with a carbon-carbon bond length of 0.142 nm. It is the basic building block of carbon materials starting from which all other dimensional structures such as 3D graphite, 2D CNT and 0D fullerene can be produced by stacking, rolling and wrapping a graphene nanosheet, respectively [9].

Graphene exhibits remarkable properties such as high theoretical surface area of 2630 m^2/g, outstanding electrical conductivity of 6000 S/cm, electron mobility of 200,000 cm^2/V s, superior thermal conductivity of 5300 W/mK and high Young's modulus of 1 TPa [27]. Graphene is a zero band-gap material with an overlap of the valence and conduction band and the charge carriers act as massless relativistic particles giving rise to a host of interesting properties such as anomalous quantum Hall effect [28]. To exploit these properties for energy storage applications, various synthesis routes have been adopted to prepare graphene and its derivatives that include mechanical cleavage of graphite [26], unzipping of CNT [29], chemical vapor deposition (CVD) [30], epitaxial growth on SiC [31] and chemical exfoliation of graphite [32]. Graphene obtained by mechanical exfoliation and CVD method is of high quality, but the yield is very low. In particular, chemical exfoliation of graphite is considered as a desirable alternative for large-scale production of graphene. In this method, graphite is first oxidized according to Hummers' method or the modified Hummers' method to graphite oxide followed by exfoliation in suitable solvents to obtain a stable suspension of few-layered sheets termed as graphene oxide (GO) [33]. Then, reduction of GO is performed using different reducing agents such as hydrazine hydrate [34], sodium borohydride [35], vitamin C [36], hydroiodic acid [37] etc. to produce reduced graphene oxide (ReGO). ReGO is chemically derived bi-layered or multi-layered graphene, which consists of many oxygen functionalities in its structure that provide opportunities for tuning its electrical and optical properties [38]. The very large electrochemically active surface of ReGO with functional groups and flexibility makes it a suitable candidate for energy storage and can greatly influence the electrochemical performance. Moreover, ReGO can significantly improve the conductivity of the nanocomposite, which is beneficial for superior electrochemical performance.

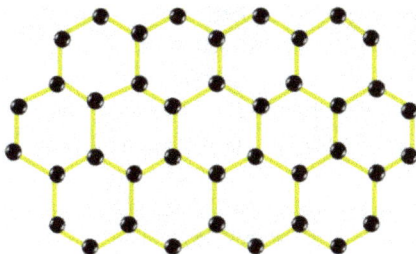

Fig. 4. Graphene: a 2D single sheet of carbon.

The applicability of ReGO as electrodes was first reported by Ruoff et al. in 2008, which delivered a specific capacitance of 135 F/g and 99 F/g in aqueous and organic electrolyte, respectively [39]. However, the experimentally obtained specific capacitance and surface area of ReGO was much lower in comparison to the large theoretical surface area. ReGO prepared by hydrothermal method also exhibited a very low SSA and specific capacitance of 14.9 m^2/g and 66 F/g, respectively in aqueous electrolyte [40]. This is due to the restacking or agglomeration of ReGO sheets after chemical reduction owing to the van der Waals interaction ($\approx$5.9 kJ/mol) between them leading to reduced surface area, which significantly hampers their application as electrodes. To prevent the re-stacking of ReGO sheets as well as to exploit its unique properties, different materials like conducting polymers [41], metal oxides [42], metals [43] etc. have been hybridized with ReGO or GO sheets to form nanocomposites, which act as effective spacers between the sheets. The advantage with GO is that it is hydrophilic due to the presence of polar oxygen functionalities like epoxide, hydroxyl, carboxylic and carbonyl groups in its basal planes and edges, which improves its dispersion in various solvents including water and eases its processability [44]. Till date, ReGO based nanocomposites have been extensively explored for supercapacitor applications and have demonstrated improved electrochemical performance [45-47].

4. Defect generation techniques in graphene-based materials

Considerable efforts have been devoted to increase the capacitive performance of graphene by combining with conducting polymers, metal oxides etc. [48, 49]. However, the capacitance values are found to be low for potential applications because not all the surface area of the electrode material could be accessed by the electrolyte ions and therefore the experimental value of specific capacitance was found to be lower than the theoretical value. This problem could be addressed by modifying the electrode surface by introducing appropriate amount of defects in graphene. Defects have a significant recognition in graphene. Different properties such as mechanical, morphological, electrical, magnetic and electrochemical properties of graphene-based materials can be tailored through defect engineering [50-53]. Defected graphene consisting of nanopores can be useful for DNA sequencing [54] or molecular sieves [55], which will allow only small molecules to pass through it. Defects can open the band gap in graphene [56]. Introduction of appropriate amount of defects has been reported to improve the capacitance of supercapacitor electrodes [57]. The electrochemical properties of a supercapacitor are significantly controlled by the microstructure, defect density, porosity and SSA of the electrode. A developed SSA with well-distributed pores or defects can promote the penetration of electrolyte ions to the bulk of the electrode and facilitate charge transport through interconnected network thereby improving the electrochemical performance. However, formation of well-distributed defects or pores in graphene with a tunable size and density is a challenge. Till date, different techniques have been developed to introduce defects in graphene-based materials, which are described below.

4.1 Doping by heterogeneous atoms

Doping graphene by heterogeneous atoms has drawn remarkable attention because it provides a way to tune and improve their structural and chemical properties that can broaden their potential application for supercapacitors. In doping, a carbon atom is replaced by a heteroatom in the hexagonal graphitic lattice that can be achieved by various synthesis procedures. Doping graphene and graphene-based derivatives (ReGO, GO etc.) by atoms such as nitrogen (N), boron (B), phosphorus (P) and sulfur (S) generates charged or active sites on the graphene surface, which offers possibilities to modify their energy storage properties. In supercapacitors that operate by charging and discharging, substitutional doping can enhance their surface properties and therefore has been widely explored. Studying the electrochemical performance of doped graphene-based materials is a trending research area for which the investigations carried out are usually a combination of experimental analysis and theoretical simulations. Graphene-based derivatives such as ReGO, GO etc. consist of a band gap, which lowers their performance as supercapacitor electrodes. Doping in ReGO and GO is advantageous as it can reduce their band gap and improve their electrical properties. N doping is easier as compared to other elemental doping and has received great attention because it can modify the local electronic structure and improve the device performance. The modified local electronic structure increases the binding of N-doped graphene with electrolyte ions, which is beneficial for high-performance supercapacitors. N-doping can increase the density of states (DOS) giving rise to increase in net specific capacitance, power and energy densities. N-doping has the ability to create holes on graphene nanosheets that increase electrolyte penetration and charge transfer rate [53]. Various chemical routes have been followed for the synthesis of N-doped graphene, which includes chemical vapour deposition (CVD), chemical synthesis, arc discharge, electrochemical reaction etc. [58-60]. N-doping by CVD method gives rise to three different localized sites of N atoms: graphitic N, pyrrolic N sharing bond with two C atoms and contributing two π electrons to the π system and pyridinic N bonded with two C atoms and donating one p electron to the aromatic π system, respectively [61]. The schematic configurations are shown in Fig. 5. Synthesis of N-doped graphene by arc discharge results a large yield of sample. However, the process requires controlled purification procedure as large amount of unwanted products are released during the process. Chemical synthesis of N-doped graphene is much preferred due to its easy route, low cost and massive production of sample. Moreover, the bonding configurations of nitrogen can be tailored in the synthesis process. Crumpled N-doped graphene have been synthesized by chemical route, which exhibited a pore volume of 3.42 cm^3/g, C_{sp} of 226 F/g at 10 A/g current density and electrochemical reversibility of 83% after 5000 cycles [62]. Porous N-doped graphene aerogels synthesized by hydrothermal reaction demonstrated a SSA of 510 m^2/g, C_{sp} of 176 F/g at 10 A/g current density and cycle life of 92% after 2000 cycles [63]. Haque et al. reported N-doped graphene prepared by thermal treatment that delivered a C_{sp} of 210 F/g at 1 A/g current density and cycle life of 90% after 5000 cycles [64]. Various reports on synthesis procedures and electrochemical performances of graphene materials doped by N are presented in Table 1.

Table 1. *Synthesis procedures and electrochemical performances of different N-doped graphene materials*

Materials	N content [%]	Specific Capacitance [F g^{-1}]	Scan rate/Current density	Cycle Life	Electrolyte	Reference
N-doped graphene	10	226	10 A/g	83% after 5000 cycles	1 M [Bu$_4$N]BF$_4$ acetonitrile	[62]
N-doped graphene aerogels	8.4	176	10 A/g	92% after 2000 cycles	1 M H$_2$SO$_4$	[63]
N-doped graphene	6	210	1 A/g	90% after 5000 cycles	0.5 M H$_2$SO$_4$	[64]
N-doped graphene	2.64	324	0.1 A/g	--	6 M KOH	[65]
N-doped graphene	3.97	246	1 A/g	89% after 2000 cycles	6 M KOH	[66]
N-doped graphene	4.5	459	1 mA/cm^2	--	1 M H$_2$SO$_4$	[67]
N-doped graphene	0.48	280	5 mV/s	99% after 40000 cycles	1 M H$_2$SO$_4$	[68]
N-doped graphene	10.85	313	0.1 A/g	85% after 3000 cycles	6 M KOH	[69]
N-doped graphene	6.85	242	1 A/g	97% after 5000 cycles	1 M H$_2$SO$_4$	[70]
N-doped graphene	15.2	348	5 mV/s	97% after 1000 cycles	1 M H$_2$SO$_4$	[71]

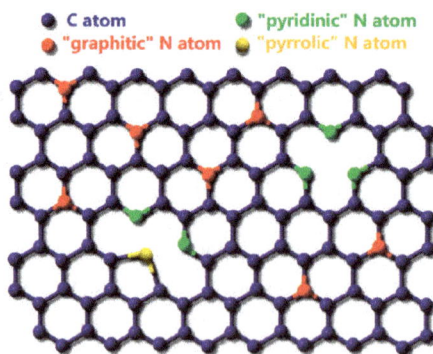

Fig. 5. Schematic of different configurations of N-doped atoms in graphene. [Reprinted with permission from Ref. [61]. Copyright (2009) American Chemical Society].

B-doping is harder as compared to N-doping and is comparatively less studied. Boron containing an electron less than carbon can create a defect by bonding with carbon atoms in the graphene lattice and therefore creates asymmetric charge distribution, which can modify the electrochemical properties. B serves as a p-type dopant in graphene and binds by sp^2 hybridization with the carbon atoms. The stability of substitutional B-doping in the same plane is more than out-of-plane bonding and hence the electrochemical performance of the resulting material depends on the method of B-doping. Numerous synthesis strategies have been carried out to dope B in graphene including CVD, arc discharge, thermal and solvothermal treatments etc. [72,59]. B-doping in graphene achieved via thermal treatment introduced a boron content of 4.7%, which delivered a C_{sp} of 172 F/g at 0.5 A/g current density and cyclic life of 96% after 5000 cycles [73]. The capacitance of B-doped graphene was enhanced by 80% than that of pristine graphene, which is attributed to the presence of oxygen functional groups in the form of BC_2O/BCO_2 bonding configurations after boron doping. In another work, B-doped graphene with porous morphologies were synthesized using 'fried-ice' concept [74]. The boron content and the SSA of the porous sample could be tailored by controlling the reaction temperature. The B-doped porous graphene exhibited a SSA of 622 m^2/g and C_{sp} of 281 F/g at 1 A/g current density. The doped B atoms as well as the oxygen functional groups contribute to pseudocapacitance, which enhances the capacitive performance of the material. Han et al. achieved 1 atom% B-doping in graphene nanoplatelets via solution processing and obtained a C_{sp} of 200 F/g along with a cycle life of 95% after 4500 cycles [75]. This result indicates that even a minimal B-doping in amounts of 1 atom% could enhance the specific capacitance. However, it is still questionable and unclear as to why the capacitive properties increase upon introducing a heteroatom into the graphene lattice. A study by Ambrosi et al. states that the increase in capacitance is not due to doping, but due to the structural defects induced due to doping into the graphitic lattice [76]. More in-depth investigations are therefore required to have a detailed understanding of the relation between doping and capacitive performance. Various reports on synthesis procedures and electrochemical performances of B-doped, S-doped and P-doped graphene materials are summarized in Table 2.

Table 2. *Synthesis procedures and electrochemical performances of different B-doped, S-doped and P-doped graphene materials*

Materials	B content [%]	Specific Capacitance [F g^{-1}]	Scan rate/Current density	Cycle Life	Electrolyte	Refer ence
B-doped graphene	4.7	172	0.5 A/g	96% after 5000 cycles	6 M KOH	[73]
B-doped graphene	3	281	1 A/g	100% after 4000 cycles	2 M H$_2$SO$_4$	[74]
B-doped graphene	1	200	0.1 A/g	95% after 4500 cycles	6 M KOH	[75]
B-doped graphene	2.56	113	1 A/g	--	0.5 M H$_2$SO$_4$	[82]
B-doped graphene	5	16.5 mF/cm^2	0.05 mA/cm^2	90% after 12000 cycles	poly(vinyl alcohol) + H$_2$SO$_4$	[83]
B-doped graphene	6	448	10 mV/s	100% after 3000 cycles	6 M KOH	[84]
S-doped graphene	3.47	320	3 A/g	92% after 1500 cycles	3 M KOH	[85]
S-doped ReGO	2	343	0.2 A/g	96% after 1000 cycles	2 M KOH	[79]
P-doped graphene	1.3	115	0.5 A/g	97% after 5000 cycles	1 M H$_2$SO$_4$	[80]
P-doped graphene	--	367	5 mV/s	96% after 5000 cycles	1 M H$_2$SO$_4$	[81]

Phosphorus and sulphur doping in graphene for supercapacitor applications is relatively less explored and represents an expanding research field in materials science. The concept of doping carbons with sulphur atoms through covalent bonding developed only a few years ago in 2011 [77]. Upon doping carbon atoms with sulphur, the charge of the neighbouring carbon atom changes to positive. Reports on the capacitive performance of S-doped carbon systems suggest that the existence of oxidized sulphur groups such as sulfones and sulfoxides play a role in enhancing the specific capacitance [78]. Sulphur-doped porous reduced graphene oxide hollow nanosphere frameworks have been synthesized by annealing graphene oxide incorporated SiO$_2$ nanoparticles with dibenzyl disulfideand etching the obtained product with hydrofluoric acid [79]. With S content of 2% in the graphene lattice, the material exhibited a SSA of 496 m^2/g, C_{sp} of 343 F/g at 0.2 A/g current density and a coulombic efficiency of 100%. This improvement is attributed to the porous hollow structure of the S-doped ReGO nanospheres, which facilitate the ion diffusion and transfer of charges across electrode/electrolyte interface. Phosphorus doping in graphitic lattice can induce structural deformation by transforming sp^2

hybridized C atoms to sp^3 hybridized state, which creates a defect-induced active surface. Few reports are available on P-doping of graphene for supercapacitor electrodes that demonstrates stable capacitive performance. P-doping level of 1.3 atomic% in graphene achieved by annealing treatment delivered a higher C_{sp} of 115 F/g than that of 29 F/g for undoped graphene at a current density of 0.5 A/g [80]. In another report, phosphorus doping was reported to induce micropores and functional groups in ReGO, which improved the electrochemical performance of the material [81]. The defect density was found to increase upon doping, which is reported to facilitate trapping of ions from the electrolyte.

4.2   Chemical functionalization

Chemical functionalization of graphene is a popular method for introducing defect sites in graphene. Functionalization can be divided into two categories: covalent and non-covalent. Covalent functionalization is achieved through covalent bonding between the guest functional group and atoms in graphene, whereas in non-covalent functionalization, the guest functional group and the graphene atoms are bonded via π-bonds i.e. by physical interaction. Functionalization through both the methods result in change in properties of graphene, however covalent functionalization appears to be more effective. Various methods have been implemented so far for incorporation of different functional groups in graphene based materials, which includes diazonium coupling, amidation, esterization, substitution etc. Such kind of functionalization is reported to change the electrical properties of graphene by inducing defects in its structure, which alters the electrochemical performance. Covalent functionalization can be achieved chemically or electrochemically that occurs at an atomic or molecular range. Various changes in electrical, thermal, mechanical and electrochemical properties of graphene based materials have been reported through covalent functionalization [86-88]. Mishra et al. have reported graphene functionalization with polyaniline and metal oxide nanoparticles (RuO_2, TiO_2, Fe_3O_4) by chemical method [89]. An increase in capacitance was obtained after functionalization which is attributed to the improved pseudocapacitance due to the increase in polar oxygen functionalities and modification of pores in graphene. Functionalization of graphene-polyaniline nanocomposite by p-aniline delivered a C_{sp} of 422 F/g, which was much greater than the capacitance value without functionalization [90]. The increased electrochemical performance of the functionalized nanocomposite is ascribed to its planar configuration with rich pores and contribution of pseudocapacitance owing to the conjugated bonding of p-aniline. Song et al. synthesized a series of diamine/triamine molecules functionalized graphene [Fig. 6] by hydrothermal route and studied their capacitive properties [91]. An enlarged interlayer spacing and SSA was obtained upon functionalization, which yielded a high specific capacitance with less than

10% capacitance decay during 10,000 charge/discharge cycles. In another work, graphene was functionalized with three phenylene diamine (PD) monomers, viz., o-phenylene diamine (OPD), m-phenylene diamine (MPD) and p-phenylene diamine (PPD) [92]. The PPD and MPD functionalized graphene network resulted in enlarged interlayer spacing and superior electrochemical performance than that of OPD functionalized graphene.

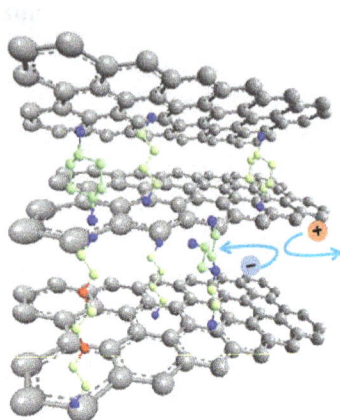

Fig. 6. Schematic diagram of graphene networks functionalized with diamine/triamine molecules. [Reprinted with permission from Ref. [91]. Copyright (2017) Elsevier]

4.3 Plasma irradiation

Plasma irradiation is one of the promising approaches to generate defects in graphene sheet with a controlled density by tailoring the plasma conditions. Plasma based irradiation is an efficient way to produce defected graphene due to its advantages like low energy intake, simple operation, high efficiency, eco-friendly nature and large-scale synthesis. Several reports have shown the formation and tailoring of pores in graphene with the help of plasma irradiation. Homogeneous defects were created in graphene by oxygen plasma irradiation, which could be controlled from a few to 10^3 μm^{-2} by tuning the power, sample temperature and time of plasma irradiation [93]. Point defects were mainly created that were further enlarged into nanopores by hydrogen plasma anisotropic etching. Various nanopores with well-defined pore size of a few nm or above could be achieved by tuning the etching time durations out of which the smallest achieved nanopores were of ≈2 nm size. Topological defects of different concentrations were introduced in graphene by Ar^+ plasma irradiation and the effect of defect concentration on the electric double layer capacitance was studied by controlling the plasma exposure time

Materials Research Forum LLC

https://doi.org/10.21741/9781644900550-8

[94]. The defect concentration was enhanced upto $3\pm0.4\times 10^{11}$cm^{-2}, which was around 30 fold increase as compared to pristine graphene. The topological defects were found to improve the density of states, which significantly affect the quantum capacitance and Helmholtz capacitance resulting in the modification of EDL capacitance. First-principles calculations and the tight-binding method estimated that the most probable defect formed during Ar$^+$ plasma irradiation is the Stone-Wales type topological defect, which resulted in the enhancement of quantum capacitance. Jingyi Zhu et al. reported Ar$^+$ plasma etching to induce defects in few layer graphene structures [95]. The defects were quantified in terms of the normalized intensity ratio (I_D /I_G) of the D-band of graphene at 1350 cm^{-1} to the graphitic or G-band at 1580 cm^{-1}. An increase in I_D/I_G ratio has been reported with the increase in plasma power. The specific capacitance increased in defected graphene, which is attributed to defect induced increase in the density of states. Jeong et al. reported nitrogen plasma irradiation of graphene to produce N-doped graphene for high-performance ultracapacitors [96]. Raman spectra of N-doped graphene exhibits a higher D/G ratio than that of reduced GO indicating the higher concentration of defects generated during the plasma treatment. Around four fold increase in capacitance of doped graphene has been observed as compared to that of pristine graphene based counterparts upon plasma irradiation. A C_{sp} of 280 F/g with high cycle life of 99% after 10000 cycles has been obtained for the plasma modified electrode. It has been reported that increase in capacitance likely resulted from the N-doped sites at basal planes, as investigated by the XPS as well as the ionic binding energy calculations.

4.4 Swift heavy ion irradiation

A unique route to tailor the electrode surface and introduce defects in a controlled way is swift heavy ion (SHI) irradiation. Irradiation with energetic ions can produce cylindrical disordered zones along each ion impact site, which provides a way to induce randomly distributed defects on the electrode surface that influence the electrochemical performance. Different SHI beams with varied energies and fluences has been utilized by researchers to tune the structure, morphology, optical, electrical and sensing properties of materials for a variety of applications [97-99]. The modifications takes place due to varied energy deposition processes of the incident ions traversing through the material. When ions traverse through a solid material, it loses energy by two independent processes during their traversal [100] as described below:

Nuclear energy loss (S_n): When the incident ions lose their energy by elastic collisions with the nuclei of the target material leading to atomic displacement, it is known as nuclear energy loss. The displacement occurs as soon as the energy transferred by the colliding particle is greater than certain displacement threshold energy (E_d) of the target atom. Ions with energy typically of few tens of keV/amu undergo nuclear energy loss [101]. It is derived by taking into consideration the momentum transferred from the incident ion to the target atom and the inner atomic potential between the charged ion and the target atom.

Electronic energy loss (S_e): In electronic energy loss, the penetrating ions lose their energy by inelastic collisions with the atoms leading to electronic excitation and ionization. This energy loss is dominant at high energies of 1 MeV/nucleon or more, where the velocity of the impinging heavy ions becomes equal to or greater than the Bohr velocity of the electron [98]. Electronic energy loss takes place by two mechanisms. One mechanism is glancing collision, which arises by distant resonant collisions with small momentum transfer and low energy loss (< 100 eV) and the other one is knock-on collision where close collisions occur with large momentum transfer and high energy loss (> 100 eV) [102]. In both the collisions, the energy transfer occurs through electronic excitation and ionization of the target material. While glancing collisions are quite frequent, knock-on collisions are very infrequent. For films sufficiently thinner than that of the stopping range of the incident ions, the electronic energy deposition is reasonably uniform throughout the film thickness.

The magnitudes of both the electronic and nuclear energy losses are decided by the incident ion mass and its energy. Heavy ions with energies so high that move at a velocity comparable to the Bohr velocity of the electron are referred to as swift heavy ions (SHI) [100]. For swift heavy ions, the electronic energy loss is dominant, where elastic collisions are insignificant. Materials modification by low energy ions (tens of keV to a few MeV) occurs through ion beam assisted deposition, implantation, doping and new phase formation, where the incident ions get embedded in the material and cause modification due to the collision cascade produced by the elastic collisions. Whereas in SHI irradiation (few tens of MeV and higher), the ions do not get implanted and pass through the film owing to their large range (few tens of μm or more), where the modifications are caused by electronic energy deposition in the material during their passage [101].

4.5   Other techniques

In addition to the above mentioned techniques, various methods such as electron beam irradiation, chemical activation, microwave irradiation have been developed to introduce defects in graphene for supercapacitor property enhancement. Electron beam irradiation is reported to reduce graphene oxide as well as to tailor the pore structure of reduced graphene oxide (ReGO). Controlled defects were introduced in ReGO by varying the dose of electron beam irradiation. Higher irradiation dose reduced the oxygen content,

Materials Research Forum LLC
https://doi.org/10.21741/9781644900550-8

increased the specific surface area and the number of micropores of ReGO, which are beneficial for supercapacitors. The electrochemical performance of ReGO based electrodes increased upon an irradiation dose of 50 to 200 kGy from 156 to 206 F/g at 0.2 A/g current density, but decreased with further increase in fluence [103]. Activation of graphene quantum dots (GQDs) with KOH has been reported to create pores and activated edges at the free ends by Hassan et al. Intensified D/G ratio was obtained from the Raman spectra. The KOH activation creates nanoscale pores on the surface of graphene sheets, which anchored ionic charges resulting in improved electrochemical performance. The activated GQDs exhibited about 6 factor enhancements in BET surface area and two factor improvements in specific capacitance than that of the non-activated GQDs [104]. Microwave irradiation have been used to generate defects on graphene nanoplatelets followed by anchoring of palladium nanoparticles on these defects, and subsequent growth of carbon nanotubes using an ionic liquid to obtain 3D graphene-nanotube-palladium nanostructures [Fig. 7] [105]. The defects such as dislocations and point defects are inferred to initiate the growth of palladium nanoparticles by acting as nucleation sites followed by the vertical growth of carbon nanotubes (CNTs) induced by palladium nanoparticles on graphene sheets. Direct bonding between graphene and CNTs has been obtained through the defect-based growth mechanism. The vertically arranged CNTs on graphene nanosheets improves its SSA as well as prevents its restacking by acting as spacers between them, which facilitates electrolyte ion diffusion paths resulting in improved electrochemical properties of the nanostructures. A C_{sp} of 700 F/g has been achieved at 100 mV/s scan rate along with a capacitive retention of more than 100% upto 1000 cycles.

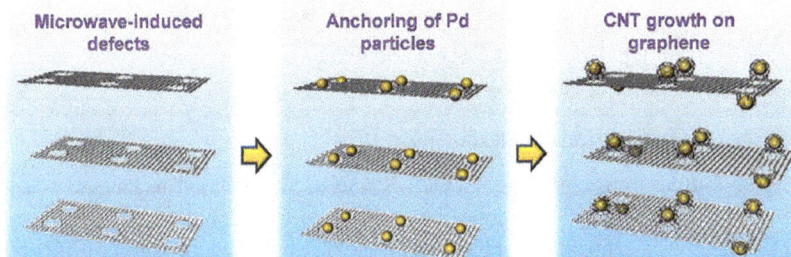

Fig. 7. Schematic mechanism of carbon nanotubes grown vertically on graphene sheets. [Reprinted with permission from Ref. [105]. Copyright (2012) American Chemical Society].

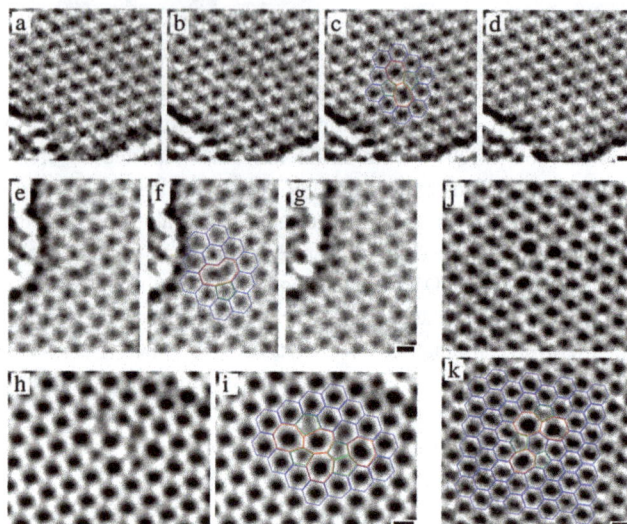

Fig. 8. (a) pristine lattice, (b, c) Stone Wales defect and atomic configuration, (d) disappearance of SW defect after 4 s, (e, f) vacancy defect and atomic configuration; green color represents pentagon. (g) disappearance of vacancies after 4 s, (h, i) Defect containing four pentagons-heptagons and atomic configuration (red indicates heptagons), (j, k) Defect containing three pentagons-heptagons and atomic configuration. [Reprinted with permission from Ref. [106]. Copyright (2008) American Chemical Society]

5.   Techniques to investigate the nature of defects in materials

The nature and type of defects in materials can be investigated by a number of techniques, which are discussed below:

5.1   Transmission electron microscopy

Transmission electron microscopy (TEM) is one of widely used techniques to obtain information of defects on an atomic scale. Defects in materials can be categorised as point defects, dislocations and grain boundaries. The presence of defects in a material results in TEM image with detectable contrasts, which have to be analysed to identify the nature of defects. High resolution transmission electron microscopy (HRTEM) provides images with high accuracy of about 0.01 nm, which allows direct detection of defect formation. Chen et al. studied the nature of defects created by Ar^+ plasma treated graphene sheet by TEM and observed the formation of heptagon-pentagon defect pairs in the graphene structure besides a few octagons, which is known as Stone-Wales defect [94]. Direct imaging of defects on an atomic scale by HRTEM in single layer graphene membranes have been demonstrated by Meyer et al. [106]. Individual carbon atoms in the graphene lattice with a contrast of about 6% were detected by adjusting the HRTEM parameters and the signal to noise ratio. A single Stone-Wales defect was observed during 1st electron beam exposure of 1 s, which disappeared in the next exposure after 4 s (Fig. 8a-8d). Defect structure consisting of several five- and seven-membered carbon rings were then generated, which remained stable up to an electron beam exposure of 20 s. All defect structures were observed to disappear with increased exposure and consisted of equal number of pentagon-heptagon configuration, which does not contain any dislocation type of defect. In addition to these defects, reconstructed vacancy configuration involving a pentagon was observed (Fig. 8e-8f), which also disappeared after a few seconds. In-situ defect formation in single graphene layer has been studied by HRTEM, where topological defects could be seen clearly in a plan view of graphene layer [107]. A missing row of zig-zag carbon chain was clearly observed in the graphene network after several tens of seconds of irradiation with an electron beam.

5.2   Raman spectroscopy

Raman spectroscopy is a significant tool for investigating the defect density of graphene sheets. The defect density can be analyzed in terms of the intensity ratio of the D and G bands of graphene that appear at 1355 cm^{-1} and 1595 cm^{-1}, respectively. The origin of these bands is due to the different phonon scattering processes occurring at different symmetry points in the first Brillouin zone of graphene. The G band appears due to in-plane vibration of sp^2 carbon atoms and the D band originates from a ring breathing mode of sp^2 atoms. The D Band known as the disorder band or the defect band becomes active only when the carbon ring is adjacent to a graphene edge or defect and its appearance indicates the presence of defects or disordered carbon in the graphene material [108]. The intensity of the D band is directly proportional to the level of defects in the graphene layer. The ratio of intensities of the D and G peaks (I_D/I_G) is therefore termed as the disorder parameter and is used to investigate defect density in graphene. Proper quantification of defect density in graphitic materials by using Raman spectroscopy has been achieved through the evaluation of the size (L_a) of in-plane crystallites formed by a number of carbon rings and this has been linked to the disorder parameter (I_D/I_G). The relation between L_a and (I_D/I_G) is given by Eq. 3 [109].

$$L_a = (2.4 \times 10^{-10})\lambda_{laser}^4 \left(\frac{I_D}{I_G}\right)^{-1} \tag{3}$$

where λ_{laser} is the laser excitation wavelength. Another band known as D' band appears for graphene at 1621 cm^{-1} due to the presence of defects and arises via an intra-valley double resonance transition. The ratio of peak intensities $I_D/I_{D'}$ can be used to characterize the nature of defects present in graphene layer. A maximum value of $I_D/I_{D'}$ (~13) indicates defects associated with sp^3 hybridization and a minimum value (~3.5) represents boundaries. The $I_D/I_{D'}$ value close to 7 suggests the presence of vacancy-like defects. A study by Jeng et al. investigates various types of defects present in graphene by different values of $I_D/I_{D'}$ [110]. A detailed study of (I_D/I_G) and ($I_{D'}/I_G$) ratios with increase in defect density in graphene sheets have been demonstrated by Zhou et al. [111]. It was observed that the D band appeared immediately after the generation of defects, the intensity of which increased with increasing amount of defects. With increase in defect density, the intensity ratio of D peak to G peak (I_D/I_G) increase first and then decrease. However, the intensity ratio of D' peak to G peak ($I_D/I_{D'}$) increased first and then saturated with increasing amount of defects.

5.3   Atomic force microscopy

Atomic force microscopy (AFM) provides non-destructive atomic-resolution imaging of sample surfaces by which the topographical structure of atomic scale defects can be studied. Defects could be observed as dark contrasts with respect to the background in an AFM image, which is due to the missing atoms from the surfaces. The distance between two point defects and the position change of atomic defects can also be studied using AFM. A quantitative study of defects can be done by calculating the power spectra of the AFM maps, which gives a measure of the root mean square roughness of the films. The power spectra could be calculated by performing fast Fourier transform of each map and plotting the square of amplitude with respect to the spatial frequency. A study of graphene morphology with increasing defect density by AFM shows changes in surface corrugation and root mean square roughness of graphene [109]. Folding characteristics of functionalized graphene sheets (FGS) have been studied by AFM, where flat monolayer sheets are changed into a folded configuration with the application of lateral forces applied by AFM tip on the sample [112]. The sheets are observed to fold and unfold reversibly several times and the folding is seen to occur always at the same position. The conclusion drawn was that the folding and bending characteristics of graphene sheets are dominated by pre-existing fault lines consisting of functional groups and/or defects and the wrinkles present in the FGS might exhibit such preferred defect structure. AFM is also reported to distinguish between unreduced and reduced GO nanosheets [113]. The difference was investigated in terms of phase values, line profiles and root mean square roughness of the sheets. The measured phase of the unreduced sheets was higher as compared to that of the substrate, while the reduced graphene sheets exhibited same phase as that of the substrate. The root mean square roughness of the unreduced sheets was found to be slightly larger (0.108 nm) than that of the reduced nanosheets (0.095 nm).

5.4   X-ray diffraction

X-ray diffraction (XRD) is a commonly used non-destructive technique to obtain information about the crystal structure, phase composition, strain and crystallinity of materials. The presence of defects can be analyzed quantitatively using XRD by determination of crystallinity, strain and crystallite size of the material. The crystallinity percentage can be determined in terms of X-ray scattering from both the crystalline and amorphous phases in the material. The difference between scattering from crystalline and amorphous phases depends on the ordering of the material. The presence of defects may result in lower ordering and increased amorphicity of the material, which reduces the crystallinity. Crystallinity percentage from XRD patterns is obtained by calculating the ratio of the areas of the crystalline peaks to the total area under the diffractogram using Eq. 4.

$$X_C = \frac{A_c}{A_c + A_a} \times 100 \% \tag{4}$$

where, A_c represents total area of the crystalline region, A_a is the area of the amorphous region and $(A_c + A_a)$ represents the total area under the XRD diffractogram. The degree of disorder of a material increases with increase in defect density, which can be investigated in terms of crystallite or domain size and strain. The crystallite size (D) decreases with the growth of defects, while the strain (ε) of the material increases with increasing defect density, which can be evaluated from Eq. 5 and Eq. 6 [114].

$$D = \frac{\lambda}{\beta_C^f Cos\theta} \tag{5}$$

$$\varepsilon = \frac{\beta_G^f}{4\,tan\theta} \tag{6}$$

Here β_C and β_G corresponds to the Cauchy and Gaussian components of integral breadth of single Bragg peak. The crystallite size as evaluated from XRD for nitrogen doped graphene has been found to decrease than that of pristine graphene due to formation of defects during N-doping [115].

5.5   Positron annihilation spectroscopy

Positron annihilation spectroscopy (PAS) is one of unique techniques for the detection of vacancy defects and free voids in a material at the nanometre level. It is a non-destructive spectroscopy that makes use of positrons to identify material defects, voids or free volume and can provide quantitative information about vacancy defects in the concentration range of 10^{15}-10^{20} cm^{-3}. In this method, the positrons are implanted into the target material and their decay rate is determined. The principle is based on the correlation of the lifetime of the implanted positrons with the pore size of the target material. If a positron is trapped at an empty void or defect, the lifetime of the positron increases due to reduction in the local electron density. Thus longer positron lifetimes will correspond to defects or voids with larger sizes and the number of positrons with longer lifetimes will determine the defect concentration. This measurement provides accurate information about defects as the positron lifetimes are sensitive to even very small free voids. Graphene layers with a thickness of 2-3 nm have been characterized using low energy positron spectroscopy (2 eV-20 keV) and the gamma-ray spectra was analyzed based on W (wing)-parameter at different positron energies [116]. The W-parameter measures the fraction of positrons annihilating with high momentum electrons and its decrease implies the presence of defects in the material. Composites of graphene oxide and polyaniline characterized by PAS showed long positron lifetimes, which suggested the presence of large sized vacancy defects within the material [117]. The positron lifetime component was observed to increase with increasing polyaniline content and increase in reduction of GO. It was attributed that the reduction of GO shranked its volume giving rise to free volume or voids that contributed to the defect concentration thereby increasing the intensity of the positron lifetime component.

6.   SHI Irradiation induced defect engineering in ReGO-PPrNTs nanocomposite system

Defect engineering of reduced graphene oxide-polypyrrole nanotubes (ReGO-PPrNTs) nanocomposite electrodes by swift heavy ion irradiation studied by the authors have been discussed in the following sub-sections. The study has been focused on the defect induced modifications in the morphology, structure and electrochemical properties of the nanocomposites with an aim to improve their potential application as supercapacitor electrodes.

6.1   Synthesis of ReGO-PPrNTs nanocomposites

The nanocomposites have been prepared chemically by in-situ reduction method of graphene oxide (GO) in presence of PPrNTs. The in-situ reduction method was chosen in order to prevent the agglomeration of graphene sheets after reduction. The synthesis of PPrNTs has been carried out by a reactive methyl orange-$FeCl_3$ template method [118]. For nanocomposite preparation, 40 mg of GO was dispersed in deionized water via ultrasonication. Modified Hummers' method was used to synthesize GO [33]. In the dispersed GO solution, 100 mg of pre-synthesized PPrNTs were added followed by stirring and ultrasonication for 3 h. Subsequently, the obtained GO-PPrNTs nanocomposite was mixed with hydrazine hydrate under stirring for 1 h at 95 °C, which reduces the GO to ReGO in the solution. The final product was washed thoroughly with ethanol and deionized water and dried at 60 °C for 12 h. SHI irradiation of the nanocomposite films was carried out at IUAC, New Delhi, India under ultrahigh vacuum ($\approx 10^{-6}$ Torr), where the ions were generated by the 15 UD tandem pelletron accelerator. The films were exposed to C^{6+} ions of 85 MeV energy over an area of 1 cm^2 at varied ion doses of 6×10^{10} (6E10), 3.6×10^{11} (3.6E11), 2.2×10^{12} (2.2E12) and 1.3×10^{13} (1.3E13) ions/cm^2 [119]. For film preparation, the nanocomposite was mixed with nafion binder and carbon black in 85:5:10 ratio to form a paste and casted on glass slides coated with ITO. The energy of the irradiating ions was determined such that the incident C^{6+} ions were not implanted and passed through the films.

6.2 Morphological studies

The modification induced by SHI irradiation of different fluences on the morphology of ReGO-PPrNTs nanocomposite has been studied by SEM and HRTEM and presented in Fig. 9 and 10, respectively. The HRTEM image of the pristine nanocomposite shows both ReGO and PPrNTs indicating formation of the nanocomposite, where the average outer diameter of PPrNTs is measured to be about 134 nm. Upon irradiation, formation and enlargement of pores on the nanocomposite surface are observed. The pore size does not vary greatly up to an irradiation dose of 6E10 ions/cm^2. However for fluences larger than 6E10 ions/cm^2, a porosity distribution from 0.41-2.18 µm is observed, which indicates the increase in pore size. The increase in pore size may be attributed to the huge electronic ionization by SHI in the target material giving rise to the generation of defects and structural fragmentation [120]. A decrease in nanotube diameter is also recorded from Fig. 9(g) at the highest ion fluence of 1.3E13 ions/cm^2 than that of pristine nanocomposite as shown in Fig. 9(f). The average diameter of PPrNTs as calculated from Fig. 10(a) and 10(b) is found to decrease from 134 nm for unirradiated to 117 nm for the nanocomposite exposed to 2.2E12 ions/cm^2 C^{6+} ions. Broken portions of nanotubes are also visible at an ion dose of 2.2E12 ions/cm^2 that indicates SHI induced fragmentation. However, with increase in SHI fluence to 1.3E13 ions/cm^2, severe damage of NTs is observed from HRTEM images though FESEM image displays a porous morphology at this fluence. This is due to the formation of cylindrical disordered zones along the path of

SHI that results in fragmentation of the target material [100]. It appears that PPrNTs are stable enough under SHI irradiation up to an ion dose of 2.2E12 ions/cm^2, after which they are completely damaged.

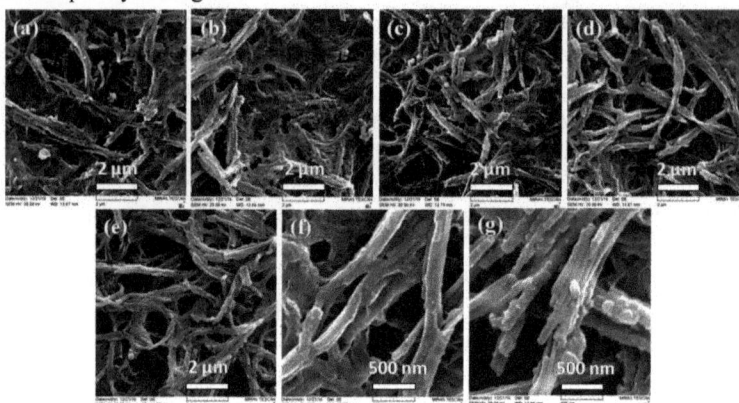

Fig. 9. FESEM micrographs of ReGO-PPrNTs nanocomposites (a) unirradiated, and irradiated with increasing ion dose of (b) 6E10, (c) 3.6E11, (d) 2.2E12, (e) 1.3E13 ions/cm^2 at 25 kx magnification; FESEM micrographs of ReGO-PPrNTs nanocomposites (f) unirradiated, and irradiated with ion dose of (g) 1.3E13 ions/cm^2 at 100 kx magnification. [Reprinted from Ref. [119] with permission from Elsevier]

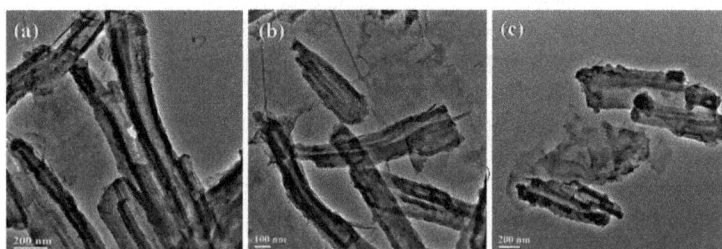

Fig. 10. HRTEM images of ReGO-PPrNTs nanocomposites (a) unirradiated, and irradiated with increasing ion dose of (b) 2.2E12 and (c) 1.3E13 ions/cm^2. [Reprinted from Ref. [119] with permission from Elsevier]

6.3   Structural analysis

Micro-Raman spectra have been recorded to investigate the defect density induced by SHI irradiation at different fluences in ReGO-PPrNTs nanocomposite and are shown in Fig. 11(i). The μ-Raman spectra of the pristine and irradiated nanocomposites depict peaks at 1595, 1355, 982 and 922 cm^{-1}, respectively. The absorption bands of doped states of PPr appear at 982 and 922 cm^{-1}. The peaks of PPrNTs attributed to the ring stretching vibration at 1382 cm^{-1} and π conjugated structure at 1588 cm^{-1} overlap with the D and G peaks of ReGO at 1355 and 1595 cm^{-1}, respectively [118]. The most prominent bands observed in the μ-Raman spectra are the D and G bands. The G band appears due to in-plane vibration of sp^2 C atoms and the D band originates from a ring breathing mode of sp^2 atoms, which is activated only in the presence of structural imperfections and defects in a carbon material [108]. Another peak named as D' band appears at 1621 cm^{-1} at the highest fluence of irradiation, which arises due to the presence of defects via an intra-valley double resonance transition. The ratio of intensities of the D and G bands (I_D/I_G), termed as the disorder parameter, provides information about defects density in a carbon nanostructure and can determine the ion irradiation induced damage in both ReGO and PPrNTs [110,121]. The plot of disorder parameter with respect to the ion dose is presented in Fig. 11(ii). Upon irradiation, the disorder parameter is lowered at an initial ion dose of 6E10 ions/cm^2 from 0.81 to 0.78 than that of the pristine nanocomposite, which indicates less disorder and is a signature of annealing or re-arrangement of carbon atoms at moderate temperatures generated by the electronic energy deposition [122]. The re-arrangement of carbon atoms is corroborated by the HRTEM micrograph that displays no major change in morphology at this fluence. On further increase in fluence, the disorder parameter increases to 0.82, 0.87 and 0.91 at a fluence of 3.6E11, 2.2E12 and 1.3E13 ions/cm^2, respectively. This implies that SHI no longer anneal the defects beyond a certain fluence and results in growth of defect sites in the nanocomposite. This occurs as the electronic energy loss required for defect creation starts dominating over that required for defect annealing in the target material [122]. The passage of SHI through the target material creates a highly ionized cylindrical zone of positive ions within a time period of 10^{-12} s due to the SHI induced excitation. As the mobility of the conduction electrons in the material is low enough to neutralize the charged cylindrical zone within a time interval of 10^{-12} s, it explodes radially under Coulomb repulsive forces [100]. This results in the generation of cylindrical shock waves through the target material leading to the formation of various types of defects such as point, columnar defects and a continuous amorphized zone. Therefore, a maximum value of $I_D/I_G = 0.91$ at the dose of 1.3E13 ions/cm^2 indicates a dramatic increase in defects and structural disorder in the nanocomposite leading to amorphous transformation, which results in the appearance of the defect induced band D' at 1621 cm^{-1}.

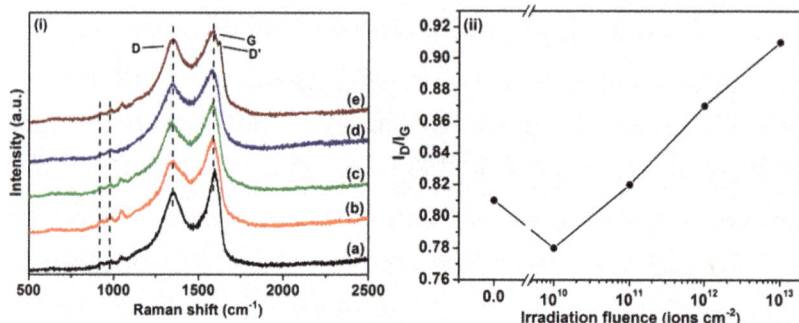

Fig. 11. (i) μ-Raman spectra of (a) unirradiated, and irradiated ReGO-PPrNTs nanocomposites with increasing ion dose of (b) 6E10, (c) 3.6E11, (d) 2.2 E12 and (e) 1.3E13 ions/cm^2; (ii) Plot of disorder parameter (I_D/I_G) versus ion dose. [Reprinted from Ref. [119] with permission from Elsevier]

6.4   Surface properties study

To investigate the change in the specific surface area (SSA) of the nanocomposite before and after SHI irradiation, N_2 adsorption-desorption data have been collected at liquid N_2 temperature of 77 K and the isotherms are shown in Fig. 12. The SSA has been calculated using the Brunauer-Emmett-Teller (BET) method in the relative pressure (P/P_0) region of 0.05-0.35. The BET results show that the unirradiated ReGO-PPrNTs nanocomposite exhibit a SSA of 208.4 m^2/g [Fig. 12(a)]. Upon irradiation, the SSA of the nanocomposite increase to 251.5 and 263.8 m^2 g^{-1} with increase in SHI dose to 2.2E12 and 1.3E13 ions/cm^2 [Fig. 12 (b) and (c)]. The volume of the pores are calculated to be 0.377, 0.491 and 0.531 cm^3 g^{-1} for the unirradiated, 2.2E12 and 1.3E13 ions/cm^2 irradiated nanocomposite, respectively. The porosity variation curves have been estimated by Barrett-Joyner-Halenda (BJH) method and are shown in insets of Fig. 12. It is observed that the pore size of the unirradiated nanocomposite ranges from 1-9 nm, which increases upon irradiation to 4-30 and 6-50 nm for the 2.2E12 and 1.3E13 ions/cm^2 irradiated nanocomposite indicating the presence of micro as well as macropores in the nanocomposite films. The increased SSA and porous structure achieved upon irradiation might be ascribed to the SHI induced cylindrical damage zones in the nanocomposite films as observed from FESEM images.

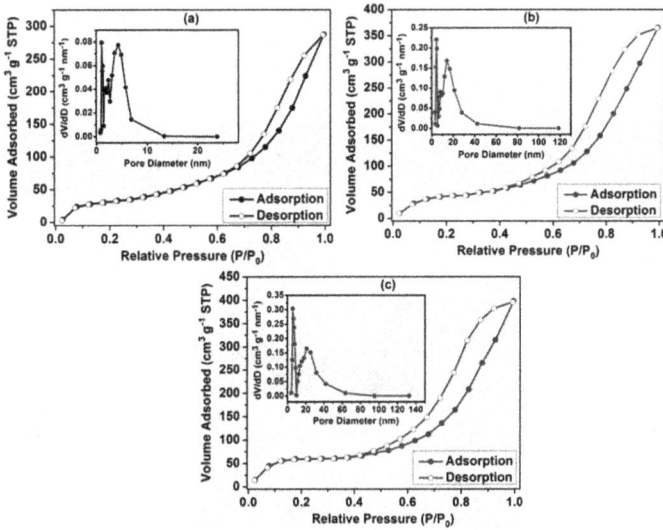

Fig. 12. N_2 adsorption-desorption data of (a) unirradiated, and irradiated ReGO-PPrNTs nanocomposites with increasing ion dose of (b) 2.2E12 and (c) 1.3E13 ions/cm^2. The pore size variation curves are presented in insets. [Reprinted from Ref. [119] with permission from Elsevier]

Surface wettability of electrode is an essential parameter influencing the electrode-electrolyte reactivity. With a view to study the degree of interaction between the electrode surface and an aqueous solvent, water contact angle measurements of the ReGO-PPrNTs nanocomposites before and after exposure to SHI ions at different doses have been conducted and displayed in Fig. 13. The water contact angle of pristine ReGO-PPrNTs nanocomposite is measured to be 83°. After SHI irradiation, a decrease in the water contact angle of the nanocomposites with increasing fluence is observed. The decrease in contact angle of the irradiated nanocomposites than that of pristine nanocomposite implies increase in wettability of the electrode surface upon irradiation. This indicates that SHI irradiation results in an increase in surface wettability of the nanocomposite electrode, which may result from enhanced SSA and pore size of the nanocomposites with increase in SHI dose.

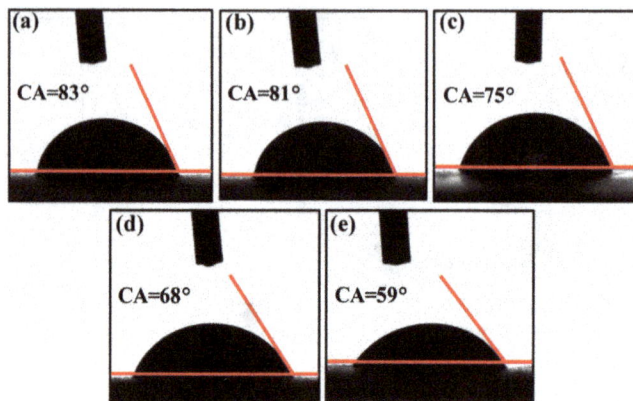

Fig. 13. Water contact angles of (a) unirradiated, and irradiated ReGO-PPrNTs nanocomposites with increasing ion dose of (b) 6E10, (c) 3.6E11, (d) 2.2E12 and (e) 1.3E13 ions/cm².

6.5   Electrochemical properties study

The changes brought by 85 MeV C^{6+} irradiation on the electrochemical performance of ReGO-PPrNTs nanocomposite has been analyzed by CV, GCD and EIS in 1 M KCl electrolyte using a three electrode system. The CV curves of unirradiated and irradiated nanocomposites have been recorded at varied scan rates in the potential window of –0.2-0.8 V and shown in Fig. 14. The cyclic voltammogram of the unirradiated nanocomposite displays a quasi-rectangular shape, which indicates that the charge storage mechanism in the electrode is a contribution of both EDLC and pseudocapacitance. The EDL behavior is exhibited by ReGO and the pseudocapacitance contributed by PPrNTs arises from the rapid and reversible faradaic reactions across the interface [118]. Upon irradiation, it is observed that the current response and integral area of the CV curves increases with ion fluence up to 2.2E12 ions/cm², followed by a drop at the highest ion dose. In addition, modifications in the shape of the cyclic voltammograms could be noted at different fluences of irradiation. Upon irradiation at an ion dose of 6E10 ions/cm², no significant change in the CV shape occurs [Fig. 14(b)]. However, as the fluence increases to 3.6E11 ions/cm², the cyclic voltammogram deviates from quasi-rectangular shape as observed in Fig. 14(c). With further increase in fluence, the non-rectangular shape of CV enlarges and becomes most prominent at the ion dose of 2.2E12 ions/cm² [Fig. 14(d)]. At the highest fluence of 1.3E13 ions/cm², the CV curve again encloses a quasi-rectangular shape [Fig. 14(e)]. The enhanced charge storage behavior upon irradiation is attributed to the increased surface area and pore size of the nanocomposite electrode upon irradiation as evident from BET and BJH measurements. The increase in pore size generates more electrochemical active sites in the electrode that allows enhanced charge storage in the irradiated nanocomposite than that of pristine [123]. The transition of CV curve to non-rectangular shape is due to the enhanced pseudocapacitive contribution from the increased electroactive sites of PPrNTs than that of ReGO since PPrNTs are in higher amount in the nanocomposite as compared to ReGO. Therefore, an enhanced non-rectangular CV area is observed up to an ion dose of 2.2E12 ions/cm^2. However, heavy irradiation of 1.3E13 ions/cm^2 leads to severe damage of PPrNTs and it could no longer contribute to the faradaic redox reactions arising from its active sites resulting in complete decline of pseudocapacitance. Therefore, the CV curve encloses a quasi-rectangular area indicating EDLC as the dominant charge storage mechanism at that fluence. The electrochemical performance of the unirradiated and irradiated nanocomposites has been further studied by GCD measurements at 0.5 A/g current density in the potential window of 0-0.8 V. Fig. 15(a) shows the first charge-discharge responses of ReGO-PPrNTs nanocomposites before and after irradiation with varied ion doses. The specific capacitances (C_{sp}) and the coulombic efficiency η have been calculated from the GCD curves using Eq. 7 and Eq. 8.

$$C_{sp} = \frac{I \times \Delta t_d}{m \times \Delta V} \tag{7}$$

$$\eta = \frac{\Delta t_d}{\Delta t_c} \times 100\% \tag{8}$$

where, ΔV, I, Δt_c, Δt_d and m are the potential and current during discharge, charge time, discharge time and electrode mass, respectively. The GCD curves of the pristine and irradiated nanocomposites reveal a linear and symmetric discharge. It is observed that SHI irradiation results in a longer discharge time of the nanocomposite electrode with increase in SHI dose up to 2.2E12 ions/cm^2 signifying improved charge storage and specific capacitance upon irradiation. The obtained values of C_{sp} and η of the nanocomposites before and after irradiation are presented in Table 3. The C_{sp} is highest upon irradiation at an SHI dose of 2.2E12 ions/cm^2 and its charge-discharge curves are displayed in Fig. 15(b). The improved supercapacitor performance upon irradiation may be ascribed to increased SSA of the nanocomposite, which contributes to the available electroactive sites to be accessible by the electrolyte and facilitates the ion transport throughout the electrode. Increase in pore size also allows the penetration of solvated ions into the bulk of the electrode thereby improving the capacitance. In addition, the presence of defects in a minimal amount induced by irradiation may also lead to improved capacitance. Defects create charged dangling bonds locally that are reactive towards adsorbates and can promote the diffusion of ions from the electrolyte to the electroactive surfaces [124]. However, this seems applicable till an ionic dose of 2.2E12 ions/cm^2, after which the discharge duration and specific capacitance decreases. This could be attributed to the chain scissioning and structural breakdown of PPrNTs that limits its charge transfer reactions at an ionic dose of 1.3E13 ions/cm^2, though an increase in surface area is recorded from BET measurements. The defected structure of ReGO at this fluence as observed from Raman measurements may also degrade the electrochemical performance. The highest specific energy and specific power achieved are 28.51 W h/kg and 192.39 W/kg, respectively for 2.2E12 ions/cm^2 irradiated nanocomposite.

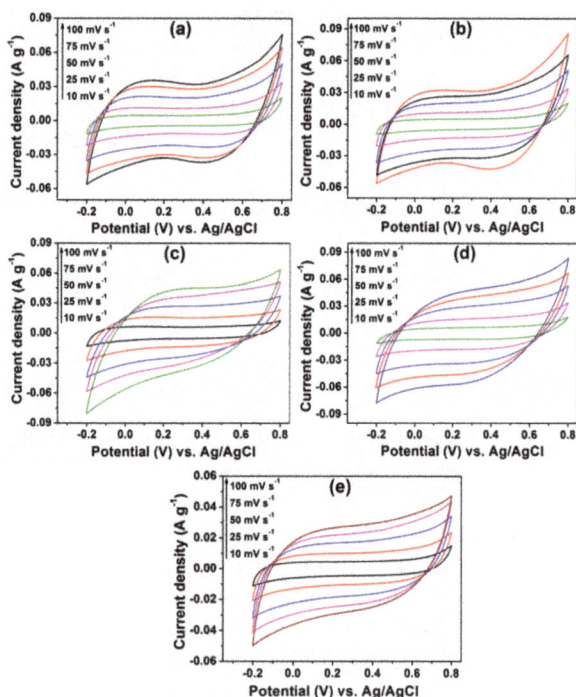

Fig. 14.Cyclic voltammograms of (a) unirradiated, and irradiated ReGO-PPrNTs nanocomposites with increasing ion dose of (b) 6E10, (c) 3.6E11, (d) 2.2E12 and (e) 1.3E13 ions/cm^2. [Reprinted from Ref. [119] with permission from Elsevier].

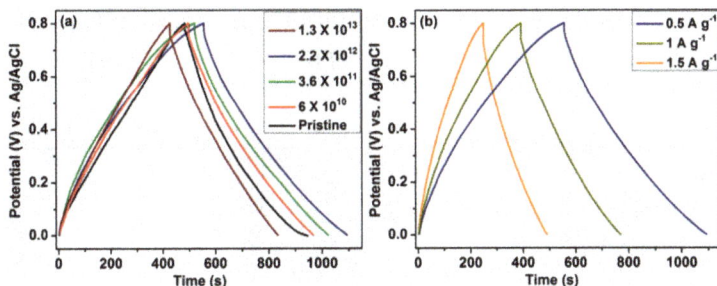

Fig. 15: *(a) GCD curves of ReGO-PPrNTs before and after irradiation with varied SHI doses; (b) Charge-discharge curves of 2.2E12 ions/cm^2 irradiated ReGO-PPrNTs nanocomposite at varied current densities.*

Table 3. *Sp. capacitance (C_{sp}), coulombic efficiency (η), specific energy (E) and specific power (P) of ReGO-PPrNTs nanocomposites before and after irradiation with varied SHI doses*

Sample	C_{sp} [F/g]	E [W h/kg]	P [W/kg]	η
Pristine	299.81	24.68	192.49	96 %
6E10	308.64	25.41	192.35	96 %
3.6E11	325.07	26.76	192.32	96 %
2.2E12	346.28	28.51	192.39	96 %
1.3E13	268.79	20.99	187.22	94 %

The cycle life is a significant parameter for electrodes that are used in supercapacitors. The cycle life of ReGO-PPrNTs nanocomposites before and after exposure to SHI has been evaluated for 1000 cycles by CV measurements at 10 mV s^{-1} scan rate. As presented in Fig. 16(i), the capacitive retention of unirradiated nanocomposite is calculated to be 77% after 1000 consecutive cycles, which increases up to 89% with increase in irradiation fluence. The highest capacitive retention with a negligible capacitance loss of about 8% after 1000 cycles is recorded for the 2.2E12 ions/cm^2 irradiated nanocomposite. Fig. 16(ii) shows the CV curves of the nanocomposite exposed to an ion dose of 2.2E12 ions/cm^2 with increasing cycle number up to 1000 cycles. It is observed that the shape of the CV curve suffers no major change with increasing cycles indicating its improved electrochemical stability. This may be ascribed to the increased surface area and electroactive sites in the irradiated nanocomposites, which increase charge storage and preserves the capacitive performance with increasing cycle number. For highest ionic dose of 1.3E13 ions/cm^2, the cyclic stability drops to 74% due to the high concentration of defect sites and structural degradation of the nanocomposites.

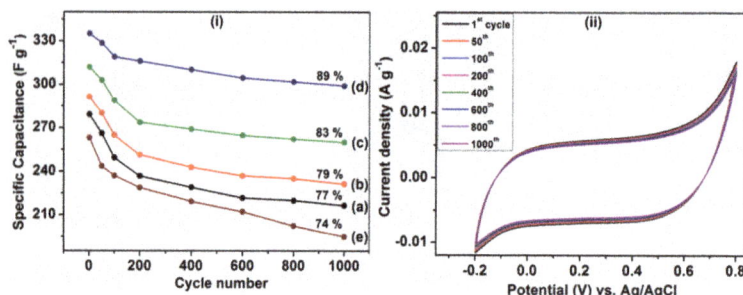

Fig. 16. (i) Plot of sp. capacitance with repeated CV cycle of (a) unirradiated, and irradiated ReGO-PPrNTs nanocomposites with increasing ion dose of (b) 6E10, (c) 3.6E11, (d) 2.E12 and (e) 1.3E13 ions/cm^2; (ii) Cyclic voltammograms of 2.2E12 ions/cm^2 irradiated ReGO-PPrNTs nanocomposite with cycle number.

Conclusions

The growing demand for power generation, hybrid vehicles and portable electronic devices and the increasing threat of global warming due to extensive use of fossil fuels has raised unprecedented requirements for clean, sustainable and secured renewable energy sources. The reliable functioning of renewable energy sources needs efficient and low cost energy storage technologies that can deliver energy efficiently on demand. Supercapacitors are receiving increasing importance as energy storage devices over batteries owing to their high rate capability, excellent cyclic stability and environmental benignity. In the past decade, graphene have received tremendous interest as supercapacitor electrodes due to its outstanding electrical, thermal and mechanical properties. Different materials viz. conducting polymers, metal nanoparticles, metal oxides have been composited with graphene derivatives to enhance their properties. Despite the many advances made, the capacitance values are not high for potential applicability. The electrochemical performance of graphene based electrodes could be modified by introducing appropriate amount of defects. A variety of studies report the modification of mechanical, morphological, electrical, magnetic and electrochemical properties of graphene based materials through defect engineering. Different defect-engineered methods namely heterogeneous atom doping, chemical functionalization,

Materials Research Foundations **64** (2020) 209-256   https://doi.org/10.21741/9781644900550-8

plasma irradiation, swift heavy ion irradiation etc. are found to display unique and improved results that can boost the development of graphene based electrodes for supercapacitor applications. Heterogeneous atom doping is reported to generate active or charged sites on the graphene surface, which offers possibilities to modify their energy storage properties. There are many works that have investigated N-doped graphene for supercapacitors, but a few reports amongst them have focused on the capacitance studies of other hetero-atom doped graphene and therefore deserve attention. Chemical functionalization is another method to introduce defects in graphene and can be achieved by covalent or non-covalent bonding between the guest functional group and atoms in graphene. Defected graphene with a controlled defect density can be achieved through plasma irradiation by tuning its power, irradiation time and sample temperature. Another method to modify the electrode surface and introduce defects in a controlled way is swift heavy ion (SHI) irradiation. Randomly distributed pores could be created in graphene based electrodes by irradiating with energetic ions, which influence the electrochemical performance. The modifications occur owing to the different energy deposition processes of the incident ions penetrating through the solid target sample, namely the nuclear energy loss and electronic energy loss. Electronic energy loss with energy deposition of more than 1 MeV/nucleon plays the dominant role in modifying the target material through swift heavy ion irradiation. In regard to this, SHI irradiation of reduced graphene oxide based polypyrrole nanocomposite electrodes with different fluences has been discussed. The passage of SHI through the target material forms cylindrical disordered zones in its path giving rise to randomly distributed pores with pore size distribution around 0.41-2.18 μm. An increase in disorder parameter (I_D/I_G) obtained from Micro-Raman studies from a SHI dose of 6E10 ions/cm^2 indicate increasing defect sites in the nanocomposites with increasing fluence. The irradiated nanocomposites exhibit improved electrochemical performance as compared to that of pristine due to the presence of pores that generates more electrochemical active sites as well as enables fast diffusion of electrolyte ions across the interface. Defects are generally considered as limiters of a material performance. However, the studies discussed above contradicts the fact and demonstrates that introduction of appropriate amount of defects in graphene could actually improve the energy storage properties.

References

[1] D.G. Nocera, Living healthy on a dying planet, Chem. Soc. Rev. 38 (2009) 13-15. https://doi.org/10.1039/B820660K

[2] K. Kaygusu, Energy for sustainable development: A case of developing countries, Renew. Sust. Energy Rev. 16 (2012) 1116-1126. https://doi.org/10.1016/j.rser.2011.11.013

[3] Y. Zhu, S. Murali, M.D. Stoller, K.J. Ganesh, W. Cai, P.J. Ferreira, A. Pirkle, R.M. Wallace, K.A. Cychosz, M. Thommes, D. Su, E.A. Stach, R.S. Ruoff, Carbon-based supercapacitors produced by activation of graphene, Science 332 (2011) 1537-1541. https://doi.org/10.1126/science.1200770

[4] H. Wang, J. Deng, Y. Chen, F. Xu, Z. Wei, Y. Wang, Hydrothermal synthesis of manganese oxide encapsulated multiporous carbon nanofibers for supercapacitors, Nano Res. 9 (2016) 2672-2680. https://doi.org/10.1007/s12274-016-1154-2

[5] B. Saravanakumar, K.K. Purushothaman, G. Muralidharan, Fabrication of two-dimensional reduced graphene oxide supported V_2O_5 networks and their application in supercapacitors, Mater. Chem. Phys. 170 (2016) 266-275. https://doi.org/10.1016/j.matchemphys.2015.12.051

[6] D. Sun, L. Jin, Y. Chen, J.R. Zhang, J.J. Zhu, Microwave-assisted in situ synthesis of graphene/PEDOT hybrid and its application in supercapacitors, ChemPlusChem 78 (2013) 227-234. https://doi.org/10.1002/cplu.201200206

[7] S.K. Simotwo, C. DelRe, V. Kalra, Supercapacitor electrodes based on high-purity electrospun polyaniline and polyaniline-carbon nanotube nanofibers, ACS Appl. Mater. Interfaces 8 (2016) 21261-21269. https://doi.org/10.1021/acsami.6b03463

[8] S. Pei, H.-M. Cheng, The reduction of graphene oxide, Carbon 50 (2012) 3210-3228. https://doi.org/10.1016/j.carbon.2011.11.010

[9] L.L. Zhang, R. Zhou, X.S. Zhao, Graphene-based materials as supercapacitor electrodes, J. Mater. Chem. 20 (2010) 5983-5992. https://doi.org/10.1039/C000417K

[10]Z. Yu, L. Tetard, L. Zhai, J. Thomas, Supercapacitor electrode materials: nanostructures from 0 to 3 dimensions, Energy Environ. Sci. 8 (2015) 702-730. https://doi.org/10.1039/C4EE03229B

[11]M. Vangari, T. Pryor, L. Jiang, Supercapacitors: review of materials and fabrication methods, J. Energy Eng. 139 (2012) 72-79. https://doi.org/10.1061/(ASCE)EY.1943-7897.0000102

[12]J. Yan, Q. Wang, T. Wei, Z. Fan, Recent advances in design and fabrication of electrochemical supercapacitors with high energy densities, Adv. Energy Mater. 4 (2014) 1300816. https://doi.org/10.1002/aenm.201300816

[13]L. Dong, C. Xu, Y. Li, Z.-H. Huang, F. Kang, Q.-H. Yang, X. Zhao, Flexible electrodes and supercapacitors for wearable energy storage: a review by category, J. Mater. Chem. A 4 (2016) 4659-4685. https://doi.org/10.1039/C5TA10582J

[14]Y. Zhang, H. Feng, X. Wu, L. Wang, A. Zhang, T. Xia, H. Dong, X. Li, L. Zhang, Progress of electrochemical capacitor electrode materials: A review, Int. J. Hydrogen Energy 34 (2009) 4889-4899. https://doi.org/10.1016/j.ijhydene.2009.04.005

[15]X. Li, B. Wei, Supercapacitors based on nanostructured carbon, Nano Energy 2 (2013) 159-173. https://doi.org/10.1016/j.nanoen.2012.09.008

[16]A.K. Shukla, A. Banerjee, M.K. Ravikumar, A. Jalajakshi, Electrochemical capacitors: technical challenges and prognosis for future markets, Electrochim. Acta 84 (2012) 165-173. https://doi.org/10.1016/j.electacta.2012.03.059

[17]E. Frackowiak, Carbon materials for supercapacitor application, Phys. Chem. Chem. Phys. 9 (2007) 1774-1785. https://doi.org/10.1039/B618139M

[18]S. Bose, T. Kuila, A.K. Mishra, R. Rajasekar, N.H. Kim, J.H. Lee, Carbon-based nanostructured materials and their composites as supercapacitor electrodes, J. Mater. Chem. 22 (2012) 767-784. https://doi.org/10.1039/C1JM14468E

[19]G. Yu, X. Xie, L. Pan, Z. Bao, Y. Cui, Hybrid nanostructured materials for high-performance electrochemical capacitors, Nano Energy 2 (2013) 213-234. https://doi.org/10.1016/j.nanoen.2012.10.006

[20]B.E. Conway, V. Birss, J. Wojtowicz, The role and utilization of pseudocapacitance for energy storage by supercapacitors, J. Power Sources 66 (1997) 1-14. https://doi.org/10.1016/S0378-7753(96)02474-3

[21]G.A. Snook, P. Kao, A.S. Best, Conducting-polymer-based supercapacitor devices and electrodes, J. Power Sources 196 (2011) 1-12. DOI: 10.1016/j.jpowsour.2010.06.084

[22]J. Libich, J. Máca, J. Vondrák, O. Čech, M. Sedlaříková, Supercapacitors: properties and applications, J. Energy Storage 17 (2018) 224-227. https://doi.org/10.1016/j.est.2018.03.012

[23]C. Zhong, Y. Deng, W. Hu, J. Qiao, L. Zhang, J. Zhang, A review of electrolyte materials and compositions for electrochemical supercapacitors, Chem. Soc. Rev. 44 (2015) 7484-7539. https://doi.org/10.1039/C5CS00303B

[24]Z.S. Wu, W. Ren, D.-W. Wang, F. Li, B. Liu, H.-M. Cheng, High-energy MnO_2 nanowire/graphene and graphene asymmetric electrochemical capacitors, ACS Nano 4 (2010) 5835-5842. https://doi.org/10.1021/nn101754k

[25] D.P. Dubal, O. Ayyad, V. Ruiz, P.G.-Romero, Hybrid energy storage: the merging of battery and supercapacitor chemistries, Chem. Soc. Rev. 44 (2015) 1777-1790. https://doi.org/10.1039/C4CS00266K

[26] K.S. Novoselov, A.K. Geim, S.V. Morozov, D. Jiang, Y. Zhang, S.V. Dubonos, I.V. Grigorieva, A.A. Firsov, Electric field effect in atomically thin carbon films, Science 306 (2004) 666-669. https://doi.org/10.1126/science.1102896

[27] X. Huang, X. Qi, F. Boey, H. Zhang, Graphene-based composites, Chem. Soc. Rev. 41 (2012) 666-686. https://doi.org/10.1039/C1CS15078B

[28] Y. Zhang, Y.-W. Tan, H.L. Stormer, P. Kim, Experimental observation of the quantum Hall effect and Berry's phase in graphene, Nature 438 (2005) 201. https://doi.org/10.1038/nature04235

[29] D.V. Kosynkin, A.L. Higginbotham, A. Sinitskii, J.R. Lomeda, A. Dimiev, B.K. Price, J.M. Tour, Longitudinal unzipping of carbon nanotubes to form graphene nanoribbons, Nature 458 (2009) 872. https://doi.org/10.1038/nature07872

[30] K.S. Kim, Y. Zhao, H. Jang, S.Y. Lee, J.M. Kim, K.S. Kim, J.H. Ahn, P. Kim, J.Y. Choi, B.H. Hong, Large-scale pattern growth of graphene films for stretchable transparent electrodes, Nature 457 (2009) 706. https://doi.org/10.1038/nature07719

[31] K.V. Emtsev, A. Bostwick, K. Horn, J. Jobst, G.L. Kellogg, L. Ley, J.L. McChesney, T. Ohta, S.A. Reshanov, J. Röhrl, E. Rotenberg, A.K. Schmid, D. Waldmann, H.B. Weber, T. Seyller, Towards wafer-size graphene layers by atmospheric pressure graphitization of silicon carbide, Nat. Mater. 8 (2009) 203. https://doi.org/10.1038/nmat2382

[32] V.C. Tung, M.J. Allen, Y. Yang, R.B. Kaner, High-throughput solution processing of large-scale graphene, Nat. Nanotechnol. 4 (2009) 25. https://doi.org/10.1038/nnano.2008.329

[33] D.C. Marcano, D.V. Kosynkin, J.M. Berlin, A. Sinitskii, Z. Sun, A. Slesarev, L.B. Alemany, W. Lu, J.M. Tour, Improved synthesis of graphene oxide, ACS Nano 4 (2010) 4806-4814. https://doi.org/10.1021/nn1006368

[34] H.A. Becerril, J. Mao, Z. Liu, R.M. Stoltenberg, Z. Bao, Y. Chen, Evaluation of solution-processed reduced graphene oxide films as transparent conductors, ACS Nano 2 (2008) 463-470. https://doi.org/10.1021/nn700375n

[35] H.J. Shin, K.K. Kim, A. Benayad, S.M. Yoon, H.K. Park, I.S. Jung, M.H. Jin, H.K. Jeong, J.M. Kim, J.-Y. Choi, Y.H. Lee, Efficient reduction of graphite oxide by

sodium borohydride and its effect on electrical conductance, Adv. Funct. Mater. 19 (2009) 1987-1992. https://doi.org/10.1002/adfm.200900167

[36] V. Dua, S.P. Surwade, S. Ammu, S.R. Agnihotra, S. Jain, K.E. Roberts, S. Park, R.S. Ruoff, S.K. Manohar, All-organic vapor sensor using inkjet-printed reduced graphene oxide, Angew. Chem. 122 (2010) 2200-2203. https://doi.org/10.1002/anie.200905089

[37] S. Pei, J. Zhao, J. Du, W. Ren, H.-M. Cheng, Direct reduction of graphene oxide films into highly conductive and flexible graphene films by hydrohalic acids, Carbon 48 (2010) 4466-4474. https://doi.org/10.1016/j.carbon.2010.08.006

[38] M. Pumera, Electrochemistry of graphene, graphene oxide and other graphenoids, Electrochem. Commun. 36 (2013) 14-18. https://doi.org/10.1016/j.elecom.2013.08.028

[39] M.D. Stoller, S. Park, Y. Zhu, J. An, R.S. Ruoff, Graphene-based ultracapacitors, Nano Lett. 8 (2008) 3498-3502. https://doi.org/10.1021/nl802558y

[40] Y. Liu, Y. Ying, Y. Mao, L. Gu, Y. Wang, X. Peng, CuO nanosheets/rGO hybrid lamellar films with enhanced capacitance, Nanoscale 5 (2013) 9134-9140. https://doi.org/10.1039/C3NR02737F

[41] J. Zhang, P. Chen, B.H. Oh, M.B. Chan-Park, High capacitive performance of flexible and binder-free graphene-polypyrrole composite membrane based on in situ reduction of graphene oxide and self-assembly, Nanoscale 5 (2013) 9860-9866. https://doi.org/10.1039/C3NR02381H

[42] Z.S. Wu, G. Zhou, L.C. Yin, W. Ren, F. Li, H.-M. Cheng, Graphene/metal oxide composite electrode materials for energy storage, Nano Energy 1 (2012) 107-131. https://doi.org/10.1016/j.nanoen.2011.11.001

[43] Y. Shao, H. Wang, Q. Zhang, Y. Li, High-performance flexible asymmetric supercapacitors based on 3D porous graphene/MnO_2 nanorod and graphene/Ag hybrid thin-film electrodes, J. Mater. Chem. C 1 (2013) 1245-1251. https://doi.org/10.1039/C2TC00235C

[44] D.R. Dreyer, S. Park, C.W. Bielawski, R.S. Ruoff, The chemistry of graphene oxide, Chem. Soc. Rev. 39 (2010) 228-240. DOI: 10.1039/B917103G

[45] Z.S. Wu, D.W. Wang, W. Ren, J. Zhao, G. Zhou, F. Li, H.M. Cheng, Anchoring hydrous RuO_2 on graphene sheets for high-performance electrochemical capacitors, Adv. Funct. Mater. 20 (2010) 3595-3602. https://doi.org/10.1002/adfm.201001054

[46] Q. Wang, J. Yan, Z. Fan, T. Wei, M. Zhang, X. Jing, Mesoporous polyaniline film on ultra-thin graphene sheets for high performance supercapacitors, J. Power Sources 247 (2014) 197-203. https://doi.org/10.1016/j.jpowsour.2013.08.076

[47] S. Ye, J. Feng, Self-assembled three-dimensional hierarchical graphene/polypyrrole nanotube hybrid aerogel and its application for supercapacitors, ACS Appl. Mater. Interfaces 6 (2014) 9671-9679. https://doi.org/10.1021/am502077p

[48] M.A. Memon, W. Bai, J. Sun, M. Imran, S.N. Phulpoto, S. Yan, Y. Huang, J. Geng, Conjunction of conducting polymer nanostructures with macroporous structured graphene thin films for high-performance flexible supercapacitors, ACS Appl. Mater. Interfaces 8 (2016) 11711-11719. https://doi.org/10.1021/acsami.6b01879

[49] Y. Haldorai, Y.S. Huh, Y.-K. Han, Surfactant-assisted hydrothermal synthesis of flower-like tin oxide/graphene composites for high-performance supercapacitors, New J. Chem. 39 (2015) 8505-8512. https://doi.org/10.1039/C5NJ01442E

[50] Y.-W. Son, M.L. Cohen, S.G. Louie, Half-metallic graphene nanoribbons, Nature 444 (2006) 347. https://doi.org/10.1038/nature05180

[51] D.W. Boukhvalov, M.I. Katsnelson, Chemical functionalization of graphene with defects, Nano Lett. 8 (2008) 4373-4379. https://doi.org/10.1021/nl802234n

[52] J. Lahiri, Y. Lin, P. Bozkurt, I.I. Oleynik, M. Batzill, An extended defect in graphene as a metallic wire, Nat. Nanotechnol. 5 (2010) 326. https://doi.org/10.1038/nnano.2010.53

[53] G. Luo, L. Liu, J. Zhang, G. Li, B. Wang, J. Zhao, Hole defects and nitrogen doping in graphene: implication for supercapacitor applications, ACS Appl. Mater. Interfaces 5 (2013) 11184-11193. https://doi.org/10.1021/am403427h

[54] G.F. Schneider, S.W. Kowalczyk, V.E. Calado, G. Pandraud, H.W. Zandbergen, L.M.K Vandersypen, C. Dekker, DNA translocation through graphene nanopores, Nano Lett. 10 (2010) 3163-3167. https://doi.org/10.1021/nl102069z

[55] S. Blankenburg, M. Bieri, R. Fasel, K. Müllen, C.A. Pignedoli, D. Passerone, Porous graphene as an atmospheric nanofilter, Small 6 (2010) 2266-2271. https://doi.org/10.1002/smll.201001126

[56] A. Nourbakhsh, M. Cantoro, T. Vosch, G. Pourtois, F. Clemente, M.H. Veen, J. Hofkens, M.M. Heyns, S.D. Gendt, B.F. Sels, Bandgap opening in oxygen plasma-treated graphene, Nanotechnology 21 (2010) 435203. DOI: 10.1088/0957-4484/21/43/435203

[57] J. Yan, J. Liu, Z. Fan, T. Wei, L. Zhang, High-performance supercapacitor electrodes based on highly corrugated graphene sheets, Carbon 50 (2012) 2179-2188. https://doi.org/10.1016/j.carbon.2012.01.028

[58] Z. Jin, J. Yao, C. Kittrell, J.M. Tour, Large-scale growth and characterizations of nitrogen-doped monolayer graphene sheets, ACS Nano 5 (2011) 4112-4117. https://doi.org/10.1021/nn200766e

[59] L.S. Panchakarla, K.S. Subrahmanyam, S.K. Saha, A. Govindaraj, H.R. Krishnamurthy, U.V. Waghmare, C.N.R. Rao, Synthesis, structure, and properties of boron-and nitrogen-doped graphene, Adv. Mater. 21 (2009) 4726-4730. https://doi.org/10.1002/adma.200901285

[60] V. Thirumal, A. Pandurangan, D. Jayakumar, R. Ilangovan, Modified solar power: electrochemical synthesis of nitrogen doped few layer graphene for supercapacitor applications, J. Mater. Sci.: Mater. Electron. 27 (2016) 3410-3419. https://doi.org/10.1007/s10854-015-4173-y

[61] D. Wei, Y. Liu, Y. Wang, H. Zhang, L. Huang, G. Yu, Synthesis of N-doped graphene by chemical vapor deposition and its electrical properties, Nano Lett. 9 (2009) 1752-1758. https://doi.org/10.1021/nl803279t

[62] Z. Wen, X. Wang, S. Mao, Z. Bo, H. Kim, S. Cui, G. Lu, X. Feng, J. Chen, Crumpled nitrogen-doped graphene nanosheets with ultrahigh pore volume for high-performance supercapacitor, Adv. Mater. 24 (2012) 5610-5616. https://doi.org/10.1002/adma.201201920

[63] Z.-Y. Sui, Y.-N. Meng, P.-W. Xiao, Z.-Q. Zhao, Z.-X. Wei, B.-H. Han, Nitrogen-doped graphene aerogels as efficient supercapacitor electrodes and gas adsorbents, ACS Appl. Mater. Interfaces 7 (2015) 1431-1438. https://doi.org/10.1021/am5042065

[64] E. Haque, M.M. Islam, E. Pourazadi, M. Hassan, S.N. Faisal, A.K. Roy, K. Konstantinov, A.T. Harris, A.I. Minett, V.G. Gomes, Nitrogen doped graphene via thermal treatment of composite solid precursors as a high performance supercapacitor, RSC Adv. 5 (2015) 30679-30686. https://doi.org/10.1039/C4RA17262K

[65] H. Jin, X. Wang, Z. Gu, Q. Fan, B. Luo, A facile method for preparing nitrogen-doped graphene and its application in supercapacitors, J. Power Sources 273 (2015) 1156-1162. https://doi.org/10.1016/j.jpowsour.2014.10.010

[66] D. Li, C. Yu, M. Wang, Y. Zhang, C. Pan, Synthesis of nitrogen doped graphene from graphene oxide within an ammonia flame for high performance supercapacitors, RSC Adv. 4 (2014) 55394-55399. https://doi.org/10.1039/C4RA10761F

[67] M.P. Kumar, T. Kesavan, G. Kalita, P. Ragupathy, T.N. Narayanan, D.K. Pattanayak, On the large capacitance of nitrogen doped graphene derived by a facile route, RSC Adv. 4 (2014) 38689-38697. https://doi.org/10.1039/C4RA04927F

[68] C. Li, Y. Hu, M. Yu, Z. Wang, W. Zhao, P. Liu, Y. Tong, X. Lu, Nitrogen doped graphene paper as a highly conductive, and light-weight substrate for flexible supercapacitors, RSC Adv. 4 (2014) 51878-51883. https://doi.org/10.1039/C4RA11024B

[69] Y. Lu, Y. Huang, F. Zhang, L. Zhang, X. Yang, T. Zhang, K. Leng, M. Zhang, Y. Chen, Functionalized graphene oxide based on p-phenylenediamine as spacers and nitrogen dopants for high performance supercapacitors, Chinese Sci. Bull. 59 (2014) 1809-1815. https://doi.org/10.1007/s11434-014-0297-3

[70] D. Wang, Y. Min, Y. Yu, B. Peng, A general approach for fabrication of nitrogen-doped graphene sheets and its application in supercapacitors, J. Colloid Interface Sci. 417 (2014) 270-277. https://doi.org/10.1016/j.jcis.2013.11.021

[71] V. Sahu, S. Grover, B. Tulachan, M. Sharma, G. Srivastava, M. Roy, M. Saxena, N. Sethy, K. Bhargava, D. Philip, H. Kim, G. Singh, S.K. Singh, M. Das, R.K. Sharma, Heavily nitrogen doped, graphene supercapacitor from silk cocoon, Electrochim. Acta 160 (2015) 244-253. https://doi.org/10.1016/j.electacta.2015.02.019

[72] T. Wu, H. Shen, L. Sun, B. Cheng, B. Liu, J. Shen, Nitrogen and boron doped monolayer graphene by chemical vapor deposition using polystyrene, urea and boric acid, New J. Chem. 36 (2012) 1385-1391. https://doi.org/10.1039/C2NJ40068E

[73] L. Niu, Z. Li, W. Hong, J. Sun, Z. Wang, L. Ma, J. Wang, S. Yang, Pyrolytic synthesis of boron-doped graphene and its application as electrode material for supercapacitors, Electrochim. Acta 108 (2013) 666-673. https://doi.org/10.1016/j.electacta.2013.07.025

[74] Z. Zuo, Z. Jiang, A. Manthiram, Porous B-doped graphene inspired by fried-ice for supercapacitors and metal-free catalysts, J. Mater. Chem. A 1 (2013) 13476-13483. https://doi.org/10.1039/C3TA13049E

[75] J. Han, L.L. Zhang, S. Lee, J. Oh, K.-S. Lee, J.R. Potts, J. Ji, X. Zhao, R.S. Ruoff, S. Park, Generation of B-doped graphene nanoplatelets using a solution process and

their supercapacitor applications, ACS Nano 7 (2012) 19-26.
https://doi.org/10.1021/nn3034309

[76] A. Ambrosi, H.L. Poh, L. Wang, Z. Sofer, M. Pumera, Capacitance of p-and n-doped graphenes is dominated by structural defects regardless of the dopant type, ChemSusChem 7 (2014) 1102-1106. https://doi.org/10.1002/cssc.201400013

[77] J.P. Paraknowitsch, A. Thomas, J. Schmidt, Microporous sulfur-doped carbon from thienyl-based polymer network precursors, Chem. Commun. 47 (2011) 8283-8285. https://doi.org/10.1039/C1CC12272J

[78] X. Zhao, Q. Zhang, C.-M. Chen, B. Zhang, S. Reiche, A. Wang, T. Zhang, R. Schlögl, D.S. Su, Aromatic sulfide, sulfoxide, and sulfone mediated mesoporous carbon monolith for use in supercapacitor, Nano Energy 1 (2012) 624-630. https://doi.org/10.1016/j.nanoen.2012.04.003

[79] X. Chen, X. Chen, X. Xu, Z. Yang, Z. Liu, L. Zhang, X. Xu, Y. Chen, S. Huang, Sulfur-doped porous reduced graphene oxide hollow nanosphere frameworks as metal-free electrocatalysts for oxygen reduction reaction and as supercapacitor electrode materials, Nanoscale 6 (2014) 13740-13747. https://doi.org/10.1039/C4NR04783D

[80] Y. Wen, B. Wang, C. Huang, L. Wang, D. Hulicova-Jurcakova, Synthesis of phosphorus-doped graphene and its wide potential window in aqueous supercapacitors, Chem.: Eur. J. 21 (2015) 80-85. https://doi.org/10.1002/chem.201404779

[81] P. Karthika, N. Rajalakshmi, K.S. Dhathathreyan, Phosphorus-doped exfoliated graphene for supercapacitor electrodes, J. Nanosci. Nanotechnol. 13 (2013) 1746-1751. https://doi.org/10.1166/jnn.2013.7112

[82] V. Thirumal, A. Pandurangan, R. Jayavel, R. Ilangovan, Synthesis and characterization of boron doped graphene nanosheets for supercapacitor applications, Synth. Met. 220 (2016) 524-532. https://doi.org/10.1016/j.synthmet.2016.07.011

[83] Z. Peng, R. Ye, J.A. Mann, D. Zakhidov, Y. Li, P.R. Smalley, J. Lin, J.M. Tour, Flexible boron-doped laser-induced graphene microsupercapacitors, ACS Nano 9 (2015) 5868-5875. https://doi.org/10.1021/acsnano.5b00436

[84] D.-Y. Yeom, W. Jeon, N.D.K. Tu, S.Y. Yeo, S.-S. Lee, B.J. Sung, H. Chang, J.A. Lim, H. Kim, High-concentration boron doping of graphene nanoplatelets by simple

thermal annealing and their supercapacitive properties, Sci. Rep. 5 (2015) 9817.
https://doi.org/10.1038/srep09817

[85] N. Parveen, M.O. Ansari, S.A. Ansari, M.H. Cho, Simultaneous sulfur doping and
exfoliation of graphene from graphite using an electrochemical method for
supercapacitor electrode materials, J. Mater. Chem. A 4 (2015) 233-240.
https://doi.org/10.1039/C5TA07963B

[86] E. Bekyarova, M.E. Itkis, P. Ramesh, C. Berger, M. Sprinkle, W.A. Heer, R.C.
Haddon, Chemical modification of epitaxial graphene: spontaneous grafting of aryl
groups, J. Am. Chem. Soc. 131 (2009) 1336-1337. https://doi.org/10.1021/ja8057327

[87] J. Zhang, Y. Xu, L. Cui, A. Fu, W. Yang, C. Barrow, J. Liu, Mechanical properties
of graphene films enhanced by homo-telechelic functionalized polymer fillers via π-π
stacking interactions, Compos. Part A: Appl. Sci. Manuf. 71 (2015) 1-8.
https://doi.org/10.1016/j.compositesa.2014.12.013

[88] Z. Lei, J. Zhang, L.L. Zhang, N.A. Kumar, X.S. Zhao, Functionalization of
chemically derived graphene for improving its electrocapacitive energy storage
properties, Energy Environ. Sci. 9 (2016) 1891-1930.
https://doi.org/10.1039/C6EE00158K

[89] A.K. Mishra, S. Ramaprabhu, Functionalized graphene-based nanocomposites for
supercapacitor application, J. Phys. Chem. C 115 (2011) 14006-14013.
https://doi.org/10.1021/jp201673e

[90] Z. Gao, F. Wang, J. Chang, D. Wu, X. Wang, X. Wang, F. Xu, S. Gao, K. Jiang,
Chemically grafted graphene-polyaniline composite for application in
supercapacitor, Electrochim. Acta 133 (2014) 325-334.
https://doi.org/10.1016/j.electacta.2014.04.033

[91] B. Song, J. Zhao, M. Wang, J. Mullavey, Y. Zhu, Z. Geng, D. Chen, Y. Ding, K.-S.
Moon, M. Liu, C.-P. Wong, Systematic study on structural and electronic properties
of diamine/triamine functionalized graphene networks for supercapacitor
application, Nano Energy 31 (2017) 183-193.
https://doi.org/10.1016/j.nanoen.2016.10.057

[92] B. Song, J.I. Choi, Y. Zhu, Z. Geng, L. Zhang, Z. Lin, C.-C. Tuan, K.-S. Moon, C.-P.
Wong, Molecular level study of graphene networks functionalized with
phenylenediamine monomers for supercapacitor electrodes, Chem. Mater. 28 (2016)
9110-9121. https://doi.org/10.1021/acs.chemmater.6b04214

[93] G. Xie, R. Yang, P. Chen, J. Zhang, X. Tian, S. Wu, J. Zhao, M. Cheng, W. Yang, D. Wang, C. He, X. Bai, D. Shi, G. Zhang, A general route towards defect and pore engineering in graphene, Small 10 (2014) 2280-2284. https://doi.org/10.1002/smll.201303671

[94] J. Chen, Y. Han, X. Kong, X. Deng, H.J. Park, Y. Guo, S. Jin, Z. Qi, Z. Lee, Z. Qiao, R.S. Ruoff, H. Ji, The origin of improved electrical double-layer capacitance by inclusion of topological defects and dopants in graphene for supercapacitors, Angew. Chem. Int. Ed. 55 (2016) 13822-13827. https://doi.org/10.1002/anie.201605926

[95] J. Zhu, A.S. Childress, M. Karakaya, S. Dandeliya, A. Srivastava, Y. Lin, A.M. Rao, R. Podila, Defect-engineered graphene for high-energy-and high-power-density supercapacitor devices, Adv. Mater. 28 (2016) 7185-7192. https://doi.org/10.1002/adma.201602028

[96] H.M. Jeong, J.W. Lee, W.H. Shin, Y.J. Choi, H.J. Shin, J.K. Kang, J.W. Choi, Nitrogen-doped graphene for high-performance ultracapacitors and the importance of nitrogen-doped sites at basal planes, Nano Lett. 11 (2011) 2472-2477. https://doi.org/10.1021/nl2009058

[97] S. Arif, M.S. Rafique, F. Saleemi, F. Naab, O. Toader, R. Sagheer, S. Bashir, R. Zia, K. Siraj, S. Iqbal, Surface topographical and structural analysis of Ag^+-implanted polymethylmethacrylate, Nucl. Instrum. Methods Phys. Res. B 381 (2016) 114-121. https://doi.org/10.1016/j.nimb.2016.05.028

[98] D.K Avasthi, Modification and characterisation of materials by swift heavy ions, Def. Sci. J. 59 (2009) 401-412. https://doi.org/10.14429/dsj.59.1540

[99] S.B. Kadam, K. Datta, P. Ghosh, A.B. Kadam, P.W. Khirade, V. Kumar, R.G. Sonkawade, A.B. Gambhire, M.K. Lande, M.D. Shirsat, Improvement of ammonia sensing properties of poly (pyrrole)-poly (n-methylpyrrole) composite by ion irradiation, Appl. Phys. A 100 (2010) 1083-1088. https://doi.org/10.1007/s00339-010-5705-1

[100] D. Kanjilal, Swift heavy ion-induced modification and track formation in materials, Curr. Sci. 80 (2001) 1560-1566.

[101] D.K Avasthi, Some interesting aspects of swift heavy ions in materials science, Curr. Sci. 78 (2000) 1297-1302.

[102] E.H. Lee, Ion-beam modification of polymeric materials-fundamental principles and applications, Nucl. Instrum. Methods Phys. Res. B 151 (1999) 29-41. https://doi.org/10.1016/S0168-583X(99)00129-9

[103] M. Kang, D.H. Lee, Y.-M. Kang, H. Jung, Electron beam irradiation dose dependent physico-chemical and electrochemical properties of reduced graphene oxide for supercapacitor, Electrochim. Acta 184 (2015) 427-435. https://doi.org/10.1016/j.electacta.2015.10.053

[104] M. Hassan, E. Haque, K.R. Reddy, A.I. Minett, J. Chen, V.G. Gomes, Edge-enriched graphene quantum dots for enhanced photo-luminescence and supercapacitance, Nanoscale 6 (2014) 11988-11994. https://doi.org/10.1039/C4NR02365J

[105] V. Sridhar, H.-J. Kim, J.-H. Jung, C. Lee, S. Park, I.-K. Oh, Defect-engineered three-dimensional graphene-nanotube-palladium nanostructures with ultrahigh capacitance, ACS Nano 6 (2012) 10562-10570. https://doi.org/10.1021/nn3046133

[106] J.C. Meyer, C. Kisielowski, R. Erni, M.D. Rossell, M.F. Crommie, A. Zettl, Direct imaging of lattice atoms and topological defects in graphene membranes, Nano Lett. 8 (2008) 3582-3586. https://doi.org/10.1021/nl801386m

[107] A. Hashimoto, K. Suenaga, A. Gloter, K. Urita, S. Iijima, Direct evidence for atomic defects in graphene layers, Nature 430 (2004) 870. https://doi.org/10.1038/nature02817

[108] I. Calizo, A.A. Balandin, W. Bao, F. Miao, C.N. Lau, Temperature dependence of the Raman spectra of graphene and graphene multilayers, Nano Lett. 7 (2007) 2645-2649. https://doi.org/10.1021/nl071033g

[109] G. Compagnini, F. Giannazzo, S. Sonde, V. Raineri, E. Rimini, Ion irradiation and defect formation in single layer graphene, Carbon 47 (2009) 3201-3207. https://doi.org/10.1016/j.carbon.2009.07.033

[110] J. Zeng, H.J. Yao, S.X. Zhang, P.F. Zhai, J.L. Duan, Y.M. Sun, G.P. Li, J. Liu, Swift heavy ions induced irradiation effects in monolayer graphene and highly oriented pyrolytic graphite, Nucl. Instrum. Methods Phys. Res. B 330 (2014) 18-23. https://doi.org/10.1016/j.nimb.2014.03.019

[111] Y.B. Zhou, Z.M. Liao, Y.-F. Wang, G.S. Duesberg, J. Xu, Q. Fu, X.S. Wu, D.-P. Yu, Ion irradiation induced structural and electrical transition in graphene, J. Chem. Phys. 133 (2010) 234703. https://doi.org/10.1063/1.3518979

[112] H.C. Schniepp, K.N. Kudin, J.-L. Li, R.K. Prud'homme, R. Car, D.A. Saville, I.A. Aksay, Bending properties of single functionalized graphene sheets probed by atomic force microscopy, ACS Nano 2 (2008) 2577-2584. https://doi.org/10.1021/nn800457s

[113] J.I. Paredes, S. Villar-Rodil, P. Solís-Fernández, A. Martínez-Alonso, J.M.D. Tascon, Atomic force and scanning tunneling microscopy imaging of graphene nanosheets derived from graphite oxide, Langmuir 25 (2009) 5957-5968. https://doi.org/10.1021/la804216z

[114] T.H.D. Keijser, J.I. Langford, E.J. Mittemeijer, A.B.P. Vogels, Use of the Voigt function in a single-line method for the analysis of X-ray diffraction line broadening, J. Appl. Crystallogr. 15 (1982) 308-314. https://doi.org/10.1107/S0021889882012035

[115] D. Geng, S. Yang, Y. Zhang, J. Yang, J. Liu, R. Li, T.-K. Sham, X. Sun, S. Ye, S. Knights, Nitrogen doping effects on the structure of graphene, Appl. Surf. Sci. 257 (2011) 9193-9198. https://doi.org/10.1016/j.apsusc.2011.05.131

[116] V.A. Chirayath, M.D. Chrysler, A.D. McDonald, R.W. Gladen, A.J. Fairchild, A.R. Koymen, A.H. Weiss, Investigation of graphene using low energy positron annihilation induced Doppler broadening spectroscopy, J. Phys. Conf. Ser. 791 (2017) 012032. https://doi.org/10.1088/1742-6596/791/1/012032

[117] U. Rana, P.M.G. Nambissan, S. Malik, K. Chakrabarti, Effects of process parameters on the defects in graphene oxide-polyaniline composites investigated by positron annihilation spectroscopy, Phys. Chem. Chem. Phys. 16 (2014) 3292-3298. https://doi.org/10.1039/C3CP54032D

[118] M. Devi, A. Kumar, In-situ reduced graphene oxide nanosheets-polypyrrole nanotubes nanocomposites for supercapacitor applications, Synth. Met. 222 (2016) 318-329. https://doi.org/10.1016/j.synthmet.2016.11.004

[119] M. Devi, A. Kumar, 85 MeV C^{6+} swift heavy ion irradiation of in-situ reduced graphene oxide-polypyrrole nanotubes nanocomposite films for supercapacitor electrodes, Electrochim. Acta 261 (2018) 1-13. https://doi.org/10.1016/j.electacta.2017.12.106

[120] D.K. Avasthi, S. Ghosh, S.K. Srivastava, W. Assmann, Existence of transient temperature spike induced by SHI: evidence by ion beam analysis, Nucl. Instrum. Methods Phys. Res. B 219 (2004) 206-214. https://doi.org/10.1016/j.nimb.2004.01.055

[121] V. Kumar, Y. Ali, K. Sharma, V. Kumar, R.G. Sonkawade, A.S. Dhaliwal, H.C. Swart, Swift heavy ions induced surface modifications in Ag-polypyrrole composite films synthesized by an electrochemical route, Nucl. Instrum. Methods Phys. Res. B 323 (2014) 7-13. https://doi.org/10.1016/j.nimb.2014.01.009

[122] S. Kumar, A. Tripathi, F. Singh, S.A. Khan, V. Baranwal, D.K. Avasthi, Purification/annealing of graphene with 100-MeV Ag ion irradiation, Nanoscale Res. Lett. 9 (2014) 126. https://doi.org/10.1186/1556-276X-9-126

[123] E. Frackowiak, S. Delpeux, K. Jurewicz, K. Szostak, D. Cazorla-Amoros, F. Beguin, Enhanced capacitance of carbon nanotubes through chemical activation, Chem. Phys. Lett. 361 (2002) 35-41. https://doi.org/10.1016/S0009-2614(02)00684-X

[124] J.M. Carlsson, M. Scheffler, Structural, electronic, and chemical properties of nanoporous carbon, Phys. Rev. Lett. 96 (2006) 046806. https://doi.org/10.1103/PhysRevLett.96.046806

Materials Research Forum LLC
https://doi.org/10.21741/9781644900550-9

Chapter 9

Future Prospects and Challenges of Graphene-Based Supercapacitors

Paul Thomas[1], Nelson Pynadathu Rumjit1,[2], Chin Wei Lai[1]*, Mohd Rafie Bin Johan[1]

[1] Nanotechnology & Catalysis Research Centre (NANOCAT), Institute for Advanced Studies (IAS), University of Malaya (UM), Level 3, Block A, 50603 Kuala Lumpur, Malaysia

[2] Department of Environmental and Water Resources Engineering, VIT University, Vellore 632014, Tamil Nadu, India

*cwlai@um.edu.my

Abstract

According to the statement made by IUPAC (International Union of Pure and Applied Chemistry), graphene is distinguished as a carbon monolayer of the graphite structure. It can be described by analogy to a polycyclic aromatic hydrocarbon of quasi-infinite size. Over the past few years, graphene-based materials have gained immense attraction as next-generation material inherited with its better mechanical characteristics, excellent electrical conductivity and better thermal conductivity as contrasted with other similar materials. In the subsequent section, the new aspects of next-generation of supercapacitors with modified graphene materials including metal oxide/graphene composites, activated graphene, reduced graphene, doped graphene, and polymer/graphene composites will be reviewed and discussed. Moreover, advantages and drawbacks of these resultant composite materials applied in supercapacitor application are summarized in this chapter.

Keywords

Graphene, Supercapacitor, Electrical Conductivity, Thermal Conductivity, Mechanical Properties, Composite Materials

Contents

1.   Introduction

Graphene-based materials with various compositions and structural modifications have gained popularity in recent years and have broad applications in energy storage systems such as supercapacitors, batteries and fuel cells [1]. Currently, the demand of cost-effective, environmental friendly and better energy efficient systems has increased [2]. Among diverse energy storage devices, the supercapacitors have gained much popularity attributable to their high specific energy density of 10 W h kg^{-1} and power greater than 10 kW kg^{-1}. These properties make the graphene material as a potential candidate for various portable energy storage applications. Graphene is an atomically thick, two-dimensional sheet composed of sp^2-carbon atoms in a honeycomb structure of six-member rings [3]. This is the primary fundamental unit of graphitic materials in diverse structures identical to three, one and zero-dimensional. Additionally, the graphene has superior optical adsorption capabilities along with excellent electrical and mechanical properties [4]. The graphene with unique specific surface area of more than 2600 m^2g^{-1} demonstrates ample porosity with excellent electrical conductivity [5].

Numerous reports on the graphene-based electrode materials, have been published but very few research investigations were reported on modifications of graphene for next-generation applications [6]. This book chapter briefly focuses on the new aspects of next-generation supercapacitors also outlined suitability of modified graphene materials including metal oxide/graphene composites, activated graphene, reduced graphene, doped graphene and polymer/ graphene composites. This chapter also summarizes the advantages and drawbacks of these materials.

2.   Overview of graphene as electrodes

Several studies have been reported in the last few years regarding the suitability of graphene on supercapacitors and various energy storage devices [7–9]. For the EDLC applications, the activated graphene is comprehensively employed owing to the facts such as the presence of high specific surface area and cost-effectiveness. The activated graphene using potassium hydroxide and steam demonstrated better performance and cycling characteristics in aqueous media due to development of meso and micropores [10]. The activated graphene sheets in EDLC application have been found to improve the specific capacitance by 35% as compared to traditional graphene nanosheets [11]. The microwave exfoliation of graphene oxide trailed by chemical activation enhanced the surface area to 3100 m^2g^{-1} along with significant amount of wide pores ranging from 0.6 to 5 nm present over in sp^2-bonded carbon walls. Therefore, the electrostatic double-layer capacitors assembled using activated graphene sheets electrodes features outstanding electrochemical capabilities.

2.1 Heteroatom doping- graphene

The heteroatom doping of the carbon materials displays unique physiochemical characteristics resulting in better electrochemical performances due to the transformations in electronic states of graphene layer and pseudo-capacitance. Nitrogen, boron, sulphur and phosphorous are commonly employed for doping [12–14]. Though many studies have been reported regarding nitrogen doping but only a few reports were available on the doping of sulphur and phosphorous due to their larger atomic sizes that restricts their occupancy on the edges or surface of graphene as the surface functional groups.

2.1.1 Boron Doping

Boron doping attained by blending it with carbon sources exposed to an extreme temperature level of 1800 to 3000 °C to attain doping on the graphene structure attributable by boron diffusion at extreme temperature level [14]. The high thermal environmental condition results in the development of substituted boron, B_2O_3 and B_4C structures [15]. Carbonization of precursors utilizing inorganic or organic templates for boron doped carbon contributes to the development of B-O, B-O-C and B-C [16]. Boron has been observed to block the contraction of the templates to maintain the mesoporous structure with fine specific surface area especially for the self-assembly systems[13]. The extreme temperature conditions minimize the boron doping contribution (2.5%), whereas boron impregnation during carbonization involving templates contributes a high level of boron contribution (14%) [17]. The boron in graphene framework enhances the number of holes leading to the improvement in the density of states which further enhances the capacitance of the energy storage devices [18]. Density functional theory studies were performed to understand the outcome of boron presence on double layer capacitance and its potency on the charge carrier's potential to adsorb onto the boron doped carbons. As boron doping improves the concentration level of charge carriers, besides, the intrinsic electrical conductivity also improves which results in faster cycling performances [19]. The graphitic crystallinity was observed to increase as born doping level increases, and electrical conductivity also gets enhanced. However, the introduction of boron as flaws in the ordered porous structure increases resistance in charge mobility for the electrolytes in porous surface, so it is essential to optimize the dosage level of boron for better conductivity and performance. The boron doping aids to bring new heteroatoms (nitrogen or oxygen) from the air at high-temperature results in broad redox peak responses at lower potentials [16,20]. The oxygen functional groups provide pseudocapacitance, suppress the oxidation process and stabilize the capacitance characteristics at higher potentials, The presence of boron- carbon bonds minimize the Femi level and enhance chemisorption of oxygen [21]. The restriction to the oxidation depends on position and level of boron atoms being introduced. At low boron concentration, doping demonstrates the catalytic effect whereas inhibiting impact prominent as the concentration level of boron gets intensifies [21,22]. However, the oxygen chemisorption is less preferred as the process is site dependent and complicated [23]. Droplet contact tests and water vapor adsorption isotherms have proved that boron-doping resulted in enhancing of pseudocapacitance and thus demonstrating superior performances whereas nitrogen doping only contributes to better wettability [24].

The probability is to develop boron doped graphene for electrostatic double-layer capacitors by combining the unique properties of graphene with the advantages of boron doping makes it a promising candidate for future electrostatic double-layer capacitors. Besides, the boron doping aids in the introduction of new atoms into the graphene structure as the temperature of the thermal treatment gets elevated, the co-doping of phosphorous and boron or nitrogen and boron exhibits superior electrochemical performance [17,20,25]. The boron-nitrogen bonding is the most stable energy state which provides pseudocapacitance through redox reactions with lithium ion, sodium ion and even with protons [17]. Co-doping of phosphorous and boron introduces more oxygen functional groups resulting in enhancing the pseudocapacitance.

2.1.2 Nitrogen doping

Nitrogen doping is another promising technique utilized to amplify the electrochemical characteristics of the supercapacitor containing using graphene. The nitrogen doping can be carried out by direct carbonization, ammoxidation, chemical vapor deposition and solvothermal or hydrothermal treatments of nitrogen-containing precursors [26–32]. There are main four categories of nitrogen-containing functional groups found on graphene layer to improvise the electrochemical performance; Pyrodine-N-oxide, pyridinic N, quaternary N and pyridine N [12,33,34]. The presence of imperfections in the graphene layers on the basal planes leads in the evolution of pyridine N and pyridinic N on the flaws of the basal planes but pyridine-N-oxide and quaternary N located on the edges. Similar to the boron doping, the introduction of nitrogen modifies the electrochemical characteristics of graphene. The density functional theory study exhibited the strong interactions of the proton with nitrogen-doped edges compared to the oxygen or boron doped edges [35]. Density functional theory study also demonstrates that the positions of nitrogen-containing functional groups have a crucial role in ion adsorption on the surface.

The nitrogen doping tunes the electrical conductivity of carbon, the presence of pyridine-N-oxide and quaternary-N-oxide enhances the electron mobility maintains superior capacitance with fast charging and discharging cycles [36,37]. Figure 2 exhibits the structural and morphological modifications of graphene on doping with nitrogen. The nitrogen-doping plays the crucial role in electrical conductivity, as the concentration exceeds to 12 %, the electrical conductivity observes to declined due to the changes graphitic structure [29,38,39]. The contribution of pseudocapacitance is another main feature of the nitrogen doping. Both pyrrolic N and pyridinic N provide pseudocapacitance through redox reactions as protons involve in redox reactions the pseudocapacitance observed to be limited positive potentials in acidic electrolytes [40]. The introduction of nitrogen-containing functional group introduced pseudocapacitance exhibits better stability and cycling performance. Nitrogen doping in graphene provides high specific surface area, superior capacitance performance, long-term stability and energy capability compared to activated carbon in non-aqueous and aqueous electrolytes.

2.1.3 Sulphur and Phosphorous Doping

The phosphorous doping carried out by the addition of phosphoric acid serves as an activation agent for graphite along with it also observes the development of phosphate groups on the graphene surface [41,42]. The evolution of pyrophosphates on the graphene layer surface displays positive influence on specific capacitance via redox reactions, while the metaphosphates improvised the capacitance and electrochemical characteristics at high scan rates also the metaphosphates aid to the development of phosphate bridges [43]. The bonding structures of P-N and P=N enhanced the capacitive performances, however similar to the boron doping, the phosphorous doping brings more oxygen-containing functional groups resulting to the faster degradation of capacitance retention during fast charging and discharging cycles [44]. Phosphorus-doped carbon demonstrated more stability during the cycling process due to the blockage of hydrogen evolution and oxygen oxidation by forming a protective oxidation layer on the surface. Sulphur doping has not been studied extensively as compared to other doping materials. The incorporation of sulphur with graphene exhibits improvement in the polarity on graphene surface but reduces the resistance in charge mobility [45]. Further studies need to be carried out to understand the prospects of sulphur doping in graphene.

3.   Electrode -graphene/ metal oxide composites

During the synthesis approach, the superior characteristics of the graphene get curtailed due to the occurrence of chemical moieties. The intrinsic specific surface area of graphene composites is minimized by stacking and aggregation due to strong π-π-interactions. The conductivity of the graphene declines as the resistance between intersheets contact increases. Studies have reported the actual capacitive performance lower compared to that of anticipated values estimated through the ultrahigh theoretical surface area due to restacking and aggregation of graphene sheets as a result of the strong van der Waals interactions [46]. This phenomenon is crucial and needs to inhibit the restacking adequately; many researchers tried to add extra additives during synthesis. The additives dispersed on the graphene sheets surfaces act as a nano spacer by restricting the restacking of graphene sheets during the reduction of graphene oxide. Uniform dispersion of additives helps to make effective utilization of graphene electrode materials. Graphene sheets longer than five μm exhibits superior conductivity by providing a two-dimensional pathway towards rapid energy delivery and storage. The new configuration of hierarchical graphene sheet composites is promising for future high-performance energy storage devices as it provides effective utilization of electroactive species, the mobility of charge, better cycling performance and restacking of graphene sheets. Table 1 lists electrochemical and structural characteristics of various metal oxide/graphene composites utilized for energy applications. Metal oxides in particular SnO_2, RuO_2, TiO_2, Mn_3O_4, $NiCo_2O_4$, Fe_3O_4 and Zn_2SnO_4 were used along with graphene to improve the electrochemical characteristics [60]. RuO_2 based graphene sheets exhibit high specific capacitance of 220 Fg^{-1} compared to other metal oxides, due to reversible Faradic reactions and excellent electrical conductivity [61,62]. Similarly, the graphene-ZnO metal oxides demonstrated better energy density (4.8 Wh kg^{-1}) and capacitance (61 Fg^{-1}) compared to graphene- SnO_2 with good cycling capabilities. Q. Yan et al. developed Fe_3O_4/ reduced graphene oxide electrode which displayed a power density of 5.5 $kWkg^{-1}$; energy density of 67 $Whkg^{-1}$ and specific capacitance of 480 Fg^{-1} [63]. Graphene composites and layered double hydroxides have gained the attention of scientific community as the evolution of layered double hydroxides forbids the restacking of graphene nanosheets; the concentration of layered double hydroxide ions influence the development of complex structures. These composites with specific capacitance of 1200 F g^{-1} have better conductivity and high specific surface area can be effectively utilizes as layered double hydroxide structure [64,65]. The combination of cobalt oxides with graphene features high-performance electrodes for supercapacitor applications. The fabricated electrodes displays superior capacitance of 1100 F g^{-1} at current density of 10 Ag-^{1} [66,67]. Fabrication of graphene/copper composites using rotary chemical vapor deposition followed by vacuum hot pressing is shown in Figure 1; the same procedure can be utilized for fabrication of other metal oxides. Cobalt-based metal oxides/ graphene is most promising for the future energy storage devices owing to its high specific capacitance, cheapness and environmentally friendly nature [68]. Nickel hydroxide encapsulated in carbon nanotubes serves as a backbone for graphene sheets could enhances electrochemical performance by exhibiting high specific capacitance of 1235 F g^{-1} at a current density of 1 Ag^{-1} [69]. The electrochemical performance of supercapacitor also depends on the microstructure, morphology and mass ratio composition of composites.

Table 1. *Reports on electrochemical and structural attributes of graphene/metal oxide composite used for supercapacitor and lithium-ion battery applications.*

Structural models	Composites	Morphology of metal oxides	Application	Specific Capacity (mA hg^{-1})	No of Cycles	References
Encapsulated	Fe$_3$O$_4$/G	Nanoparticles	Lithium-Ion Battery	650	100	[48]
	G/Fe$_3$O$_4$	Hollow Nanoparticles	Lithium-Ion Battery	900	50	[49]
2D sandwich	G/Co$_3$O$_4$	Nanosheets	Lithium-Ion Battery	915	30 (84 % retention)	[52]
Anchored	RuO$_2$/G	Nanoparticles	Supercapacitor	570	1000 (97.9 % retention)	[55]
	G/Co$_3$O$_4$	Nanoparticles	Supercapacitor	243.2	2000(95.6% retention)	[56]
	G/MnO$_2$	Nanoflowers	Supercapacitor	315	5000(95% retention)	[57]
Layered	MnO$_2$/G	Nanotube	Lithium-Ion Battery	495	40	[58]
	MnO$_2$/G	Rod-shaped	Lithium-Ion Battery	1105	15	[59]

Materials Research Forum LLC
https://doi.org/10.21741/9781644900550-9

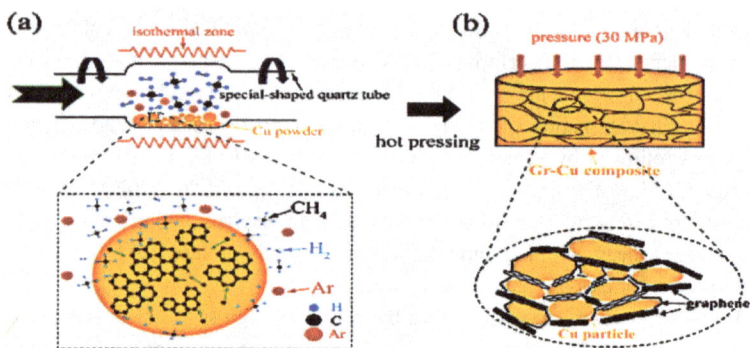

Figure 1. Fabrication of graphene-copper composites a) Synthesis using rotary chemical vapor deposition b) Fabrication by vacuum hot pressing [70].

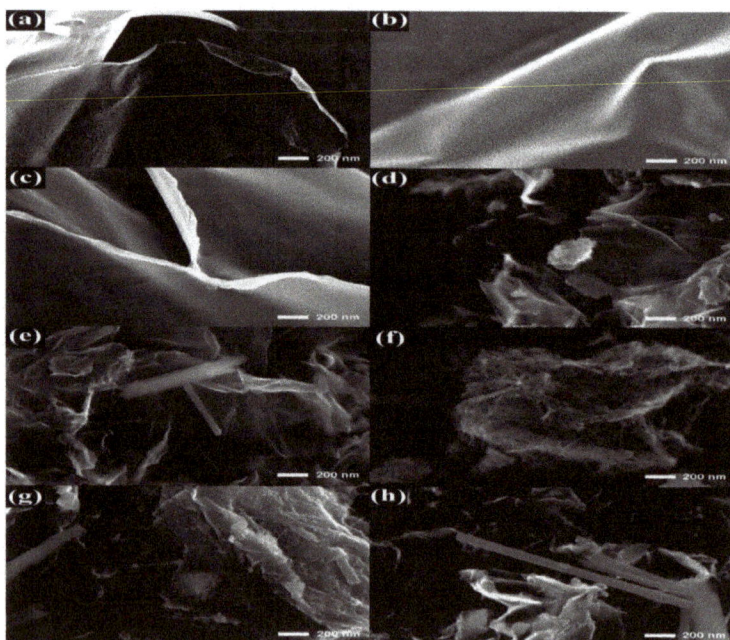

Figure 2. SEM micrographs of (a) Graphite oxide, (b) Graphene (c) Nitrogen doped graphene, (d) Nitrogen doped graphene manganese (NGM), [(e),(f),(g),(h)] NGM powders [71]

4. Electrode -graphene/ polymer composites

Polymer composites have gained wide attraction owing to their superior strength and excellent conductivity-graphene materials blended with polymers to develop composites for the fabrication of flexible energy storage devices. Studies on successful fabrication of graphene/polymer composite electrodes for numerous applications such as counter electrodes for sensitized solar cells, organic solar cell, fuel cells and flexible supercapacitors have been reported in literature [72]. Graphene/ polyaniline composites have shown capacitance of 233 F g^{-1} at scan rate of 2mVs-^{1} with good cycling performances [73], Similarly, study also exhibited the specific capacitance of 555 F g^{-1} at current density of 0.2 Ag^{-1}, better cycling stability in H$_2$SO$_4$ to make it an appropriate choice for supercapacitor applications [74] as graphene employed in composites the conductivity of polyaniline fibers was enhanced up to 44 % [75]. The performance and characteristics of various graphene/polymer composites are listed in Tables 2 and 3.

As commented above the π- π interactions between graphene sheets result in the lower specific surface area compared with theoretical estimation results. The blending of polymer with graphene is essential to develop new technologies for the homogeneous blending of the polymer matrix into single layer graphene sheets. The fabrication of three-dimensional graphene architectures in the composite can overcome these drawbacks.

Table 2. Graphene/Polyaniline capacitive properties based on various morphological structures

Materials	Preparation method	Capacitance (F g^{-1})	Current Density (A g^{-1})	Performance	References
PANI/G nanorods	Interfacial polymerisation	497	0.2	After 2000cycle 5.7% capacitance loss	[76]
G/PANI nanofibers Vacuum filtration	Filtration mediated by vacuum.	210	0.3	-	[73]
PANI/G	In-situ polymerisation	480	0.1	-	[77]
PANI/G hollow spheres	-	614	0.1	-	[78]
Flakes of G/PANI	-	764	0.1	-	[79]

Table 3*. Graphene/Polymer composite synthesis procedures and electrical characteristics*

Composites	Procedure	Conductivity (S m^{-1})	References
RGO/CNT/PPy	Mixing in solution	1530	[80]
GO/ PANI	In-situ polymerisation in solution	1000	[81]
RGO/PP	In-situ polymerisation in solution	685	[82]
RGO/PANI nanofibers	Mixing in solution	550	[73]

6. Summary and future outlooks

To meet the growing energy demands, scientists are looking forwards for new emerging technology. Graphene arises as an attractive option for energy storage as it constitutes of numerous advantages such as interconnected porous structure, high specific surface area, good wettability, matching pore size with ions of electrolyte ions and better electrical conductivity. However, the pristine graphene has lower capacitance, and the electrochemical characteristics possibly improved by linking double-layer capacitance with reliable and rapid reversible pseudocapacitance. However, utilization of graphene composites in an ionic liquid of durable electrochemical window of four volts improvised the range of working potential for each electrode triggering in an extensive potential window. The superior cell voltage enhanced power and energy densities. To meet the demands of future the reliability, power and energy characteristics of graphene-based supercapacitors needs to be improved. The next generation research hoped to be focus on the electrodes with better electrochemical performances and minimum equivalent series resistance.

Acknowledgements

This research work was financially supported by the University Malaya Research Grant (NO. RP045B-17AET), Impact-Oriented Interdisciplinary Research Grant (No. IIRG018A-2019) and Global Collaborative Programme - SATU Joint Research Scheme (No. ST012-2019).

References

[1] R.R. Salunkhe, Y.-H. Lee, K.H. Chang, J.M. Li, P. Simon, J. Tang, N.L. Torad, C.C. Hu, Y. Yamauchi, Nanoarchitectured graphene-based supercapacitors for next-generation energy-storage applications, Chem. Eur. J. 20 (2014) 13838–13852. https://doi.org/10.1002/chem.201403649.

[2] Q. Ke, J. Wang, Graphene-based materials for supercapacitor electrodes–A review, J. Materiomics 2 (2016) 37–54. https://doi.org/10.1016/J.JMAT.2016.01.001.

[3] M. Gocyla, M. Pisarek, M. Holdynski, M. Opallo, Electrochemical detection of graphene oxide, Electrochem. Commun. 96 (2018) 77-82. https://doi.org/10.1016/J.ELECOM.2018.10.004.

[4] M. Ye, Z. Zhang, Y. Zhao, L. Qu, Graphene platforms for smart energy generation and storage, Joule 2 (2018) 245–268. https://doi.org/10.1016/J.JOULE.2017.11.011.

[5] H. Zhang, H. Guo, A. Li, X. Chang, S. Liu, D. Liu, Y. Wang, F. Zhang, H. Yuan, High specific surface area porous graphene grids carbon as anode materials for sodium ion batteries, J. Energy Chem. 31 (2019) 159-166. https://doi.org/10.1016/J.JECHEM.2018.06.002.

[6] J. Park, Y.S. Cho, S.J. Sung, M. Byeon, S.J. Yang, C.R. Park, Characteristics tuning of graphene-oxide-based-graphene to various end-uses, Energy Storage Mater. 14 (2018) 8–21. https://doi.org/10.1016/J.ENSM.2018.02.013.

[7] C. (John) Zhang, V. Nicolosi, Graphene and MXene-based transparent conductive electrodes and supercapacitors, Energy Storage Mater. 16 (2019) 102–125. https://doi.org/10.1016/J.ENSM.2018.05.003.

[8] H.G. Kang, J. Jeong, S.B. Hong, G.Y. Lee, D.H. Kim, J.W. Kim, B.G. Choi, Scalable exfoliation and activation of graphite into porous graphene using microwaves for high–performance supercapacitors, J. Alloys Compd. 770 (2019) 458–465. https://doi.org/10.1016/J.JALLCOM.2018.08.042.

[9] S. Sheng, W. Liu, K. Zhu, K. Cheng, K. Ye, G. Wang, D. Cao, J. Yan, Fe_3O_4 nanospheres in situ decorated graphene as high-performance anode for asymmetric supercapacitor with impressive energy density, J. Colloid Interface Sci. 536 (2019) 235–244. https://doi.org/10.1016/J.JCIS.2018.10.060.

[10] F.C. Wu, R.L. Tseng, C.C. Hu, C.C. Wang, The capacitive characteristics of activated carbons—comparisons of the activation methods on the pore structure and effects of the pore structure and electrolyte on the capacitive performance, J. Power Sources 159 (2006) 1532–1542. https://doi.org/10.1016/J.JPOWSOUR.2005.12.023.

[11] Y. Li, M. van Zijll, S. Chiang, N. Pan, KOH modified graphene nanosheets for supercapacitor electrodes, J. Power Sources 196 (2011) 6003–6006. https://doi.org/10.1016/J.JPOWSOUR.2011.02.092.

[12] H.M. Jeong, J.W. Lee, W.H. Shin, Y.J. Choi, H.J. Shin, J.K. Kang, J.W. Choi, Nitrogen-doped graphene for high-performance ultracapacitors and the importance of

nitrogen-doped sites at basal planes, Nano Lett. 11 (2011) 2472–2477. https://doi.org/10.1021/nl2009058.

[13] X. Zhao, A. Wang, J. Yan, G. Sun, L. Sun, T. Zhang, Synthesis and electrochemical performance of heteroatom-incorporated ordered mesoporous carbons, Chem. Mater. 22 (2010) 5463–5473. https://doi.org/10.1021/cm101072z.

[14] X. Zhao, Q. Zhang, C.M. Chen, B. Zhang, S. Reiche, A. Wang, T. Zhang, R. Schlögl, D. Sheng Su, Aromatic sulfide, sulfoxide, and sulfone mediated mesoporous carbon monolith for use in supercapacitor, Nano Energy 1 (2012) 624–630. https://doi.org/10.1016/J.NANOEN.2012.04.003.

[15] S. Shiraishi, M. Kibe, T. Yokoyama, H. Kurihara, N. Patel, A. Oya, Y. Kaburagi, Y. Hishiyama, Electric double layer capacitance of multi-walled carbon nanotubes and B-doping effect, Appl. Phys. A 82 (2006) 585–591. https://doi.org/10.1007/s00339-005-3399-6.

[16] X. Zhai, Y. Song, J. Liu, P. Li, M. Zhong, C. Ma, H. Wang, Q. Guo, L. Zhi, In-situ preparation of boron-doped carbons with ordered mesopores and enhanced electrochemical properties in supercapacitors, J. Electrochem. Soc. 159 (2012) E177–E182. https://doi.org/10.1149/2.047212jes.

[17] T. Tomko, R. Rajagopalan, P. Aksoy, H.C. Foley, Synthesis of boron/nitrogen substituted carbons for aqueous asymmetric capacitors, Electrochim. Acta 56 (2011) 5369–5375. https://doi.org/10.1016/J.ELECTACTA.2011.03.112.

[18] P.N. Vishwakarma, S.V. Subramanyam, Hopping conduction in boron doped amorphous carbon films, J. Appl. Phys. 100 (2006) 113702. https://doi.org/10.1063/1.2372585.

[19] H. Konno, T. Ito, M. Ushiro, K. Fushimi, K. Azumi, High capacitance B/C/N composites for capacitor electrodes synthesized by a simple method, J. Power Sources 195 (2010) 1739–1746. https://doi.org/10.1016/J.JPOWSOUR.2009.09.072.

[20] H. Konno, T. Nakahashi, M. Inagaki, T. Sogabe, Nitrogen incorporation into boron-doped graphite and formation of B–N bonding, Carbon 37 (1999) 471–475. https://doi.org/10.1016/S0008-6223(98)00215-2.

[21] D. Zhong, H. Sano, Y. Uchiyama, K. Kobayashi, Effect of low-level boron doping on oxidation behavior of polyimide-derived carbon films, Carbon 38 (2000) 1199–1206. https://doi.org/10.1016/S0008-6223(99)00245-6.

[22] L.R. Radovic, M. Karra, K. Skokova, P.A. Thrower, The role of substitutional boron in carbon oxidation, Carbon 36 (1998) 1841–1854. https://doi.org/10.1016/S0008-

6223(98)00156-0.

[23] X.W. and, L.R. Radovic, Ab initio molecular orbital study on the electronic structures and reactivity of boron-substituted carbon, J. Phys. Chem. A 108 (2004) 9180-9187 (2004). https://doi.org/10.1021/JP048212W.

[24] T. Kwon, H. Nishihara, H. Itoi, Q.-H. Yang, T. Kyotani, Enhancement mechanism of electrochemical capacitance in nitrogen-/boron-doped carbons with uniform straight nanochannels, Langmuir 25 (2009) 11961–11968. https://doi.org/10.1021/la901318d.

[25] M. Wu, Y. Ren, N. Guo, S. Li, X. Sun, M. Tan, D. Wang, J. Zheng, N. Tsubaki, Hydrothermal co-doping of boron and phosphorus into porous carbons prepared from petroleum coke to improve oxidation resistance, Mater. Lett. 82 (2012) 124–126. https://doi.org/10.1016/J.MATLET.2012.05.080.

[26] D. Usachov, O. Vilkov, A. Grüneis, D. Haberer, A. Fedorov, V.K. Adamchuk, A.B. Preobrajenski, P. Dudin, A. Barinov, M. Oehzelt, C. Laubschat, D. V. Vyalikh, Nitrogen-doped graphene: efficient growth, structure, and electronic properties, Nano Lett. 11 (2011) 5401–5407. https://doi.org/10.1021/nl2031037.

[27] M. Rybin, A. Pereyaslavtsev, T. Vasilieva, V. Myasnikov, I. Sokolov, A. Pavlova, E. Obraztsova, A. Khomich, V. Ralchenko, E. Obraztsova, Efficient nitrogen doping of graphene by plasma treatment, Carbon 96 (2016) 196–202. https://doi.org/10.1016/J.CARBON.2015.09.056.

[28] H. Xu, L. Ma, Z. Jin, Nitrogen-doped graphene: Synthesis, characterizations and energy applications, J. Energy Chem. 27 (2018) 146–160. https://doi.org/10.1016/J.JECHEM.2017.12.006.

[29] R. Yadav, C.K. Dixit, Synthesis, characterization and prospective applications of nitrogen-doped graphene: A short review, J. Sci. Adv. Mater. Devices 2 (2017) 141–149. https://doi.org/10.1016/J.JSAMD.2017.05.007.

[30] B. Jiang, C. Tian, L. Wang, L. Sun, C. Chen, X. Nong, Y. Qiao, H. Fu, Highly concentrated, stable nitrogen-doped graphene for supercapacitors: Simultaneous doping and reduction, Appl. Surf. Sci. 258 (2012) 3438–3443. https://doi.org/10.1016/J.APSUSC.2011.11.091.

[31] Y.J. Kim, Y. Abe, T. Yanagiura, K.C. Park, M. Shimizu, T. Iwazaki, S. Nakagawa, M. Endo, M.S. Dresselhaus, Easy preparation of nitrogen-enriched carbon materials from peptides of silk fibroins and their use to produce a high volumetric energy density in supercapacitors, Carbon 45 (2007) 2116–2125.

https://doi.org/10.1016/J.CARBON.2007.05.026.

[32] N.D. Kim, W. Kim, J.B. Joo, S. Oh, P. Kim, Y. Kim, J. Yi, Electrochemical capacitor performance of N-doped mesoporous carbons prepared by ammoxidation, J. Power Sources 180 (2008) 671–675. https://doi.org/10.1016/J.JPOWSOUR.2008.01.055.

[33] T.E. Rufford, D. Hulicova-Jurcakova, Z. Zhu, G.Q. Lu, Nanoporous carbon electrode from waste coffee beans for high performance supercapacitors, Electrochem. Commun. 10 (2008) 1594–1597. https://doi.org/10.1016/J.ELECOM.2008.08.022.

[34] S. Liu, D. Lentz, C.C. Tzschucke, Conversion of pyridine N-oxides to tetrazolopyridines, J. Org. Chem. 79 (2014) 3249–3254. https://doi.org/10.1021/jo500231m.

[35] T. Liao, C. Sun, A. Du, Z. Sun, D. Hulicova-Jurcakova, S. Smith, Charge carrier exchange at chemically modified graphene edges: A density functional theory study, J. Mater. Chem. 22 (2012) 8321. https://doi.org/10.1039/c2jm30387f.

[36] R. Rajagopalan, A. Balakrishnan, Innovations in engineered porous materials for energy generation and storage applications, CRC Press, 2018.

[37] K.N. Wood, R. O'hayre, S. Pylypenko, Recent progress on nitrogen/carbon structures designed for use in energy and sustainability applications, Energy Environ. Sci. 7 (2014) 1212-1249. https://doi.org/10.1039/c3ee44078h.

[38] L. Zheng, H. Zheng, D. Huo, F. Wu, L. Shao, P. Zheng, Y. Jiang, X. Zheng, X. Qiu, Y. Liu, Y. Zhang, N-doped graphene-based copper nanocomposite with ultralow electrical resistivity and high thermal conductivity, Sci. Rep. 8 (2018) 9248. https://doi.org/10.1038/s41598-018-27667-9.

[39] N.E. Derradji, M.L. Mahdjoubi, H. Belkhir, N. Mumumbila, B. Angleraud, P.Y. Tessier, Nitrogen effect on the electrical properties of CN_x thin films deposited by reactive magnetron sputtering, Thin Solid Films 482 (2005) 258–263. https://doi.org/10.1016/J.TSF.2004.11.137.

[40] J. Biemolt, I.M. Denekamp, T.K. Slot, G. Rothenberg, D. Eisenberg, Boosting the supercapacitance of nitrogen-doped carbon by tuning surface functionalities, ChemSusChem 10 (2017) 4018–4024. https://doi.org/10.1002/cssc.201700902.

[41] M. An, C. Du, L. Du, Y. Sun, Y. Wang, C. Chen, G. Han, G. Yin, Y. Gao, Phosphorus-doped graphene support to enhance electrocatalysis of methanol oxidation reaction on platinum nanoparticles, Chem. Phys. Lett. 687 (2017) 1–8.

https://doi.org/10.1016/J.CPLETT.2017.08.058.

[42] Y. Wen, B. Wang, C. Huang, L. Wang, D. Hulicova-Jurcakova, Synthesis of phosphorus-doped graphene and its wide potential window in aqueous supercapacitors, Chem. Eur. J. 21 (2015) 80–85. https://doi.org/10.1002/chem.201404779.

[43] M.Y. Rekha, N. Mallik, C. Srivastava, First Report on High Entropy alloy nanoparticle decorated graphene, Sci. Rep. 8 (2018) 8737. https://doi.org/10.1038/s41598-018-27096-8.

[44] L. Sun, J. Liu, Z. Liu, T. Wang, H. Wang, Y. Li, Sulfur-doped mesoporous carbon via thermal reduction of CS_2 by Mg for high-performance supercapacitor electrodes and Li-ion battery anodes, RSC Adv. 8 (2018) 19964-19970. https://doi.org/10.1039/c8ra01729h.

[45] B. Mensah, S.I. Kang, W. Wang, C. Nah, Effect of graphene on polar and nonpolar rubber matrices, Mech. Adv. Mater. Mod. Process 4 (2018) 1. https://doi.org/10.1186/s40759-017-0034-0.

[46] A.K. Farquhar, P.A. Brooksby, A.J. Downard, Controlled spacing of few-layer graphene sheets using molecular spacers: capacitance that scales with sheet number, ACS Appl. Nano Mater. 1 (2018) 11420-11429. https://doi.org/10.1021/acsanm.8b00280.

[47] W. Lv, F. Sun, D.-M. Tang, H.-T. Fang, C. Liu, Q.H. Yang, H.M. Cheng, A sandwich structure of graphene and nickel oxide with excellent supercapacitive performance, J. Mater. Chem. 21 (2011) 9014-9019. https://doi.org/10.1039/c1jm10400d.

[48] J.Z. Wang, C. Zhong, D. Wexler, N.H. Idris, Z.-X. Wang, L.-Q. Chen, H.-K. Liu, Graphene-encapsulated Fe_3O_4 nanoparticles with 3D laminated structure as superior anode in lithium ion batteries, Chem. Eur. J. 17 (2011) 661–667. https://doi.org/10.1002/chem.201001348.

[49] D. Chen, G. Ji, Y. Ma, J.Y. Lee, J. Lu, Graphene-encapsulated hollow Fe_3O_4 nanoparticle aggregates as a high-performance anode material for lithium ion batteries, ACS Appl. Mater. Interfaces 3 (2011) 3078–3083. https://doi.org/10.1021/am200592r.

[50] Z.S. Wu, D.-W. Wang, W. Ren, J. Zhao, G. Zhou, F. Li, H.-M. Cheng, Anchoring hydrous RuO_2 on graphene sheets for high performance electrochemical capacitors, Adv. Funct. Mater. 20 (2010) 3595–3602. https://doi.org/10.1002/adfm.201001054.

[51] G. Wang, T. Liu, Y. Luo, Y. Zhao, Z. Ren, J. Bai, H. Wang, Preparation of Fe_2O_3/graphene composite and its electrochemical performance as an anode material for lithium ion batteries, J. Alloys Compd. 509 (2011) L216–L220. https://doi.org/10.1016/j.jallcom.2011.03.151.

[52] S. Yang, X. Feng, L. Wang, K. Tang, J. Maier, K. Müllen, Graphene-based nanosheets with a sandwich structure, Angew. Chemie Int. Ed. 49 (2010) 4795–4799. https://doi.org/10.1002/anie.201001634.

[53] S. Yang, X. Feng, K. Müllen, Sandwich-Like, Graphene-based titania nanosheets with high surface area for fast lithium storage, Adv. Mater. 23 (2011) 3575–3579. https://doi.org/10.1002/adma.201101599.

[54] Z.S. Wu, W. Ren, D.W. Wang, F. Li, B. Liu, H.M. Cheng, High-energy MnO_2 nanowire/graphene and graphene asymmetric electrochemical capacitors, ACS Nano 4 (2010) 5835–5842. https://doi.org/10.1021/nn101754k.

[55] Y. Li, H. Yadegari, X. Li, M.N. Banis, R. Li, X. Sun, Superior catalytic activity of nitrogen-doped graphene cathodes for high energy capacity sodium–air batteries, Chem. Commun. 49 (2013) 11731. https://doi.org/10.1039/c3cc46606j.

[56] X.-H. Zhang, S.-M. Wang, J. Wu, X.-J. Liu, Electropolymerization of PoPD from aqueous solutions of sodium dodecyl benzene sulfonate at conducting glass electrode, J. Appl. Polym. Sci. 104 (2007) 1928–1932. https://doi.org/10.1002/app.25877.

[57] G. Yu, L. Hu, M. Vosgueritchian, H. Wang, X. Xie, J.R. McDonough, X. Cui, Y. Cui, Z. Bao, Solution-processed graphene/MnO_2 nanostructured textiles for high-performance electrochemical capacitors, Nano Lett. 11 (2011) 2905–2911. https://doi.org/10.1021/nl2013828.

[58] A. Yu, H.W. Park, A. Davies, D.C. Higgins, Z. Chen, X. Xiao, Free-Standing layer-by-layer hybrid thin film of graphene-MnO_2 nanotube as anode for lithium ion batteries, J. Phys. Chem. Lett. 2 (2011) 1855–1860. https://doi.org/10.1021/jz200836h.

[59] C.X. Guo, M. Wang, T. Chen, X.W. Lou, C.M. Li, A Hierarchically nanostructured composite of MnO_2/conjugated polymer/graphene for high-performance lithium ion batteries, Adv. Energy Mater. 1 (2011) 736–741. https://doi.org/10.1002/aenm.201100223.

[60] M. Zheng, X. Xiao, L. Li, P. Gu, X. Dai, H. Tang, Q. Hu, H. Xue, H. Pang, Hierarchically nanostructured transition metal oxides for supercapacitors, Sci. China

Mater. 61 (2018) 185–209. https://doi.org/10.1007/s40843-017-9095-4.

[61] K.H. Chang, Y.-F. Lee, C.-C. Hu, C.-I. Chang, C.-L. Liu, Y.-L. Yang, A unique strategy for preparing single-phase unitary/binary oxides–graphene composites, Chem. Commun. 46 (2010) 7957. https://doi.org/10.1039/c0cc01805h.

[62] A.K. Mishra, S. Ramaprabhu, Functionalized graphene-based nanocomposites for supercapacitor application, J. Phys. Chem. C. 115 (2011) 14006–14013. https://doi.org/10.1021/jp201673e.

[63] W. Shi, J. Zhu, D.H. Sim, Y.Y. Tay, Z. Lu, X. Zhang, Y. Sharma, M. Srinivasan, H. Zhang, H.H. Hng, Q. Yan, Achieving high specific charge capacitances in Fe_3O_4/reduced graphene oxide nanocomposites, J. Mater. Chem. 21 (2011) 3422. https://doi.org/10.1039/c0jm03175e.

[64] S. Huang, G.-N. Zhu, C. Zhang, W.W. Tjiu, Y.-Y. Xia, T. Liu, Immobilization of Co–Al layered double hydroxides on graphene oxide nanosheets: Growth mechanism and supercapacitor studies, ACS Appl. Mater. Interfaces 4 (2012) 2242–2249. https://doi.org/10.1021/am300247x.

[65] X. Dong, L. Wang, D. Wang, C. Li, J. Jin, Layer-by-layer engineered Co–Al hydroxide nanosheets/graphene multilayer films as flexible electrode for supercapacitor, Langmuir 28 (2012) 293–298. https://doi.org/10.1021/la2038685.

[66] W. Zhou, J. Liu, T. Chen, K.S. Tan, X. Jia, Z. Luo, C. Cong, H. Yang, C.M. Li, T. Yu, Fabrication of Co_3O_4-reduced graphene oxide scrolls for high-performance supercapacitor electrodes, Phys. Chem. Chem. Phys. 13 (2011) 14462-14465. https://doi.org/10.1039/c1cp21917k.

[67] X.C. Dong, H. Xu, X.W. Wang, Y.X. Huang, M.B. Chan-Park, H. Zhang, L.H. Wang, W. Huang, P. Chen, 3D graphene–cobalt oxide electrode for high-performance supercapacitor and enzymeless glucose detection, ACS Nano 6 (2012) 3206–3213. https://doi.org/10.1021/nn300097q.

[68] G. Xu, P. Nie, H. Dou, B. Ding, L. Li, X. Zhang, Exploring metal organic frameworks for energy storage in batteries and supercapacitors, Mater. Today 20 (2017) 191–209. https://doi.org/10.1016/J.MATTOD.2016.10.003.

[69] D.G. Papageorgiou, I.A. Kinloch, R.J. Young, Mechanical properties of graphene and graphene-based nanocomposites, Prog. Mater. Sci. 90 (2017) 75–127. https://doi.org/10.1016/J.PMATSCI.2017.07.004.

[70] Q. Xiao, X. Yi, B. Jiang, Z. Qin, J. Hu, Y. Jiang, H. Liu, B. Wang, D. Yi, In-situ synthesis of graphene on surface of copper powder by rotary CVD and its application

in fabrication of reinforced Cu-matrix composites, Adv. Mater. Sci. 2 (2017). https://doi.org/10.15761/AMS.1000123.

[71] H.Y. Chiu, C.P. Cho, H.Y. Chiu, C.P. Cho, Mixed-Phase MnO_2/N-containing graphene composites applied as electrode active materials for flexible asymmetric solid-state supercapacitors, Nanomaterials 8 (2018) 924. https://doi.org/10.3390/nano8110924.

[72] J. Wu, Z. Lan, J. Lin, M. Huang, Y. Huang, L. Fan, G. Luo, Y. Lin, Y. Xie, Y. Wei, Counter electrodes in dye-sensitized solar cells, Chem. Soc. Rev. 46 (2017) 5975–6023. https://doi.org/10.1039/C6CS00752J.

[73] Q. Wu, Y. Xu, Z. Yao, A. Liu, G. Shi, Supercapacitors based on flexible graphene/polyaniline nanofiber composite films, ACS Nano 4 (2010) 1963–1970. https://doi.org/10.1021/nn1000035.

[74] D.W. Wang, F. Li, J. Zhao, W. Ren, Z.G. Chen, J. Tan, Z.S. Wu, I. Gentle, G.Q. Lu, H.M. Cheng, Fabrication of graphene/polyaniline composite paper via in situ anodic electropolymerization for high-performance flexible electrode, ACS Nano 3 (2009) 1745–1752. https://doi.org/10.1021/nn900297m.

[75] H. Wang, J. Lin, Z.X. Shen, Polyaniline (PANi) based electrode materials for energy storage and conversion, J. Sci. Adv. Mater. Devices 1 (2016) 225–255. https://doi.org/10.1016/J.JSAMD.2016.08.001.

[76] B. Ma, X. Zhou, H. Bao, X. Li, G. Wang, Hierarchical composites of sulfonated graphene-supported vertically aligned polyaniline nanorods for high-performance supercapacitors, J. Power Sources 215 (2012) 36–42. https://doi.org/10.1016/j.jpowsour.2012.04.083.

[77] K. Zhang, L.L. Zhang, X.S. Zhao, J. Wu, Graphene/polyaniline nanofiber composites as supercapacitor electrodes, Chem. Mater. 22 (2010) 1392–1401. https://doi.org/10.1021/cm902876u.

[78] W. Fan, C. Zhang, W.W. Tjiu, K.P. Pramoda, C. He, T. Liu, Graphene-wrapped polyaniline hollow spheres as novel hybrid electrode materials for supercapacitor applications, ACS Appl. Mater. Interfaces 5 (2013) 3382–3391. https://doi.org/10.1021/am4003827.

[79] F. Chen, P. Liu, Q. Zhao, Well-defined graphene/polyaniline flake composites for high performance supercapacitors, Electrochim. Acta 76 (2012) 62–68. https://doi.org/10.1016/J.ELECTACTA.2012.04.154.

[80] X. Lu, H. Dou, B. Gao, C. Yuan, S. Yang, L. Hao, L. Shen, X. Zhang, A flexible

graphene/multiwalled carbon nanotube film as a high performance electrode material for supercapacitors, Electrochim. Acta 56 (2011) 5115–5121. https://doi.org/10.1016/J.ELECTACTA.2011.03.066.

[81] H. Wang, Q. Hao, X. Yang, L. Lu, X. Wang, Graphene oxide doped polyaniline for supercapacitors, Electrochem. Commun. 11 (2009) 1158–1161. https://doi.org/10.1016/j.elecom.2009.03.036.

[82] D. Zhang, X. Zhang, Y. Chen, P. Yu, C. Wang, Y. Ma, Enhanced capacitance and rate capability of graphene/polypyrrole composite as electrode material for supercapacitors, J. Power Sources 196 (2011) 5990–5996. https://doi.org/10.1016/J.JPOWSOUR.2011.02.090.

[83] C. Zhu, J. Zhai, D. Wen, S. Dong, Graphene oxide/polypyrrole nanocomposites: one-step electrochemical doping, coating and synergistic effect for energy storage, J. Mater. Chem. 22 (2012) 6300-6306. https://doi.org/10.1039/c2jm16699b.

[84] Y. Xu, Y. Wang, J. Liang, Y. Huang, Y. Ma, X. Wan, Y. Chen, A hybrid material of graphene and poly (3,4-ethyldioxythiophene) with high conductivity, flexibility, and transparency, Nano Res. 2 (2009) 343–348. https://doi.org/10.1007/s12274-009-9032-9.

[85] P.J. Hung, K.H. Chang, Y.F. Lee, C.C. Hu, K.M. Lin, Ideal asymmetric supercapacitors consisting of polyaniline nanofibers and graphene nanosheets with proper complementary potential windows, Electrochim. Acta 55 (2010) 6015–6021. https://doi.org/10.1016/J.ELECTACTA.2010.05.058.

[86] B.G. Choi, S.-J. Chang, H.-W. Kang, C.P. Park, H.J. Kim, W.H. Hong, S. Lee, Y.S. Huh, High performance of a solid-state flexible asymmetric supercapacitor based on graphene films, Nanoscale 4 (2012) 4983-4988. https://doi.org/10.1039/c2nr30991b.

[87] H. Gao, F. Xiao, C.B. Ching, H. Duan, High-performance asymmetric supercapacitor based on graphene hydrogel and nanostructured MnO_2, ACS Appl. Mater. Interfaces 4 (2012) 2801–2810. https://doi.org/10.1021/am300455d.

Keyword Index

About the Editors

Dr. Inamuddin is currently working as Assistant Professor in the Chemistry Department, Faculty of Science, King Abdulaziz University, Jeddah, Saudi Arabia. He is a permanent faculty member (Assistant Professor) at the Department of Applied Chemistry, Aligarh Muslim University, Aligarh, India. He obtained Master of Science degree in Organic Chemistry from Chaudhary Charan Singh (CCS) University, Meerut, India, in 2002. He received his Master of Philosophy and Doctor of Philosophy degrees in Applied Chemistry from Aligarh Muslim University (AMU), India, in 2004 and 2007, respectively. He has extensive research experience in multidisciplinary fields of Analytical Chemistry, Materials Chemistry, and Electrochemistry and, more specifically, Renewable Energy and Environment. He has worked on different research projects as project fellow and senior research fellow funded by University Grants Commission (UGC), Government of India, and Council of Scientific and Industrial Research (CSIR), Government of India. He has received Fast Track Young Scientist Award from the Department of Science and Technology, India, to work in the area of bending actuators and artificial muscles. He has completed four major research projects sanctioned by University Grant Commission, Department of Science and Technology, Council of Scientific and Industrial Research, and Council of Science and Technology, India. He has published 147 research articles in international journals of repute and eighteen book chapters in knowledge-based book editions published by renowned international publishers. He has published 60 edited books with Springer (U.K.), Elsevier, Nova Science Publishers, Inc. (U.S.A.), CRC Press Taylor & Francis Asia Pacific, Trans Tech Publications Ltd. (Switzerland), IntechOpen Limited (U.K.), and Materials Research Forum LLC (U.S.A). He is a member of various journals' editorial boards. He is also serving as Associate Editor for journals (Environmental Chemistry Letter, Applied Water Science and Euro-Mediterranean Journal for Environmental Integration, Springer-Nature), Frontiers Section Editor (Current Analytical Chemistry, Bentham Science Publishers), Editorial Board Member (Scientific Reports-Nature), Editor (Eurasian Journal of Analytical Chemistry), and Review Editor (Frontiers in Chemistry, Frontiers, U.K.) He is also guest-editing various special thematic special issues to the journals of Elsevier, Bentham Science Publishers, and John Wiley & Sons, Inc. He has attended as well as chaired sessions in various international and national conferences. He has worked as a Postdoctoral Fellow, leading a research team at the Creative Research Initiative Center for Bio-Artificial Muscle, Hanyang University, South Korea, in the field of renewable energy, especially biofuel cells. He has also worked as a Postdoctoral Fellow at the Center of Research Excellence in Renewable Energy, King Fahd University of Petroleum and Minerals, Saudi Arabia, in the field of polymer electrolyte membrane fuel cells and computational fluid dynamics of polymer electrolyte membrane fuel cells. He is a life member of the Journal of the Indian Chemical Society. His research interest includes ion exchange materials, a sensor for heavy metal ions, biofuel cells, supercapacitors and bending actuators.

Dr. Rajender Boddula is currently working as Chinese Academy of Sciences-President's International Fellowship Initiative (CAS-PIFI) at National Center for Nanoscience and Technology (NCNST, Beijing). He obtained Master of Science in Organic Chemistry from Kakatiya University, Warangal, India, in 2008. He received his Doctor of Philosophy in Chemistry with the highest honours in 2014 for the work entitled "Synthesis and Characterization of Polyanilines for Supercapacitor and Catalytic Applications" at the CSIR-Indian Institute of Chemical Technology (CSIR-IICT) and Kakatiya University (India). Before joining National Center for Nanoscience and Technology (NCNST) as CAS-PIFI research fellow, China, worked as senior research associate and Postdoc at the Aligarh Muslim University (AMU, India) and National Tsing-Hua University (NTHU, Taiwan) respectively in the fields of bio-fuel and CO2 reduction applications. His academic honors include University Grants Commission National Fellowship and many merit scholarships, study-abroad fellowships from Australian Endeavour Research Fellowship, and CAS-PIFI. He has published many scientific articles in international peer-reviewed journals and has authored around twenty book chapters, and he is also serving as an editorial board member and a referee for reputed international peer-reviewed journals. He has published edited books with Springer (UK), Elsevier, Materials Research Forum LLC (USA) and CRC Press Taylor & Francis group. His specialized areas of research are energy conversion and storage, which include sustainable nanomaterials, graphene, polymer composites, heterogeneous catalysis for organic transformations, environmental remediation technologies, photoelectrochemical water-splitting devices, biofuel cells, batteries and supercapacitors.

Dr. Mohammad Faraz Ahmer is presently working as Assistant Professor in the Department of Electrical Engineering, Mewat Engineering College, Nuh Haryana, India, since 2012 after working as Guest Faculty in University Polytechnic, Aligarh Muslim University Aligarh, India, during 2009-2011. He completed M.Tech. (2009) and Bachelor of Engineering (2007) degrees in Electrical Engineering from Aligarh Muslim University, Aligarh in the first division. He obtained a Ph.D. degree in 2016 on his thesis entitled "Studies on Electrochemical Capacitor Electrodes". He has published six research papers in reputed scientific journals. He has edited two books with Materials Research Forum LLC, U.S.A. His scientific interests include electrospun nano-composites and supercapacitors. He has presented his work at several conferences. He is actively engaged in searching of new methodologies involving the development of organic composite materials for energy storage systems.

Prof. Abdullah M. Asiri is the Head of the Chemistry Department at King Abdulaziz University since October 2009 and he is the founder and the Director of the Center of Excellence for Advanced Materials Research (CEAMR) since 2010 till date. He is the Professor of Organic Photochemistry. He graduated from King Abdulaziz University (KAU) with B.Sc. in Chemistry in 1990 and a Ph.D. from University of Wales, College of Cardiff, U.K. in 1995. His research interest covers color chemistry, synthesis of novel photochromic and thermochromic systems, synthesis of novel coloring matters and dyeing of textiles, materials chemistry, nanochemistry and nanotechnology, polymers and plastics. Prof. Asiri is the principal supervisors of more than 20 M.Sc. and six Ph.D. theses. He is the main author of ten books of different chemistry disciplines. Prof. Asiri is the Editor-in-Chief of King Abdulaziz University Journal of Science. A major achievement of Prof. Asiri is the research of tribochromic compounds, a new class of compounds which change from slightly or colorless to deep colored when subjected to small pressure or when grind. This discovery was introduced to the scientific community as a new terminology published by International Union of Pure and Applied Chemistry (IUPAC) in 2000. This discovery was awarded a patent from European Patent office and from UK patent. Prof. Asiri involved in many committees at the KAU level and on the national level. He took a major role in the advanced materials committee working for King Abdulaziz City for Science and Technology (KACST) to identify the national plan for science and technology in 2007. Prof. Asiri played a major role in advancing the chemistry education and research in KAU. He has been awarded the best researchers from KAU for the past five years. He also awarded the Young Scientist Award from the Saudi Chemical Society in 2009 and also the first prize for the distinction in science from the Saudi Chemical Society in 2012. He also received a recognition certificate from the American Chemical Society (Gulf region Chapter) for the advancement of chemical science in the Kingdome. He received a Scopus certificate for the most publishing scientist in Saudi Arabia in chemistry in 2008. He is also a member of the editorial board of various journals of international repute. He is the Vice- President of Saudi Chemical Society (Western Province Branch). He holds four USA patents, more than one thousand publications in international journals, several book chapters and edited books.

www.ingramcontent.com/pod-product-compliance
Lightning Source LLC
Chambersburg PA
CBHW071334210326
41597CB00015B/1448